Environmental Toxicology and Ecosystem

This book covers varied aspects of environmental contaminants and their effects on the living organisms. It addresses the basics of ecotoxicity assessment, interaction of the abiotic or biotic factors with the novel chemical entities, and the fate of the natural organic matter upon interaction with new chemical entities. It further includes models for ecotoxicity studies and high-throughput approaches including OMICS. It provides an overview of the ecological risk assessment, regulatory toxicology guidelines, and possible roadmaps for protection of environmental health.

Features:

- Discusses environmental toxicology facets and their effects on the ecosystem.
- Provides an introduction to environmental toxicology keeping in view the paradigm shift on entry of novel materials in the environment.
- Includes bioavailability, bioconcentration, and biomagnification of trophic transfer of pollutants.
- Covers high-throughput approaches for ecotoxicity assessment.
- Explores roadmaps for environmental protection and sustainable development.

This book is geared toward graduate students and researchers in Environmental Sciences and Engineering, Toxicology, and Ecology.

Environmental Toxicology and Ecosystem

Edited by
Ashutosh Kumar

CRC Press
Taylor & Francis Group
Boca Raton London New York

CRC Press is an imprint of the
Taylor & Francis Group, an **informa** business

Designed cover image: © Shutterstock

First edition published 2023
by CRC Press
6000 Broken Sound Parkway NW, Suite 300, Boca Raton, FL 33487-2742

and by CRC Press
4 Park Square, Milton Park, Abingdon, Oxon, OX14 4RN

CRC Press is an imprint of Taylor & Francis Group, LLC

© 2023 selection and editorial matter, Ashutosh Kumar; individual chapters, the contributors

Reasonable efforts have been made to publish reliable data and information, but the author and publisher cannot assume responsibility for the validity of all materials or the consequences of their use. The authors and publishers have attempted to trace the copyright holders of all material reproduced in this publication and apologize to copyright holders if permission to publish in this form has not been obtained. If any copyright material has not been acknowledged please write and let us know so we may rectify in any future reprint.

Except as permitted under U.S. Copyright Law, no part of this book may be reprinted, reproduced, transmitted, or utilized in any form by any electronic, mechanical, or other means, now known or hereafter invented, including photocopying, microfilming, and recording, or in any information storage or retrieval system, without written permission from the publishers.

For permission to photocopy or use material electronically from this work, access www.copyright.com or contact the Copyright Clearance Center, Inc. (CCC), 222 Rosewood Drive, Danvers, MA 01923, 978-750-8400. For works that are not available on CCC please contact mpkbookspermissions@tandf.co.uk

Trademark notice: Product or corporate names may be trademarks or registered trademarks and are used only for identification and explanation without intent to infringe.

ISBN: 9781032154947 (hbk)
ISBN: 9781032154954 (pbk)
ISBN: 9781003244349 (ebk)

DOI: 10.1201/9781003244349

Typeset in Times
by Newgen Publishing UK

Contents

Preface vii
Editor Biography ix
Contributors xi

Chapter 1 Environmental Paradigm Change and Risk Assessment 1

Dhairya Mazmudar, Dixit Raiyani, and Ashutosh Kumar

Chapter 2 Environmental and Ecotoxicological Impacts of Engineered Nanomaterials 11

Neha Shrivastava, Vikas Shrivastava, Rajesh Singh Tomar, and Anurag Jyoti

Chapter 3 Toxicology of Metal Ions, Pesticides, Nanomaterials, and Microplastics 35

Kamayani Vajpayee, Krupa Kansara, and Ritesh K. Shukla

Chapter 4 Natural Organic Matter: A Ubiquitous Adsorbent in Aquatic Systems to Probe Nanoparticles Behavior and Effects in Ecosystem 59

Abdulkhaik Mansuri and Ashutosh Kumar

Chapter 5 Analysis of Emerging Contaminants in Water Samples: Advances and Challenges 85

Sivaperumal P., Nikhil P. K., Smith C., Shaikh S., Rajnish G., Rupal T., and Tejal G. M.

Chapter 6 Advanced Techniques for Detection of Environmental Pollutants and Recent Progress 121

Kriti Srivastava, Harsh Prajapati, Gajendra Singh Vishwakarma, Nidhi Verma, and Alok Pandya

Chapter 7 Impact of Heavy Metals on Different Ecosystems 139

Onila Lugun, Richa Singh, Shambhavi Jha, and Alok Kumar Pandey

Chapter 8 Metabolism of Environmental Pollutants with Specific Reference to Pesticides, Endocrine Disrupters, and Mutagenic Pollutants 165

Simran Nasra and Ashutosh Kumar

Chapter 9 Bioavailability, Bioconcentration, and Biomagnification of Pollutants 179

Shrushti Shah and Ashutosh Kumar

Chapter 10 High Throughput Approaches for Engineered Nanomaterial-Induced Ecotoxicity Assessment 215

Krupa Kansara and Dhiraj Devidas Bhatia

Chapter 11 Impact of Heavy Metals on Gut Microbial Ecosystem: Implications in Health and Disease 233

Dhirendra Pratap Singh, Ravinder Naik Dharavath, Raghunath Singh, Keya Patel, Shirali Patel, Vandana Bijalwan, Praveen Kolimi, Mahendra Bishnoi, and Santasabuj Das

Chapter 12 Roadmaps for Environmental Protection and Sustainable Development 267

Suman Mallick and Ratna Ghosal

Index 287

Preface

The applications of new materials/chemicals in diverse areas of medicine, personal care products, electronics, energy production, and many others has led to their inadvertent release in surface and subsurface environments through landfills and other waste disposal methods. It is quite likely that some of these materials/chemicals will induce toxic response in both lower and higher trophic organisms. The discipline of Environmental Toxicology has gained a lot of importance due to the release of these emerging pollutants such as new chemical entities (NCEs), genetically modified organisms, and engineered nanomaterials.

Environmental health is an intrinsic part of maintaining quality of life and protecting human health. Rapid industrial development, population growth, and urban crowding have made environmental toxicology more relevant to the developing world. The hazards posed by novel materials and enabling technologies pose risks to sustainable development. Life cycle assessment and long-term impact of these materials need to be explored well for their sustainable growth. Environmental monitoring and risk assessment depend on high-throughput hazard assessment methods and alternate models representing biota in the ecosystem. Understanding and knowledge of such tools and methods for environmental toxicology will lead to development of new guidelines for protection of human and environmental health.

This book provides an knowledge on all the methods used in environmental toxicology and their limitation for hazard identification and risk assessment. The book starts with an introduction to environmental toxicology keeping in view the paradigm shift of entry of novel materials in the environment. Further, topics like basic definitions of ecotoxicity assessment, ecotoxicological impacts of engineered nanomaterials, metal ions, pesticides, and microplastics are discussed in detail. A specific chapter focused on understanding the interaction of abiotic or biotic factors with the new chemical entities and the effects and the impacts on the natural organic matter is also included. Additionally, chapters on bioavailability, bioconcentration, and biomagnification of pollutants; metabolism of pollutants with specific reference to pesticides, endocrine disrupters, and mutagenic potential; and impact of heavy metals on gut microbial ecosystem and their implications in health and disease are also included to increase knowledge depth and understanding of the used methodologies and impact of different pollutants.

The book also attempts to address the current methodologies used for the analysis of emerging contaminants in water samples, advancements, and challenges associated with these methodologies; recent progress in the development of advanced techniques for detection of environmental pollutant; and high-throughput approaches including OMICS for ecotoxicity assessment.

Readers will also be given an overview of regulatory toxicology guidelines and possible roadmaps for environmental protection and sustainable development.

Editor Biography

Ashutosh Kumar works as Associate Professor in the Division of Biological and Life Sciences, School of Arts and Sciences, Ahmedabad University, Gujarat, India. He obtained his master's degree in Applied Microbiology from Vellore Institute of Technology, Vellore in 2008 and worked at CSIR-Indian Institute of Toxicology Research, Lucknow for his doctorate degree and was involved in understanding the fate of nanomaterials on biological systems.

Ashutosh is an exceptionally talented and motivated nanotechnology researcher who has the passion for understanding the effects of nanotechnology on the human life and environment at large. His deep understanding of the environment combined with his skills in nanotechnology and his versatile analytical understanding of different animal systems allowed him to make significant seminal contributions to the field of environmental nanotechnology. In the past, Professor Kumar's research work has been focused on development of new methods for nanomaterial safety in India. He developed a novel method for the detection of uptake of nanoparticles in live cells for several generations. Additionally, he has been involved in studying the potential effect of nanoparticles on human cells, mice, and aquatic organisms including zebrafish and tetrahymena. His understanding of the role of abiotic factors in influencing the nanotechnology outcomes of biotic community is outstanding and by far an exemplar in the field. His study makes a significant contribution to understanding the possible negative effects of novel technological solutions and the ways to mitigate them. He has supervised 27 master's theses, 5 doctoral theses, and published close to 60 international research articles, 23 book chapters, and 3 edited books. His contribution in the field has led to the publication of multiple articles in high impact factor journals. Some of his contributions have also been extensively discussed in the policy documents such as "Guidance on the safety assessment of nanomaterials in cosmetics, prepared by Scientific Committee on Consumer Safety (SCCS), European union, 2019" and in "United Nations in the Asia-Pacific research and training network in Science, Technology & Innovation (STI) policies for sustainable development."

The impact of his research and his contributions to the field are also evident from the awards won by Ashutosh at this young age. He has won the "Indian National Science Academy (INSA) Medal for Young Scientist 2014" in the area of Health Sciences while pursuing his PhD. Similarly, he won the National Academy of Sciences India (NASI) – Young Scientist Platinum Jubilee Award (2015) in the field of Bio-medical, Molecular Biology and Biotechnology, DBT Young Investigator Award and vLife Science Best Publication Award.

Contributors

Dhiraj Devidas Bhatia
Biological Engineering Discipline, Indian Institute of Technology Gandhinagar, Palaj, Gujarat, India

Vandana Bijalwan
Division of Clinical Epidemiology, ICMR-National Institute of Occupational Health, Meghani Nagar, Gujarat, India

Mahendra Bishnoi
Food and Nutrition Biotechnology Division, National Agri-food Biotechnology Institute, Sector 81, SAS Nagar, Punjab, India

Smith C.
ICMR-National Institute of Occupational Health, Meghani Nagar, Ahmedabad-16, Gujarat, India

Santasabuj Das
Department of Pharmaceutics and Drug Delivery, University of Mississippi, Oxford, MS, USA

Ravinder Naik Dharavath
Food and Nutrition Biotechnology Division, National Agri-food Biotechnology Institute, Sector 81, SAS Nagar, Punjab, India

Rajnish G.
ICMR-National Institute of Occupational Health, Meghani Nagar, Ahmedabad-16, Gujarat, India

Ratna Ghosal
Biological & Life Sciences, School of Arts & Sciences, Ahmedabad University, Central Campus, Navrangpura, Ahmedabad, Gujarat, India

Shambhavi Jha
Nanomaterial Toxicology Laboratory, Systems Toxicology & Health Risk Assessment Group, CSIR-Indian Institute of Toxicology Research (CSIR-IITR), VishvigyanBhawan, 31, Mahatma Gandhi Marg, Lucknow, Uttar Pradesh, India

Anurag Jyoti
Amity Institute of Biotechnology, Amity University Madhya Pradesh, Maharajpura Dang, Gwalior

Nikhil P. K.
ICMR-National Institute of Occupational Health, Meghani Nagar, Ahmedabad-16, Gujarat, India

Krupa Kansara
Biological Engineering Discipline, Indian Institute of Technology Gandhinagar, Palaj, Gujarat, India

Praveen Kolimi
Department of Pharmaceutics and Drug Delivery, University of Mississippi, Oxford, MS, USA

Ashutosh Kumar
Biological & Life Sciences, School of Arts & Sciences, Ahmedabad University, Central Campus, Navrangpura, Ahmedabad, Gujarat, India

Onila Lugun
Nanomaterial Toxicology Laboratory, Systems Toxicology & Health Risk

Assessment Group, CSIR-Indian
Institute of Toxicology Research
(CSIR-IITR), Vishvigyan Bhawan,
31, Mahatma Gandhi Marg, Lucknow,
Uttar Pradesh, India

Tejal G. M.
ICMR-National Institute of
Occupational Health, Meghani Nagar,
Ahmedabad-16, Gujarat, India

Suman Mallick
Biological & Life Sciences, School
of Arts & Sciences, Ahmedabad
University, Central Campus,
Navrangpura, Ahmedabad,
Gujarat, India

Abdulkhaik Mansuri
Biological & Life Sciences, School
of Arts & Sciences, Ahmedabad
University, Central Campus,
Navrangpura, Ahmedabad,
Gujarat, India

Dhairya Mazmudar
Biological & Life Sciences, School
of Arts & Sciences, Ahmedabad
University, Central Campus,
Navrangpura, Ahmedabad,
Gujarat, India

Simran Nasra
Biological & Life Sciences, School
of Arts & Sciences, Ahmedabad
University, Central Campus,
Navrangpura, Ahmedabad,
Gujarat, India

Sivaperumal P.
ICMR-National Institute of
Occupational Health, Meghani Nagar,
Ahmedabad-16, Gujarat, India

Alok Kumar Pandey
Nanomaterial Toxicology Laboratory,
Systems Toxicology & Health Risk
Assessment Group, CSIR-Indian
Institute of Toxicology Research
(CSIR-IITR), Vishvigyan Bhawan,
31, Mahatma Gandhi Marg, Lucknow,
Uttar Pradesh, India

Alok Pandya
Department of Engineering and Physical
Sciences, Institute of Advanced
Research, Gandhinagar, India

Keya Patel
Department of biological and Life
sciences, School of arts and sciences,
Ahmedabad University, Ahmedabad,
Gujarat, India

Shirali Patel
Department of biological and Life
sciences, School of arts and sciences,
Ahmedabad University, Ahmedabad,
Gujarat, India

Harsh Prajapati
Department of Engineering and Physical
Sciences, Institute of Advanced
Research, Gandhinagar, India

Dixit Raiyani
Biological & Life Sciences, School
of Arts & Sciences, Ahmedabad
University, Central Campus,
Navrangpura, Ahmedabad,
Gujarat, India

Shaikh S.
ICMR-National Institute of
Occupational Health, Meghani
Nagar, Ahmedabad-16,
Gujarat, India

Contributors

Shrushti Shah
Biological & Life Sciences, School of Arts & Sciences, Ahmedabad University, Central Campus, Navrangpura, Ahmedabad, Gujarat, India

Neha Shrivastava
Amity Institute of Biotechnology, Amity University Madhya Pradesh, Maharajpura Dang, Gwalior, India

Vikas Shrivastava
Amity Institute of Biotechnology, Amity University Madhya Pradesh, Maharajpura Dang, Gwalior, India

Ritesh K. Shukla
Biological and Life Sciences, School of Arts and Sciences, Ahmedabad University, Ahmedabad, Gujarat, India

Dhirendra Pratap Singh
Division of Toxicology, ICMR-National Institute of Occupational Health, Meghani Nagar, Gujarat, India

Raghunath Singh
Schizophrenia Division, Centre for Addiction and Mental Health, 250 College Street, Toronto, ON, Canada

Richa Singh
Nanomaterial Toxicology Laboratory, Systems Toxicology & Health Risk Assessment Group, CSIR-Indian Institute of Toxicology Research (CSIR-IITR), VishvigyanBhawan, 31, Mahatma Gandhi Marg, Lucknow, Uttar Pradesh, India

Kriti Srivastava
Department of Engineering and Physical Sciences, Institute of Advanced Research, Gandhinagar, India

Rupal T.
ICMR-National Institute of Occupational Health, Meghani Nagar, Ahmedabad-16, Gujarat, India

Rajesh Singh Tomar
Amity Institute of Biotechnology, Amity University Madhya Pradesh, Maharajpura Dang, Gwalior, India

Kamayani Vajpayee
Biological and Life Sciences, School of Arts and Sciences, Ahmedabad University, Ahmedabad, Gujarat, India

Nidhi Verma
Department of Engineering and Physical Sciences, Institute of Advanced Research, Gandhinagar, India

Gajendra Singh Vishwakarma
Department of Biological Science and Biotechnology, Institute of Advanced Research, Gandhinagar, India

1 Environmental Paradigm Change and Risk Assessment

Dhairya Mazmudar, Dixit Raiyani, and Ashutosh Kumar

CONTENTS

1.1 Introduction 1
1.2 Risk Assessment 2
1.3 Uncertainties That May Hamper the Process of Evaluation of Risk 3
1.4 Historical Perspective and Current Approaches 3
    1.4.1 General Outline of Environmental Risk Assessment in the European Union 4
    1.4.2 General Outline of Environmental Risk Assessment in United States 5
1.5 Tools Used for Environmental Risk Assessment 5
    1.5.1 European Union System for the Evaluation of Substances (EUSES) 5
    1.5.2 European Centre for Ecotoxicology and Toxicology of Chemicals (ECETOC) Targeted Risk Assessment (TRA) 6
    1.5.3 EcoFate 6
    1.5.4 Exposure Model for Soil-organic Fate and Transport (EMSOFT) 7
1.6 Future Trends and Challenges 7
Acknowledgments 8
References 8

1.1 INTRODUCTION

Risk is defined as the chances of uncertain or undesired events occurring. It includes the possible consequences of such events, and the extent to which they can occur. When we look towards various criteria to measure risk, we consider its cause – at what concentrations would a substance be harmful to the environment and surrounding life, and what would be the time of exposure. As we know, some substances are not harmful at low concentrations but may pose a risk when their exposure levels increase. In similar fashion, when exposure time of some substances is low, they may not cause damage but prolonged exposure may be harmful (Bartell, 2008).

Ecological risk is defined as the cause of undesired events that lead to the damaging of ecological systems or the disruption of balance in those ecosystems. It includes various toxic effects that may be caused by physical agents, biological agents, and other foreign particles. Such harmful agents can deteriorate entire ecosystems or cause extensive damage to biodiversity. Global warming is an example of ecological risk caused by human activities (Bartell, 2008). Risk assessment is used in several fields such as medical science, marine biology, environmental science, and engineering. It is mainly used to evaluate or to determine environmental risk, which is majorly caused by human activities (Manuilova, 2003). There are multiple approaches and methods that can be used to define specific risk. Commonly, risk is defined as the mixture of probability of occurrence of an event, frequency of occurrence, associated hazard, and magnitude of effect.

Let's look at a simple example to understand the risk associated with any substance. We consume table salt in our day-to-day life; we know that salt at low concentrations is not harmful for human life, but what if we consume spoonfuls of salt every day? It would lead to an increase in blood pressure (Maneton et al., 2005; He & MacGregor, 2009), which would lead to a higher risk of cardiovascular disease (Cook et al., 2007) and renal disease (Heeg et al., 1989). Thus, concentration or amount of any substance along with the exposure time can be used to define risk. Similarly, concentrated acids are severely corrosive and can cause possibly fatal burns to humans, but in greatly diluted forms they are benign. Exposure time is also important when assessing risk. Overall risk of a substance is assessed as a combination of probability of an exposure event, and the hazardous effect of that particular chemical or substance, the concentration which enters the environment and exposure levels in different environmental compartments (Manuilova, 2003).

1.2 RISK ASSESSMENT

Risk assessment is a scientific technique used to assess risk caused by hazardous material. The estimation of risk can be done both qualitatively and quantitatively. Risk may arise during the manufacture of chemicals or during distribution, or may be at the time of recycling or disposal. There are many possible routes from which risk may arise. The process of risk assessment is carried out to determine the effects of potentially harmful material on humans as well as ecosystems and other organisms. Risk assessment related to the environment is known as environmental or ecological risk assessment (Manuilova, 2003).

Ecological Risk Assessment (ERA) is a toxicological test that is performed to evaluate various ecological effects caused by different types of stressors. ERA mainly estimates potential adverse effects of various stressors and their risk to the ecosystem. It measures the risk level of a specific stressor and how harmful it can be for the environment. It evaluates adverse effects caused by undesirable agents by calculating its concentration, its reaction with the environment, and its exposure duration. It involves models of survival rate, growth, and reproduction of various species present in the ecosystem, and includes tests for the life expectancy of organisms. It mainly includes evaluation of risk factors and how we can manage such risks and focus on conservation of environment. ERA is a combined result of the principles of toxicology

Environmental Paradigm Change and Risk Assessment

and environmental science in which the former measures toxicity of stressors present in the environment and the latter focuses on conservation of resources present in that ecosystem. ERA is mainly designed to cover toxic chemicals present in an environment.

Several kinds of human activities result in risk for the environment and belonging species. Humans produce many toxic chemicals in industries and other sectors, and such harmful toxic substances often leach out into the environment. We know that with the increases in the world's population, demand for food has increased exponentially (Alexandratos & Bruinsma, 2012) and to meet such high demand, use of fertilizers increases (Ritchie & Roser, 2013), which pollutes soil and water bodies (Viets & Lunin, 1975). There is a likelihood that such harmful agents may enter into the food chain and lead to biomagnification. This increased use of fertilizers and pesticides to meet larger production of food can lead to degradation or disruption of the environment and such harmful effects may create risk for the ecosystem and its species.

1.3 UNCERTAINTIES THAT MAY HAMPER THE PROCESS OF EVALUATION OF RISK

There are several factors that affect the process of risk assessment such as

(1) *Availability of Information:* There is a need for primary and basic data in order to evaluate risk and to make precise assumptions. Lack of data or insufficient information may result in uncertainty.
(2) *Measurement errors:* Errors in observations, instruments used for measurements, and human handling may lower the accuracy of assessment.
(3) *Environmental conditions:* Base data may differ due to variability in climatic conditions and types of natural resources such as soil and terrain composition. Additionally, the biodiversity and structure of the ecosystem can also bring variability to the scope of the assessment. Another major cause for uncertainty is the difference between natural environment and lab conditions.
(4) *Mechanism inadequacies:* Lack of prior knowledge about any tools or models used may cause unforeseen anomalies in results and can cause uncertainty. Failing to account for all available stressors and the probable responses of different species, instability in conditions, or parameters used can also create various challenges.

Therefore, we can say that the process of risk assessment changes as we move from place to place and from environment to environment. As conditions change from field to lab, the availability of data can also change (Manuilova, 2003).

1.4 HISTORICAL PERSPECTIVE AND CURRENT APPROACHES

Ecological risk assessment is a blend of two fields, risk assessment and ecological assessment, and it is important to learn the history of both. Quantitative risk assessment developed as a base for regulatory decision-making on chemicals only about 40 years ago (Paustenbach, 1995; Van Leeuwen and Hermens, 1995). Risk assessment as an idea first developed in 17th century in England and the Netherlands. It was carried out for the insurance of merchant ships. Later on, from the context of insurance, risk assessment then spread to the safety of people, property, and also health risk assessment. In the European Union (EU), environmental policy started in 1973 with the adoption of the first five-year European Community Environmental Action.

The context of environmental or ecological risk assessment was first formed in the United States in the 1970s, which was initially developed with the aim of forming a series of environmental laws such as The Clean Air act of 1970, Federal Insecticide, Fungicide, and Rodenticide Act of 1972, and the Clean Water Act of 1977. Over the years, the approaches and regulations surrounding environmental assessment and policy have changed, with numerous new laws, amendments to existing guidelines, and standardization of procedural outlines.

Currently, two major approaches are in practice based on the region. The EU plans and carries out assessments according to the Technical Guidance Document on Risk Assessment for New and Existing Substances (TGD, 2003). However, the United States follows the guidelines defined by the United States Environmental Protection Agency (US EPA).

1.4.1 GENERAL OUTLINE OF ENVIRONMENTAL RISK ASSESSMENT IN THE EUROPEAN UNION

In the EU the risk assessment process begins with the identification of hazard and creation of a *Hazard Identification* report. This document indicates the potential adverse effects induced by the chemical. Further, a *Dose-response Assessment* is made, which is an estimate of the relationship between dosage/exposure levels and the probable severity of effect. It is very important for a hazardous material to induce a dose-dependent response in living systems. The next step in the risk assessment process is *Exposure Assessment*. Exposure assessment is a quantitative estimation of how much chemical or hazardous substance is interacting with various environmental compartments or organisms. This also determines source, route of exposure, and degradation pathways. The last step in the risk assessment process is *Risk characterization*. This is an estimation of possible effects and their severity in the event of exposure to a certain chemical in any environmental compartment. This involves the calculation of risk characterization ratio (RCR) (Verdonck et al., 2007; Valavanidis & Vlachogianni, 2015). It is the ratio of predicted environmental concentration (PEC) and predicted no-effect concentration (PNEC). PEC is the calculated concentration of a chemical in the environment. This calculation is based on modelling and third-party data (Peake et al., 2016). Subsequently, PNEC is the concentration of a pharmaceutical at which no pharmacological effect is expected to occur for a specific organism. A PNEC value is usually derived using ecotoxicity testing data. A risk characterization ratio that is ≥1 indicates potential risk of adverse effects and warrants further research, monitoring and implementation of risk reduction measures. Alternatively, a ratio less than 1 is read as no need of further research or testing (Verdonck et al., 2007).

1.4.2 GENERAL OUTLINE OF ENVIRONMENTAL RISK ASSESSMENT IN UNITED STATES

The United States follows a slightly different approach as compared to the EU. The first step in the risk assessment process is *Problem Formulation* (i.e., to define the purpose of the assessment and the problem), followed by formulation of a plan for risk characterization and analysis (National Research Council, 2014). The crux of this step is collating information on potential sources, effects, possible aggravators, and interactions with the ecosystem. This produces two endproducts – assessment endpoints and conceptual models. The next step is *Analysis* of the data, where the gathered data is evaluated to determine possible routes of exposure. This step is also known as characterization of exposure and the prediction of potential ecological effects. This step generates two major profiles, one for response to the chemical and the other for exposure.

The final step in risk assessment is *Risk characterization* (US EPA, 2014), which is very similar to the EU approach to risk assessment. Both the profiles (characterization of exposure and the prediction of potential ecological effects) generated during the previous step are integrated via a risk estimation process. The characterization process also takes into account any assumptions, scientific uncertainties, and limitations of analytical tools. The resulting end product is a complete report of profile integration along with potential ecological burdens, scientific evidence, and unavoidable uncertainties (Fowle & Dearfield, 2000).

1.5 TOOLS USED FOR ENVIRONMENTAL RISK ASSESSMENT

Researchers, industries, and governmental agencies make use of several tools and computer modelling software in order to carry out environmental risk assessment. In the following section some of the most widely used and effective tools and programs that have been developed for making risk assessments are discussed in detail.

1.5.1 EUROPEAN UNION SYSTEM FOR THE EVALUATION OF SUBSTANCES (EUSES)

EUSES is a computer program for assessment of risk posed by new and existing chemicals to the environment (Vermeire et al., 1997). It is a tool that serves as the European reference for authorities, companies, and researchers to carry out environmental exposure assessments in line with the Biocidal Products Regulation and the REACH (Registration, Evaluation, Authorisation and Restriction of Chemicals) Regulation *(European Chemicals Agency, ECHA)*. EUSES is the standard Tier I tool for exposure calculations of industrial chemicals (Spaniol et al., 2021).

EUSES requires the user to possess sufficient expertise in order to properly evaluate the input data, make appropriate data selection, to understand certain assumptions that are made and the inherent limitations of estimation methods, and to finally, interpret the results (Vermeire et al., 1997). Some of the data parameters that EUSES uses in order to perform the assessment include (Manuilova, 2003):

1. Tonnage substance produced in EU
2. Tonnage substance used in EU

3. Molecular weight
4. Vapor pressure
5. Log octanol-water partition coefficient
6. Water solubility
7. Aquatic toxicity
8. Biodegradability

Risk assessment is carried out in three steps:

1. Exposure assessment – creates estimates for doses/concentrations that different compartments can be exposed to
   a. Emission module – factors in basic properties, applications, and emission vectors to virus life cycle stages
   b. Distribution module – contains models to estimate all forms of distribution
   c. Exposure module – calculates exposure to various mammals, birds, and humans
2. Effects assessment
   a. Hazard identification – identification of inherent harmful effects of the chemical
   b. Dose-response assessment – comparison of exposure and risk levels
3. Risk characterization – estimation of the incidence and severity of effects likely to occur in environmental compartments due to exposure

1.5.2 European Centre for Ecotoxicology and Toxicology of Chemicals (ECETOC) Targeted Risk Assessment (TRA)

ECETOC TRA (Ecetoc, 2004) is a software based on the principles of EUSES and other established software. The program combines occupational, consumer, and environmental exposure aspects in a web-based tool (Urbanus et al., 2020). It brings together established exposure models, risk assessment techniques, and standard risk/safety concepts in a structured framework, making it a suitable tool for assessing a vast range of chemicals (Zaleski, 2008). It is useful for estimation of worker, consumer, and environmental exposure by means of three separate tool modules.

1.5.3 EcoFate

This is an assessment software package developed by researchers from the Environmental Toxicology Research Group at Simon Fraser University, Canada. It is a model created using Visual Basic for the Windows platform. The main purpose of the model is to predict the environmental fate and bioaccumulation on an ecosystem level (Gobas, 1995).

The model is particularly useful for ecosystem-based risk assessments of chemical emissions in freshwater and marine aquatic ecosystems. It evaluates point and non-point sources in aquatic systems such as lakes, rivers, and marine inlets. It also helps model the cumulative impact of chemical inputs by calculating contaminant concentrations in water, sediment, and biota of an entire ecosystem (Manuilova, 2003).

The overall software consists of a combination of an environmental fate, food-web bioaccumulation, toxicological hazard, and human health risk assessment models. These models are integrated to directly relate chemical emissions to concentrations, toxic effects, and human health risks.

1.5.4 Exposure Model for Soil-organic Fate and Transport (EMSOFT)

EMSOFT is a computer screening model based on the work of Jury et al. (1990) and is used to estimate the emission flux of volatile soil contaminants to ambient air. It is used to:

- Determine the amounts of contaminants that remain in the soil over a given time (initial concentration in soil is known)
- Quantify the mass flux (rate of transfer) of contaminants into the atmosphere over time
- Calculate concentrations of contaminants is air by inputting mass flux values into atmospheric dispersion models
- The EMSOFT further makes use of various input parameters. These parameters include soil fraction organic carbon (f_{oc}), temperature, porosity, and soil water content. EMSOFT uses a graphical user interface (US EPA, 2002). The graphic interface and limited number of parameters make EMSOFT easy to learn and use.

1.6 FUTURE TRENDS AND CHALLENGES

Current risk assessment approaches heavily rely on deterministic methods. The methods use "worst-case" values for exposure and toxicity, as well as comparing the resulting risk index to a standard threshold in order to deal with uncertainty and variability (Suciu et al., 2018). However, it is seen that in more complex conditions, such as environmental exposure analysis, probabilistic methods are more reliable. Deterministic methods helps to provide a more realistic basis for comparison of the accuracy of projections and alternatives (Mowrer, 2000). This is achieved by using distributions to quantify variability and uncertainty of the predicted risk (Frewer et al., 2008). One of the biggest challenges that needs to be addressed in the future is how to integrate uncertainty considerations into the assessment of exposure, effects, or both (Suciu et al., 2018).

The biggest hurdle for future developments in risk assessment is the development of approaches that can efficiently use and exploit the huge amount of existing available data of various toxicants and pollutants. The most advocated idea is a more integrated approach for risk assessment (Péry et al., 2013). The potential benefits of such an approach have been recognized for quite some time now, but no legal mandates exist yet (Suciu et al., 2018). An attempt to integrate environmental and human exposure assessment has already been made as part of the International Program on Chemical Safety (IPCS) of the World Health Organization (WHO), the European Commission (EC), the Organization for Economic Cooperation and Development (OECD), and the US Environmental Protection Agency (US EPA) (WHO Organization, 2001).

Unfortunately results haven't been as expected, with each approach (EEA and HEA) developing individually with poor linkages between them (Ciffroy et al., 2016).

It seems that integrating scientific disciplines with sociobehavioural and socio-economic lines of enquiry is the path forward (O'Brien, 2012). This may initially increase the complexity of the risk assessment process, but in the long run it will provide a better and more useful estimation of the risk. It will also reinforce transparency and drive a more efficient use of risk management resources (Calliera et al., 2016).

ACKNOWLEDGMENTS

Financial assistance by The Gujarat Institute for Chemical Technology (GICT) for the Establishment of a Facility for environmental risk assessment of chemicals and nanomaterials is acknowledged.

REFERENCES

Alexandratos, N. (n.d.). *World Agriculture towards 2030/2050: The 2012 Revision*. 154.

Bartell, S. M. (2008). Ecological Risk Assessment. In S. E. Jørgensen & B. D. Fath (Eds.), *Encyclopedia of Ecology* (pp. 1097–1101). Academic Press. https://doi.org/10.1016/B978-008045405-4.00387-6

Breure, A. M., Jager, D. T., van de Meent, D., Mulder, C., Rutgers, M., Schouten, A. J., Sterkenburg, A., Struijs, J., van Beelen, P., Vonk, M., & de Zwart, D. (n.d.). *Ecological Risk Assessment of Environmental Stress*. 15, 37–68.

Calliera, M., Marchis, A., Sacchettini, G., & Capri, E. (2016). Stakeholder consultations and opportunities for integrating socio-behavioural factors into the pesticide risk analysis process. *Environmental Science and Pollution Research International*, 23(3), 2937–2947. https://doi.org/10.1007/s11356-015-5553-9

Ciffroy, P., Péry, A. R. R., & Roth, N. (2016). Perspectives for integrating human and environmental exposure assessments. *The Science of the Total Environment*, 568, 512–521. https://doi.org/10.1016/j.scitotenv.2015.11.083

Cook, N. R., Cutler, J. A., Obarzanek, E., Buring, J. E., Rexrode, K. M., Kumanyika, S. K., Appel, L. J., & Whelton, P. K. (2007). Long term effects of dietary sodium reduction on cardiovascular disease outcomes: Observational follow-up of the trials of hypertension prevention (TOHP). *BMJ (Clinical Research Ed.)*, 334(7599), 885–888. https://doi.org/10.1136/bmj.39147.604896.55

Environmental Risk Assessment: Approaches, Experiences and Information Sources – European Environment Agency. (n.d.). Retrieved October 10, 2021, from www.eea.europa.eu/publications/GH-07-97-595-EN-C2/riskindex.html

EUSES – European Union System for the Evaluation of Substances – ECHA. (n.d.). Retrieved October 10, 2021, from https://echa.europa.eu/support/dossier-submission-tools/euses

Fowle, J. R., & Dearfield, K. L. (2000). *Risk Characterization: Handbook: (519222012-001)* [Data set]. American Psychological Association. https://doi.org/10.1037/e519222012-001

Frewer, L. J., Fischer, A. R. H., van den Brink, P. J., Byrne, P., Brock, T., Brown, C., Crocker, J., Goerlitz, G., Hart, A., Scholderer, J., & Solomon, K. (2008). Potential for the adoption of probabilistic risk assessments by end-users and decision-makers. *Human and Ecological Risk Assessment: An International Journal*, 14(1), 166–178. https://doi.org/10.1080/10807030701790355

He, F. J., & MacGregor, G. A. (2009). A comprehensive review on salt and health and current experience of worldwide salt reduction programmes. *Journal of Human Hypertension, 23*(6), 363–384. https://doi.org/10.1038/jhh.2008.144

Heeg, J. E., de Jong, P. E., van der Hem, G. K., & de Zeeuw, D. (1989). Efficacy and variability of the antiproteinuric effect of ACE inhibition by lisinopril. *Kidney International, 36*(2), 272–279. https://doi.org/10.1038/ki.1989.190

Leeuwen, C. J. van, & Hermens, J. L. M. (Eds.) (1995). *Risk Assessment of Chemicals: An Introduction.* Springer Netherlands. https://doi.org/10.1007/978-94-015-8520-0

Manuilova, A. (2003). *Methods and Tools for Assessment of Environmental Risk. 21.*

Meneton, P., Jeunemaitre, X., de Wardener, H. E., & Macgregor, G. A. (2005). Links between dietary salt intake, renal salt handling, blood pressure, and cardiovascular diseases. *Physiological Reviews, 85*(2), 679–715. https://doi.org/10.1152/physrev.00056.2003

Mowrer, H. T. (2000). Uncertainty in natural resource decision support systems: Sources, interpretation, and importance. *Computers and Electronics in Agriculture, 27*(1/3), 139–154.

National Research Council (2014). *A Framework to Guide Selection of Chemical Alternatives.* https://doi.org/10.17226/18872

O'Brien, K. (2012). Global environmental change II: from adaptation to deliberate transformation. *Progress in Human Geography, 36*(5), 667–676. https://doi.org/10.1177/0309132511425767

Parkes, M., Panelli, R., & Weinstein, P. (2003). Converging paradigms for environmental health theory and practice. *Environmental Health Perspectives, 111*(5), 669–675. https://doi.org/10.1289/ehp.5332

Paustenbach, D. J. (1995). The practice of health risk assessment in the united states (1975–1995): How the U.S. and other countries can benefit from that experience. *Human and Ecological Risk Assessment: An International Journal, 1*(1), 29–79. https://doi.org/10.1080/10807039509379983

Peake, B. M., Braund, R., Tong, A. Y. C., & Tremblay, L. A. (2016). Impact of pharmaceuticals on the environment. In *The Life-Cycle of Pharmaceuticals in the Environment* (pp. 109–152). Elsevier. https://doi.org/10.1016/B978-1-907568-25-1.00005-0

Péry, A. R. R., Schüürmann, G., Ciffroy, P., Faust, M., Backhaus, T., Aicher, L., Mombelli, E., Tebby, C., Cronin, M. T. D., Tissot, S., Andres, S., Brignon, J. M., Frewer, L., Georgiou, S., Mattas, K., Vergnaud, J. C., Peijnenburg, W., Capri, E., Marchis, A., & Wilks, M. F. (2013). Perspectives for integrating human and environmental risk assessment and synergies with socio-economic analysis. *The Science of the Total Environment, 456–457,* 307–316. https://doi.org/10.1016/j.scitotenv.2013.03.099

Ritchie, H., & Roser, M. (2013). *Fertilizers.* Our World in Data. https://ourworldindata.org/fertilizers

Spaniol, O., Bergheim, M., Dawick, J., Kötter, D., McDonough, K., Schowanek, D., Stanton, K., Wheeler, J., & Willing, A. (2021). Comparing the European Union System for the Evaluation of Substances (EUSES) environmental exposure calculations with monitoring data for alkyl sulphate surfactants. *Environmental Sciences Europe, 33*(1), 3. https://doi.org/10.1186/s12302-020-00435-1

Suciu, N. A., Panizzi, S., Ciffroy, P., Ginebreda, A., Tediosi, A., Barceló, D., & Capri, E. (2018). Evolution and future of human health and environmental risk assessment. In P. Ciffroy, A. Tediosi, & E. Capri (Eds.), *Modelling the Fate of Chemicals in the Environment and the Human Body* (Vol. 57, pp. 1–21). Springer International Publishing. https://doi.org/10.1007/978-3-319-59502-3_1

Targeted Risk Assessment (TRA). (n.d.). *Ecetoc.* Retrieved October 10, 2021, from www.ecetoc.org/tools/targeted-risk-assessment-tra/

TR 093 – Targeted Risk Assessment. (n.d.). *Ecetoc*. Retrieved October 10, 2021, from www.ecetoc.org/publication/tr-093-targeted-risk-assessment/

Urbanus, J., Henschel, O., Li, Q., Marsh, D., Money, C., Noij, D., van de Sandt, P., van Rooij, J., & Wormuth, M. (2020). The ECETOC-targeted risk assessment tool for worker exposure estimation in REACH registration dossiers of chemical substances – current developments. *International Journal of Environmental Research and Public Health*, *17*(22), 8443. https://doi.org/10.3390/ijerph17228443

U.S. EPA (2002). *Emsoft User's Guide and Modeling Software (2002 Update)*. U.S. Environmental Protection Agency, Office of Research and Development, National Center for Environmental Assessment, Washington Office, Washington, DC, NCEA/W-0073, NCEA/W-0073, 2002.

US EPA, O. (August 18, 2014). *Ecological Risk Assessment* [Data and Tools]. www.epa.gov/risk/ecological-risk-assessment

US EPA, O. (April 23, 2015). *Exposure Assessment Tools by Media – Soil and Dust* [Collections and Lists]. www.epa.gov/expobox/exposure-assessment-tools-media-soil-and-dust

Valavanidis, A., & Vlachogianni, T. (2015). *Ecotoxicity Test Methods and Ecological Risk Assessment. Aquatic and Terrestrial Ecotoxicology Tests under the Guidelines of International Organizations*. Website: www.chem-tox-ecotox.org, *1*, 1–29.

Verdonck, F., Souren, A. F. M. M., Asselt, M., Van Sprang, P., & Vanrolleghem, P. (2007). Improving uncertainty analysis in European Union Risk Assessment of Chemicals. *Integrated Environmental Assessment and Management*, *3*, 333–343. https://doi.org/10.1002/ieam.5630030304

Vermeire, T. G., Jager, D. T., Bussian, B., Devillers, J., den Haan, K., Hansen, B., Lundberg, I., Niessen, H., Robertson, S., Tyle, H., & van der Zandt, P. T. (1997). European Union System for the Evaluation of Substances (EUSES). Principles and structure. *Chemosphere*, *34*(8), 1823–1836. https://doi.org/10.1016/s0045-6535(97)00017-9

Viets, F. G., & Lunin, J. (1975). The environmental impact of fertilizers. *C R C Critical Reviews in Environmental Control*, *5*(4), 423–453. https://doi.org/10.1080/10643387509381630

Zaleski, R. (2008). ECETOC Targeted Risk Assessment Screening Tool. *Epidemiology*, *19*(6), S59. https://doi.org/10.1097/01.ede.0000339710.15816.9b

Zhang, X., & Gobas, F. A. P. C. (1995). *ECOFATE: A user-friendly environmental fate, bioaccumulation and ecological risk assessment model for contaminants in marine and freshwater aquatic ecosystems – Applications and validation* (No. CONF-9511137-). Society of Environmental Toxicology and Chemistry, Pensacola, FL (United States).

2 Environmental and Ecotoxicological Impacts of Engineered Nanomaterials

Neha Shrivastava, Vikas Shrivastava, Rajesh Singh Tomar, and Anurag Jyoti

CONTENTS

2.1 Introduction 11
2.2 Regulation of Nanomaterials Risk Assessment..13
2.3 Toxicity of Nanomaterials 14
2.4 Role of Physicochemical Properties in Nanoparticles Toxicity 14
    2.4.1 Shape 14
    2.4.2 Size 15
    2.4.3 Surface Charge 16
    2.4.4 Morphology 16
    2.4.5 Degree of Agglomeration 16
2.5 Toxicity Mechanism in Natural Ecosystem Induced by Nanoparticles.........16
2.6 Toxicity Mechanism in Plants Induced by Nanoparticles 18
    2.6.1 Effect of Nanoparticles on Crop Quality 18
2.7 Mechanism of Plant-nanoparticles Interaction...18
2.8 Nanoparticles Exposure to Humans via Potential Routes and Toxicity 20
    2.8.1 Gastrointestinal Tract (GIT) 21
    2.8.2 Respiratory Tract 22
2.9 Dermal Penetration...23
2.10 Conclusion..24
Acknowledgement 24
References...24

2.1 INTRODUCTION

Fast development of nanotechnology has significant effects on society, economy, and environment. Nanotechnology promises potential applications in nanomedicine for the detection, prevention, and treatment of disease. The United States National Nanotechnology Initiative (USNNI) defines nanotechnology as "The understanding and control of particles at one dimension in the size between1 to 100 nm." On the other hand, nanoparticles (NPs) are "nano-objects with all three dimensions in the nanoscale," where the size ranges between 1 to 100 nm (ISO, 2008) (Bawa, 2016: Technical specification, 2008). This technology has revolutionized world by introducing novel materials in industrial and consumer products in various areas (NNI, 2017; NNI, 2014). NPs exist in various shapes and sizes. They also have unique physico-chemical properties having compatibility with biological systems that are different from their bulk partners. Due to their small size and high surface area they have greater stability, strength, physical, chemical, and biological activity. Engineered nanoparticles (ENPs) have several applications in medicine (Kahan et al., 2009), cosmetics (Morganti, 2010), food (Siegrist et al., 2008; Handford et al., 2014), agriculture (Handford et al., 2014; Mukhopadhyav, 2014), electronics, and environmental safety. In medicine, NPs have shown great promise in treatment efficacy and in drug delivery systems. Inorganic ceramic NPs like alumina, silica, and titania are used for administration of drug for cancer treatment, but due to their non-biodegradable nature their applications are limited. On the other hand, various metallic NPs such as cobalt, copper, iron, zinc, etc., have applications in the food industry and medicine (Yang et al., 2014). As per the records of nanotechnology-based consumer products, there are 1814 nonmaterial-containing products from 622 companies in 32 countries, along with metal oxides (TiO_2, Ag, ZnO, SiO_2, etc.) (Vance et al., 2015). SiO_2NPs are used in the food industry, healthcare products, and in daily use products such as detergents, toothpastes, and cosmetics (Hansen et al., 2008). TiO_2 and ZnONPs are used in commercial products like food additives, sunscreen, and paints (Chen and Mao, 2007). AgNPs are employed in antibacterial products, food, water disinfectants, diagnostic biosensors, textile industries, imaging probes, and conductive inks (Marin et al., 2015). AuNPs and QDs (quantum dots) are applicable in medical, photothermal therapy, bioimaging, and drug delivery systems (Biosselier and Astruc, 2009; Zhao and Zhu, 2016).

NPs can be originated from several sources and classified as engineered and incidental. Engineered NPs are man-made, designed during the synthesis process, whereas (Junam and Lead, 2008) incidental NPs originate from various sources like volcanoes, diesel particles, forest fires, etc. (Oberdorster et al., 2005). Metal-based nanomaterials such as nanogold, nanosilver, and metal oxides (e.g., titanium oxide) are often used in cosmetics, foods, and drug-related products. Carbon-based nanomaterials such as nanotubes and fullerenes are used in coatings, films, and electronics. The dendrimers are nanopolymers used in drug delivery. Composites, used in packaging material, such as nanoclays combine one NP with other.

In spite of the enormous applications of nanomaterials, their harmful effects on humans and the environment have also been reported. Human beings are often exposed to airborne nano-sized particles. Various natural nano-sized particles like dust, storms, volcanic eruptions, etc., are present on the earth since through several natural processes. Extensive burning of fossil fuels and rapid industrial also result in increased numbers of NPs that have posed manifold risk to the environment (Handy & Owen et al., 2008). In the present decade, the most frequently used nanomaterials are fullerene, silver, zinc oxide, carbon nanotubes, silica, and titanium dioxide in household appliances, tableware, cleaning agents, children's toys, and clothing (Trojanowski and Fthenakis, 2019; Nowack and Bucheli, 2007). AgNP-based disinfectants have received much attention due to their enormous applications (Marambio-jones and Hoek, 2010). Sunscreen contains ZnO- and TiO$_2$NPs that generate free radicals in the presence of light, which are harmful for the skin (Deshmukh et al., 2019). TiO$_2$NPs are also extensively used on a daily basis in cosmetics, food products and packaging, colorants in paints, and antibacterial agents (Colvin, 2003; Frohlich and Roblegg, 2012). Annual production of these is > 200,000 t (Runa et al., 2018; Mineral and Commodity Summaries, 2015). The FDA has set the upper limit of application of these NPs in sunscreen and food products of up to 25 % (w/w) and 1 % (w/w), respectively (Jovanovic, 2015; Food and Drug Administration 1999.) NPs present in food can enter the digestive tract and lymphatic vessels as compared to bulk material. Further they can easily get spread into other organs and tissues (Hund-Rinke, 2006).

Various chemical industries are located near aquatic ecosystems. Industries discharge their effluent and hazardous substances in oceans, lakes, or rivers. A large concentration of toxic NPs like TiO2, Ag, ZnO, carbon nanotubes, titanium dioxide, SiO2, etc., are discharged into aquatic bodies each year. These NPs severely impact the water quality and pose toxicity on flora and fauna. In rural-urban settings, the contaminated water is used for irrigation purposes. This ultimately leads to the absorption of toxic NPs in soil. In addition to this, toxic NPs also get incorporated into plants. Upon consumption of these agricultural products, toxic NPs get internalized into the human body and cause severe diseases like damage of central nervous system, asthma, cardiac diseases, cancer, degenerative diseases, and many more.

The toxic effects of NPs on environment and human health are a serious problem (Hegde et al., 2016). Over the past few years a lot of emphasis has been placed on understanding the impact of toxic NPs on health and environment (Larguinho et al., 2014; Oberdorster et al., 2008, 2005). Different nanomaterials have different mechanisms and levels of toxicity. A number of factors are responsible for causing the toxicity. Particle size, surface chemistry, compatibility with biorecognition elements, etc., play major roles in the toxicity of NPs. Other parameters like zeta potential and surface charge are crucial features for providing mechanistic details of uptake and biological toxicity of NPs inside living cells. In urban settings, a major source of NPs is through road transport (60%), while 23% is derived from residential incineration and industrial processes (Oberdorster, 2010). Different trophic organisms are usually selected to evaluate the ecotoxicity and uptake of NPs, including algae, mammals, plants, crustaceans, and microbes (Hoyt and Mason, 2008).

2.2 REGULATION OF NANOMATERIALS RISK ASSESSMENT

The United States and Europe have developed various evaluation approaches for the safety regulation of nanomaterials. The United States Environmental Protection Agency (USEPA) regulates the exploitation of NPs in the United States, in cooperation with other federal government agencies. Chemical organization is ruled by the "Registration, Evaluation, Authorization and Restriction of Chemicals" (REACH) regulation in Europe (Hegde et al., 2016). Moreover, the Europe member states, in collaboration with the Chemical Agency, have published a sequence of documents regarding the use of nanomaterials in industries, enriching the registration of nanomaterials in REACH (European Commission Web Resource).

2.3 TOXICITY OF NANOMATERIALS

The increasing frequency of nano-based products in consumer goods has also bought nanotoxicity. Accumulation of NPs in different organs and unfavorable side effects have hindered their use. Nanotoxicity occurs due to generation of reactive oxygen species (ROS), resulting in the successive development of oxidative stress in cells and tissues. Oxidative stress is also produced by secretion of pro-inflammatory mediators through cascade pathways such as (PI3-K) phosphoinosotode 3- kinase, NF-kB (Nuclear Factor-kB), and (MAPK) mitogen-activated protein kinase (Pojiak-Blazi et al., 2010, Rim et al., 2013). The toxicity of NPs has been reported in various organisms such as humans, rodents, and aquatic species like catfish (Wang et al., 2011), zebrafish (Ozel et al., 2013), algae (Wang, Z et al., 2011), and macrophages (Sohaebuddin et al., 2010). Adverse effects of single-walled carbon nanotubes (SWCNTs) have been observed in human bronchi and kidney, where they were shown to stimulate cell apoptosis and reduce cell adhesion due to either upregulating genes, resulting cell death or downregulating genes related with cell proliferation (Alazzam et al., 2010). Engineered carbon and metallic NPs were observed to stimulate the *in vitro* aggregation of platelets, which further increases the vascular thrombosis in rat carotid artery (Radomski et al., 2005). Solid NPs enter into the skin through creams and lotions containing NPs. TiO_2, Al_2O_3, and ZnONPs accumulate in the brain and can cause auxiliary toxicity, interrupting normal metabolism of neurotransmitters, ultimately leading to brain damage (Poli et al., 2004).

2.4 ROLE OF PHYSICOCHEMICAL PROPERTIES IN NANOPARTICLES TOXICITY

The toxicity of NPs is enhanced by various factors. Nanomaterials use different mechanisms of toxicity, and some lack a clear mechanism (Seung et al., 2015). Different parameters such as shape, size, surface charge, degree of agglomeration, chemical properties, dose, solubility, and translocation have their role in NP toxicity (Figure 2.1) (Ferreira et al., 2012; Qiyu et al., 2010; Wu Y-N et al., 2011).

2.4.1 Shape

The shape of the NPs is an important parameter. It plays a fundamental role in the trafficking as well as uptake pathways of NPs. Different-shaped NPs have different thickness, surface area, and degradation rates (Panyam et al., 2003). Spherical NPs have advantages over non-spherical NPs (Moghimi et al., 2001). Cristina et al. reported that different shapes and diameter of nickel (Ni) NPs can cause toxicity to zebrafish embryos. Aggregated particles showed higher toxicity (LD10 and LD50) due to shape and aggregation even with a similar composition and synthesis method (Ispas et al., 2009). Chithrani et al. investigated the effect of colloidal AuNPs on the uptake of

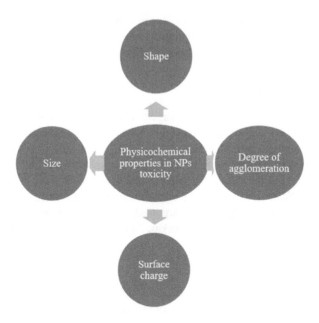

FIGURE 2.1 Parameters affecting uptake of NPs.

HeLa cells (Chithrani et al., 2006). The results demonstrated that rod-shaped AuNPs had low uptake whereas, spherical AuNPs had five-fold higher uptake. They also studied the uptake efficiency of spherical and rod-shaped transferrin-coated AuNPs on three different cell lines: HeLa cells, STO cells, and SNB19 cells (Chithrani and Chan, 2007). They observed that uptake of spherical AuNPs by all the cell lines was at a higher rate than for rod-shaped AuNPs. In another study, cell uptake efficiency was investigated by employing different shapes of NPs, disc-shaped, rod-shaped polystyrene (PS), and spherical on Caco-2 cells, and they found that disc- and rod-shaped NPs were internalized two-fold higher than spherical shape NPs (Banerjee et al., 2016).

2.4.2 Size

Size plays an important factor in determining the efficiency of cellular uptake and toxicities, when NPs interact with biological systems (Foster et al., 2001; Zhu et al., 2012 and Nel et al., 2009). Particle size is inversely proportional to the volume and surface area and enhances the interaction with biological systems. NPs between 250 nm to 3000 nm size have shown excellent *in vitro* phagocytosis, while size range between 120–150 nm are internalized through clathrin- or caveolin-mediated endocytosis (Panariti et al., 2012; Reiman et al., 2004). NP distribution is also size dependent (De Jong et al., 2008). Many studies have shown that for cellular uptake of NPs, optimum size of 50 nm of NPs is internalized more proficiently with a higher uptake rate. Uptake of NPs has been shown to be lower for small 15–30 nm size particles or about 70-240 nm for larger particles (Chithrani and Chan, 2007; Jin et al., 2009; Wang et al.,

2010). Additionally, NPs smaller than 20 nm can be easily absorbed by the small intestinal epithelial cells and further translocate to other organs by capillaries and lymphatic system within a very short period (Hoet et al., 2004).

2.4.3 SURFACE CHARGE

Based on particle kinetics, negatively charged cell membrane promotes the uptake of positively charged NPs. Positively charged NPs have higher internalization than negatively and neutrally charged NPs (Hoet et al., 2004; Marano et al., 2011). The uptake of positively charged NPs are more cytotoxic than negatively charged NPs, which may disturb the integrity of the cell and induce cell death (Schaeubl et al., 2011; Baek et al., 2011), but few scientists have proved that positively charged NPs like gold, silver, and superparamagnetic iron oxide particles are used at a more prominent level than negatively charged particles (Thorek and Tsourkas, 2008; Cho and Caruso, 2005).

2.4.4 MORPHOLOGY

The morphology of NPs is also a big concern in nanotoxicology. Many studies have been shown that carbon nanotubes are more toxic than other silica dust or ultra-fine carbon black. Inhaled fibers (e.g., asbestos) and nano-scaled fibers (e.g., carbon nanotubes) have a serious risk of lung inflammation and also delayed exposure may cause several types of cancers (Cha and Myung, 2007; Huczko et al., 2001; Firme and Bandaru, 2010). In industries most of the workers are exposed to SWCNTs beyond the acceptance criteria. Therefore, CNTs inhibit cell growth by reducing cell adhesion and may also cause death of kidney cells (Handy & Henry et al., 2008). Fullerene causes harsh lung damage upon human exposure (Baker et al., 2008; Elder et al., 2000) and also causes water flea death and damage to fish brain (Oberdorster, 2004).

2.4.5 DEGREE OF AGGLOMERATION

The chemical composition and agglomeration of NPs could be a strong inducer of inflammatory lung injury in humans (Li et al., 2007; Bantz et al., 2014). Exposure to certain types of chemicals at higher concentrations has been shown to cause severe chronic diseases like cancer and fibrosis (Donaldson et al., 2006).

2.5 TOXICITY MECHANISM IN NATURAL ECOSYSTEM INDUCED BY NANOPARTICLES

Engineered NPs are synthesized to have unique properties that are not present in bulk materials of the same sample. Metal oxide NPs are manufactured with the reaction of reducing or precipitating/oxidizing agents during their synthesis process. NP reactivity with bio-molecules is affected by several factors including NP core composition, shape, size, surface properties, stability, purity, and method of synthesis (Teske and Detweiler, 2015; Wang et al., 2016). Since NPs may preserve the unique properties of their bulk material, the efficacy of the bulk material should be considered in the study of NP interaction in the environment. It was observed that heavy metals are toxic to plant while silicon is a metalloid beneficial for plants (Tubana et al., 2016; Helaly et al., 2017). In the last decades, oxide NPs have been among the most commonly used nanomaterials due to having distinctive properties, and received considerable attention over their prospective ecological effects (Klain et al., 2008). Different industries innovate novel NPs for improvement of their products and services. Many NPs are used on a very large scale, and thus are expected to be released into the environment. Many studies have shown that certain oxide NPs such as CuO, ZnO, and TiO$_2$ (Heinlaan et al., 2008), etc., may differ in their toxicological effects, which depend on particle size and variety, test method, and test organism species. Zinc oxide nanoparticles (ZnONPs) are extensively used in textile, cosmetics, pigments, solar cells, etc. (Rancan et al., 2012) due to their attractive properties (e.g., antibacterial, antifungal, UV filtering properties, conductivity, catalytic properties, chemical stability, photonics, and optoelectronics) and low cost. ZnONPs are highly demanding in the cosmetics industry generally in facial creams and sunscreens (Nohynek et al., 2007). ZnONPs have the third highest production rates of NPs worldwide with an estimated range of about 550 to 10,000 tons/year (Wang, 2004).

Three discharge scenarios are generally considered for NP exposure to the environment (Bawa, 2016): NPs to release during the production of raw material; (Technical specification, 2008) release during use (NNI, 2017); and release after removal of NP-containing products (waste handling) (Gottschalk et al., 2009; Gottschalk et al., 2013). No standard parameter has been established for the tolerance limit of NPs in the environment. NP emissions to the environment can be directly or indirectly through effluent of wastewater treatment plants or landfills (Figure 2.2). There is improper management of industrial waste and improper discarding of products by users. ZnONPs tend to precipitate and aggregate in the environment because of their poor colloidal stability and increase with ionic strength (Liu et al., 2014) like other NPs. Reduction in electrostatic repulsion forces occurs between the NPs in high ionic strength environments, promoting sedimentation and aggregation. Natural substances present in the environment such as humic acid stabilize the ZnONPs and help their mobility, transport, and dispersion (Akhil et al., 2015). Other alterations that may occur in the environment and lead to toxic effects of ZnONPs are the redox transformations and dissolution. ZnONPs may suspend and release Zn ions, which may

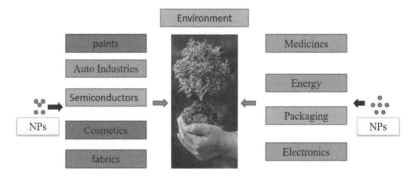

FIGURE 2.2 NPs used in different industries and their release to the environment.

cause toxic effects due to faster dissolution at nano sizes (Franklin et al., 2007; Bian et al., 2011). Redox reaction taking place on ZnONP surface may produce ROS, which are capable of oxidizing organic compounds and promoting oxidative stress. Data shows that 85% of total TiO_2 NP emissions occur via wastewater (Keller et al., 2013). TiO2NPs are accumulated in sewage sludge during wastewater treatment, ultimately incorporated onto soils. Around 30% of sewage sludge is deposited onto landfills directly or after ignition. AgNPs are predominantly emitted by use and production, and are directly deposited at landfills (Sun et al., 2016).

2.6 TOXICITY MECHANISM IN PLANTS INDUCED BY NANOPARTICLES

Metal oxide NPs such as zinc oxide, silver, titanium dioxide, copper, copper dioxide, cerium dioxide, nickel, aluminum, and iron are the most frequently used in industries. Some non-metal NPs like fullerene and SWCNTs have been well reported for their nanotoxicity mechanisms (Joner et al., 2008). In the present era, NPs have been widely used in agriculture as nanofertilizers and nanopesticides (Wang et al., 2016). NPs penetrate in the soil via agricultural application, rain erosion, surface runoff, atmospheric deposition, or other pathways. The NPs may accumulate in the soil due to their weak migration ability in soil. Much research has indicated that the NP concentration in soil is much higher than that in air and water; therefore soil might be the main source of NP emission into the environment (Jie Yang et al., 2017; Gottschalk et al., 2009). As primary producers, plants supply a possible pathway for NP transportations through the food chain and can be deposited in high trophic-level consumers (Zhu et al., 2008). NPs can have extensive toxic effects on plants, such as suppression of plant elongation, reduction in seed germination, and leaf number and even cause plant death (Lee et al., 2010).

2.6.1 Effect of Nanoparticles on Crop Quality

Previous studies on plants have found that accumulated NPs in the environment can strongly change the quality and yield of the soil-based food crop (Priester et al., 2012; Rani et al., 2016). They discovered that protein was insensible to stimulation by Ag NPs. In comparison with carbohydrates, protein content improved only at high concentrations of AgNPs (100 mg/L). ZnO stimulated the protein and starch content but decreased the micronutrient concentration (Mo and Cu) in cucumber. It was also found that CeO2NPs alter the content of fatty acids, amino acids, phenolics, and non-reducing sugars in plants (Rico et al., 2014).

2.7 MECHANISM OF PLANT-NANOPARTICLES INTERACTION

NPs in higher concentrations are toxic to plants. They damage the plasma membrane and plant cell wall, and can penetrate plant tissue with the different plant processes (Rastogi et al., 2017; Mirzajani et al., 2013), either through root or the aboveground parts including wounds and root junctions. For translocation and uptake, NPs are passed on to various physiological or chemical barriers. The plant cell wall is the first barrier to crossed NPs. The cell wall is made up of cellulose, which restricts the entry of larger particles and allows the entry of small particles. The size-barring limit for the plant cell wall is 5 and 20 nm (Dietz and Herth, 2011). NPs may translocate by endocytosis from the plant cell wall and further may travel to different plant tissues via symplastic transport (Ma et al., 2010). The study shows that magnitude, size, and zeta potential are important factors in determining the NP transport inside the plant. Entered NPs may obstruct plant metabolism in many ways, like the regulation of genes (Nair et al., 2014), by providing micronutrients (Liu et al., 2015) or interfering with different oxidative mechanisms in plants (Hossain et al., 2015) and can also interrupt the electron transport chain of chloroplast and mitochondria, ultimately resulting into oxidative burst, due to the increase in ROS concentration (Pakrashi et al., 2014; Cvjetko et al., 2017). Many published data on metal-based NP phytotoxicity suggest that NPs can stimulate oxidative stress in many plant species. Lin and Xing reported ZnO NP phytotoxicity to ryegrass; they confirmed that lipid peroxidation and particle-dependent ROS formation take place on the surface of cell membranes (Lin and Xing, 2008). The latest research found genotoxicity and phytotoxicity of AgNPs on germinating wheat seedlings. It was found that 10 mg/L AgNPs may alter all sorts of proteins that are incredibly connected to cell metabolism (Vannini et al., 2014). Hossain et al. determined the phytotoxicity of ZnONPs-, AgNPs, and Al_2O_3 in soybean seedling at the proteome level; they found significant changes in 16 common proteins of soybean leaves, which were mainly related tophoto-system and protein degradation (Hossain et al., 2016).

- Impact of silver nanoparticles (AgNP)

Among various NPs, AgNPs are used in several products because of their unique antibacterial and antimicrobial properties. Due to their extensive use, the production of AgNPs is increasing rapidly; the United States itself has been documented to fabricate 2,500 tons/year of AgNPs, of which around 80 tons end up in surface waters and 150 tons end up in sewage sludge (El-Temsah and Joner, 2012) and through surface waters and sludge AgNP may easily enter plants. It was observed that at a high concentration of 25 nm AgNPs breaks the cell wall and damages the vacuoles of root cells of Oryza sativa (Mazumdar and Ahmad, 2011). Various studies show a clear connection between the size and toxic relation of NP to the plant. Small size NP was always found to have higher toxicity to plants compared to larger size NP (Cvjetko et al., 2017). The synthesized NPs decreased the total phenol content and induced the production of carbohydrate and protein, which can be considered as a supportive effect, due to the different sizes (2–50 nm) of NPs with different entrance capacity in the highest applied (100 ppm) concentration or the different chemical property of synthesized NPs. The 200–800 nm size AgNP was observed to increase plant growth (Jasim et al., 2016) while 35–40 nm size of AgNP decreased the shoot and root growth of the different plants (Mehata et al., 2016). Major alterations in various macromolecule, protein, pectin, lipid, and cellulose were seen in *Raphanus sativus* when experimented with 2 nm AgNP with 500 mg/L concentration (Zuvera-Mena et al., 2016). The above study indicates AgNP has an impact on different aspects of plant physiology, morphology, and biochemistry, which strongly depends on the properties, size, and concentration of NPs in use.

- Impact of Copper and copper oxide nanoparticles (CuO NPs)

Copper is a fundamental micronutrient that plays an important role in plant nutrition and health because it is integrated into many enzymes and proteins. Copper nanoparticles (CuNP) are commonly used as catalysts, antimicrobial agents, gas sensors, batteries, electronics, heat transfer fluids, etc. (Kasana et al., 2017). CuONPs shows higher toxic effects than CuNP due to their oxidative property. Some scientists have reported a mutagenic, oxidatively modified DNA in different plants indicating DNA damage when treated with CuONP (Nair and Chung, 2014; Atha et al., 2012). Recent studies have shown the negative effect of CuO NP on photosynthetic activity by inactivating PS II reaction centers and indicated by inactivation of photosynthetic rate, electron transport, thylakoid number per grana, photosynthetic pigments, stomatal conductance, and transpiration rate (Perreault et al., 2014; Costa et al., 2016). It was also observed that CuONP altered phytohormones (Cvjetko et al., 2017). This study demonstrates that Cu- and CuONPs present in higher concentrations of more than 0.2 mg/L are toxic to plants and affect the physiology, growth, and biochemistry of plants (Cvjetko et al., 2017; Atha et al., 2012).

- Impact of titanium dioxide nanoparticles (TiO_2NPs)

Titanium dioxide NPs are the most commonly used NPs, and are widely used in skin-care products, cosmetic, cleaning air products, antibacterials, paints, and for corrosion of organic matter in wastewater (Clement et al., 2013). Due to photocatalytic properties, TiO_2NPs may show positive and negative effects in plants. It was observed that smaller sized TiO_2 NPs were transported by roots to different parts of plants, where as above 140 nm sized nanoparticles did not reach root parenchyma, meaning that their uptake by wheat roots is stopped by the epidermis. TiO_2NPs larger than 36 nm in diameter accumulated in wheat root parenchyma but did not translocate to the shoot (Larua et al., 2012,). Therefore, it is clear that TiO_2NPs are toxic to plants at higher concentrations, even in soil system (Rafique et al., 2014). It was also found that the phytotoxic response was similar to CuO- or AgNPs with a decrease in mitotic index, plant growth, and an increase in ROS, genotoxicity, and antioxidant activity. TiO_2NP toxicity on different plants depends on the type of treatment, size, and concentration.

2.8 NANOPARTICLES EXPOSURE TO HUMANS VIA POTENTIAL ROUTES AND TOXICITY

Due to the enormous diversity of NPs human beings and other living organisms are being exposed to NPs through their evolutionary phases. It is difficult to define how NPs are penetrating living body systems. The distribution and accumulation of a specific type of particle within a particular part of the body is totally dependent on the surface characteristic and particle size (Oberdorster et al., 2005; Pojiak-Blazi et al.,

2010). When some accumulated NPs stay in the body system due to improper excretion, severe toxicity may result. The main distribution site of NPs and target organs are not yet known, but it seems that the spleen and liver may be target organs (Handy et al., 2008; Hussain et al., 2005). NPs can be introduced into the human body through various routes such as inhalation, dermal penetration, ingestion, and blood circulation and could carry on in the system due to the inability of the macrophages to phagocytose them. These ingested, inhaled, or absorbed NPs through the skin can stimulate the formation of ROS as well as free radicals (Brown et al., 2002; Fu et al., 2014). ROS generate inflammation and oxidative stress, resulting in damage to different biological materials such as DNA, protein, etc. In addition to ROS production, other factors may influence toxicity. When these persisting NPs interact with the body system, they will release their toxic properties, which are predominantly dependent on their particle size, morphology, shape, agglomeration status, surface properties, solubility, and surface charge (Karakoti et al., 2006; Holsapple et al., 2005). As a result of their toxic properties, NPs can penetrate cell membranes and even tissue junctions where they encourage structural injure to the mitochondria (Hoshino et al., 2004; Salnikov et al., 2007). Or attack the nucleus where they cause severe DNA mutation (Donaldson et al., 2002) leading to cell death (Wilson, 2006).

2.8.1 Gastrointestinal Tract (GIT)

NPs are one of the major factors that cause gastrointestinal (GI) diseases and damage to accessory digestive organs such as the pancreas, liver, and gall bladder (Bellmann et al., 2015). The human GI tract is divided into the upper (esophagus, stomach, and duodenum) and lower tract (small intestine and large intestine) (John et al., 2014); this tract represents a mucosal barrier that specifically induces the uptake and degradation of nutrients such as peptides, carbohydrates, and fats. All of these are compulsory to maintain the immune system and physiological homeostasis of the body. A large number of engineered nanomaterials used in agri-food products can enter the GI tract (Figure 2.3).

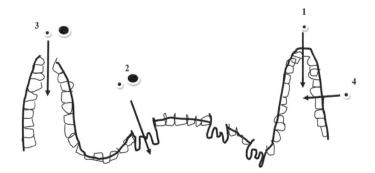

FIGURE 2.3 Particles crossing the gastrointestinal tract. 1. Small NPs (<50–100nm) cross the regular epithelial cells by endocytosis. 2. Small microparticles and large NPs cross the M-cell by transcytosis. 3. Viable channel through villous gaps of microparticles and nano. 4. Paracellular uptake of small NPs.

They can also be ingested directly by intake of food or water or through drug delivery (Martirosyan et al., 2014; Chen et al., 2014; Athinarayanan et al., 2015). A large number of inhaled NPs can easily disperse to the trachea with the help of mucociliary cells and become part of the bloodstream and translocate to other organs of the body (Jani et al., 1990). In the GI tract, there is a variable pH value inside all sections of the tract (Frohlich and Frohlich, 2016). An acidic environment in the stomach also affects the dissolution of NPs at low pH and influences their additional degradation in digestive fluids. It has been documented that the uptake of NPs smaller than 100 nm occurs mainly by endocytosis (Axson et al., 2015). Generally, endogenous and exogenous are the two main sources of NP absorption into the GI tract. ndogenous NPs naturally occur in soluble form such as phosphate and calcium secretion of intestine through NPs from water, food, cosmetics, and pharmaceuticals (Lomer et al., 2004). Engineered TiO2, aluminium silicates, or silicates have been extensively used as exogenous inorganic NPs for the food industries and in the pharmaceutical industry as additives (Kaida et al., 2004). These types of NPs that have been found in the gut are spherical in shape and around 100 nm in size and show resistivity to degradation. E171 is the commercial label for TiO_2NPs used in sweets (Weir et al., 2012) in the size range of 1–5 ug/mg. Lomer et al has reported daily ingestions of 2.5 mg for a 70 kg person in one study and 5.9 mg in another one (Lomer et al., 2000). SiO_2 NPs of different composition, labelled as E551, E554, E556, or E559, and often used as an anti-caking agent. The amount ingested daily is around 126 mg/day for a 70 kg person (Dekkers et al., 2011). The permissible limit accepted by the committee of the European Food Safety Authority is 20–50 mg SiO2NPs for a 60 kg person (Winkler et al., 2016). SiO2NPs and TiO2NPs exhibit lower genotoxicity than AgNPs of the same size (Rim et al., 2013). In a recent study, TiO2 TiO_2NPs extracted from chewing gum caused collapse of enteral epithelium, while Caco-2 cells (Caucasian colon adenocarcinoma) treated with 10 ug/ml TiO_2NPs solution showed an interruption of tight junctions and microvilli (Faust et al., 2014). Moos et al. observed a mitochondrial dysfunction when assessing the toxicity of nano (8-10 nm) and micro (<44 μm) ZnONPs on human colon carcinoma cells.

2.8.2 Respiratory Tract

The respiratory tract is the most common route of humans to airborne NPs through inhalation in the workplace and environment. The respiratory system consists of two parts, the upper (nasal cavity, the pharynx, and the larynx) respiratory tract and the lower (trachea, bronchi, and lungs) respiratory tract (Weibel, 1963; Weibel et al., 2005). The upper respiratory tract permits air passage and also protects the lower respiratory tract. The size of NPs allows them to penetrate the human respiratory system (Qiao et al., 2015; Bakand and Hayes, 2016). With a diameter between 5 and 30 μm large size NPs stay in the nasopharyngeal region, while small size NPs between 1 and 5 μm have tend to deposit in the tracheobronchial region. Whereas, NPs range between 0.1-1 um can reach the deepest region of the respiratory system in the alveolar region by Brownian diffusion and gravitational sedimentation (Siegmann et al., 1999; Witschi, 2001). Once the NPs enter in the respiratory system, body mechanisms such as coughing can promote NP motion, which is not possible for small sized NPs that have penetrated the deepest alveolar region. This results into the settlement of these nanoparticles (Muhlfeld et al., 2008). Once the smaller NPs are deposited, their elimination gets delayed and toxics effects are shown (Blank et al., 2009). Pujalte et al. conducted an experiment with male Sprague-Dawley rats exposed to ~20 nm anatase TiO$_2$NPs (15 mg/m3) for 6 h through the inhalation route. After giving treatment to the rats, they observed a high level of Ti in the lungs, as well as in the liver, kidney, and spleen (Pujalte et al., 2017). Citrate-capped et al. demonstrated that AuNPs (13 nm) show high toxicity in lung carcinoma cells, but no adverse effects were shown in liver carcinoma cells (Patra et al., 2007). Treatment of primary human bronchial epithelial cells (BEAS-2B) with ZnONPs generates oxidative stress, cytotoxicity, variation of mitochondrial membrane potential, and high intracellular Ca2+ levels (Vandebriel et al., 2012). In a recent work, the authors investigated the toxicity of inhaled ZnO < 50 nm in an in-vitro air-blood barrier (ABB) model made up of immortalized pulmonary microvascular endothelial cells (HPMEC-ST1 6R) and transwell co-culture of lunge epithelial cells (NCI-H441). The discharge of pro-inflammatory mediators (IL-6, IL-8), particularly 260 ug/ml of IL-8, were detected in treated cells against control cells followed by the secretion of dysfunction markers (sICAM-1 and sVCAM-1) (Bengalli et al., 2017). In another experiment Lee et al. exposed Adenocarcinomic human alveolar basal epithelial cells (A549) to AgNP in an increasing manner for 24 h, and investigated cell death and morphological changes in a dose-dependent approach (Lee et al., 2011). Nguyen et al. studied the dose-dependent toxicity of PVP-coated AgNPs (10, 50, and 75 nm) and uncoated (20, 40, 60, and 80 nm) in a macrophage cell line (J774A) (Nguyen et al., 2013). The effect clearly showed cell elongation due to PVP-coated AgNPs whereas cell shrinkage was evident after treatment with uncoated AgNPs. This experiment suggested that the long-term inhalational exposure to AgNPs induced inflammatory responses and alterations in lung function (Sung et al., 2008). Lin et al. reported toxicity of SiO$_2$NPs (15, 46 nm) exposed to human bronchoalveolar carcinoma- derived cell line (A549) at 10, 50, and 100 ug/ml for 48 h. The results showed SiO2NPs induce lipid peroxidation, ROS production, and loss of viability (Lin et al., 2006).

2.9 DERMAL PENETRATION

The skin is structurally divided into epidermis, dermis, and hypodermis layers. It plays different functions, such as s selective permeable barrier, UV protection, defense against foreign agents, and it also regulates human body temperature and immunological response because of the presence of Langerhans cells (Wickett and Visscher, 2006). With the increasing application of nanomaterials in the cosmetics field since 2006, it has been predicted that NP-containing cosmetic products contain ZnONPs (70%), AgNPs (20%), and TiO2NPs (70–80%) (Niska et al., 2018; Monteiro-Riviere, 2010; Piccinno et al., 2012). Many studies have shown that the penetration of NPs through the skin is influenced by size. NPs with a diameter of around 4 nm can enter integral skin, whereas with an increase in size up to 45 nm can only penetrate damaged skin (Filon et al., 2015). Meyer et al. observed mitochondrial damage in human dermal fibroblasts due to the protein p53 cell activation after 4 h exposure to 10, 50, and 100 ug/ml of 20 nm ZnONPs (Meyer et al., 2011). Rancan et al. experimented with the penetration of amorphous SiO2NPs in keratinocyte with different sizes, 291±9 to 42±3 nm, and surface charges (negative and positive). This study revealed that only smaller NPs were competent to enter the injured stratum corneum. Therefore, people with skin diseases like chronic eczema and dermatitis can have numerous toxic effects because of SiO_2NP penetration (Rancan et al., 2012). Park et al. used 20 nm and 100 nm SiO_2NPs with negative and positive charges on keratinocytes. The results demonstrated that ROS production was enhanced by small-sized NPs (Park et al., 2013). A total of 70 and 300 nm SiO_2NPs exposed to HaCaT in a size-dependent manner showed growth inhibition of HaCaT (Nabeshi et al., 2012).

2.10 CONCLUSION

The unique characteristics of NPs increases their applications in industrial and domestic processes and also contributes to enhance their toxicity to higher organisms including human, flora and fauna, and wildlife. Cellular uptake of NPs may stimulate cytotoxicity by generation of ROS. Toxicity of NPs is affected by their fundamental properties such as shape, size, charge, medium of synthesis, storage time, aggregation, stability, mobility, and reactivity. This chapter provided a better understanding of the mechanism of ecosystem, plant, and human NP interaction and the impact of different NPs such as AgNPs, CuONPs, and TiO2NPs on humans as well as flora and fauna. Dermal contact, digestion, and inhalation are the three main routes of exposure of NPs to humans.

ACKNOWLEDGEMENT

The authors wish to express our sincere thanks to Dr. Ashok Kumar Chauhan, President, RBEF parent organization of Amity University Madhya Pradesh (AUMP), Dr. Aseem Chauhan, Additional President, RBEF and chairman of AUMP, Gwalior, and Lt. Gen. V.K. Sharma, AVSM (Retd.), Vice Chancellor of AUMP, Gwalior for providing necessary facilities and their valuable support and encouragement throughout the work.

REFERENCES

Atha, D.H., Wang, H., Petersen, E.J., Cleveland, D., Holbrook, R.D., Jaruga, P., et al., 2012. Copper oxide nanoparticles mediated DNA damage in terrestrial plant models. *Environ. Sci. Technol.* 46, 1819–1827.
Athinarayanan, J., Alshatwi, A.A., Periasamy,V.S., and Al-Warthan, A.A., 2015. Identification of nanoscale ingredient in commercial food products and their induction of mitochondrially mediated cytotoxic effects on human mesenchymal stem cells. *J Food Sci.* 80, N459–64.
Akhil, K., Chandran, P., and Sudheer Khan, S., 2015. Influence of humic acid on the stability and bacterial toxicity of zinc oxide nanoparticles in water. *Journal of Photochemistry and Photobiology B: Biology* 153, 289–295.
Alazzam, A., M foumou, E., Stiharu, I., Kassab, A., Darnel, A., Yasmeen, A., et al., 2010. Identrification of deregulated genes by single wall carbon-nanotubes in human normal bronchial epithelial cells. *Nanomedicine* 6, 563–569.

Axson, J., Stark, D., Bondy, A., Capracotta, S., Maynard, A., Philbert, M., et al., 2015. Rapid Kinetics of size and pH-dependent dissolution and aggregation of silver nanoparticles in simulated gastric fluid. *J. Phys. Chem. C.* 119, 20632–20641.

Baek, M., Kimis, Yu J., Chung, H.E., Choy, J.H., et al., 2011. Effect of different forms of anionic nanoclays on Cytotoxicity. *J NanosciNanotechnol* 11, 1803–1806.

Bakand, S., and Hayes, A., 2016. Toxicological considerations, toxicity assessment, and risk managmenr of inhaled nanoparticles. *Int. J. Mol. Sci.* 17, 929.

Baker, G.L., Gupta, A., Clark, M.L., Valenzuela, B.R., Staska, L.M., Harbo, S.J., et al., 2008. Inhalation toxicity and lung toxicokinetics of C60 fullerene nanoparticles and microparticles. *Toxicol. Sci.* 101, 122–131.

Banerjee A., Qi J., Gogoi R., Wong J., and Mitragotri, 2016. Role of nanoparticles size, shape and surface chemistry in oral drug delivery. *J Control Release* 238, 176–185.

Bantz, C., Koshina, O., Lang, T., Galla, H-J., Kirkpatrick, C.J., Stauber, R.H., et al., 2014. The surface properties of nanoparticles determine the agglomeration state and the size of the particles under physiological conditions. *Beilstein J. Nanotechnol.* 5, 1774–1786.

Bawa, R., 2016. *What's In a Name? Defining "Nano" in the Context of Drug Delivery*, Bawa, R., Ed. Pan Stanford Publishing: Singapore.

Bellmann, S., Carlander, D., Fasano, A., Momcilovic, D., Scimeca, J.A., et al., 2015. Mammalian gastro intestinal tract parameters modulating the integrity, surfaces, and adsorption of food-relevant Nanomaterials. *Wiley Interdiscrip Rev Nanomed Nanobiotechnol* 7, 609–622.

Bengalli, R., Gualtieri, M., Capasso, L., Urani, C., Camatini, M., 2017. Impact of Zinc oxide nanoparticles on an in vitro model of the human air-blood barrier. *Toxicol. Lett.* 279, 22–32.

Bian, S.W., Mudunkotuwa, I.A., Rupasinghe, T., and Grassian, V.H., 2011 Aggregation and dissolution of 4 nm ZnO nanoparticles in aqueous environments: influence of pH, ionic strength, size and adsorption of humic acid. *Langmuir* 27, 6059–6068.

Biosselier, E., and Astruc, D., 2009. Gold nanoparticles in nanomedicine: preparations, imaging, diagnostics, therapies and toxicity. *Chem. Soc. Rev.* 38, 1759–1782.

Blank, F., Gehr, P., and Rutishauser, R.R., 2009. *In Vitro Human Lung Cell Culture Models to Study the Toxic Potential of Nanoparticles*. John Wiley & Sons Ltd: Chichester, UK.

Brown, J.S., Zeman, K.L., and Bennett, W.D., 2002. Ultrafine particle deposition and clearance in the healthy and obstructed lung. *Am. J. Respir. Crit. Care Med.* 166, 1240–1247.

Cha, K.E., and Myung, H., 2007. Cytotoxic effects of nanoparticles assessed in vitro and in vivo. *J. Microbiol. Biotechnol.* 17, 1573–1578.

Chen, H., Seiber, J.N., and Hotze, M., 2014. ACS selection on nanotechnology in food and agriculture: a perspective on implications and applications. *J Agric Food Chem.* 62, 1209–1212.

Chen, X., and Mao, S.S., 2007. Nanostructured titanium dioxide materials: properties, preparation and applications. *Chem. Rev.* 107, 2891–2959.

Chithrani, B.D., and Chan, W.C., 2007. Elucidating the mechanism of cellular uptake and removal of protein coated gold nanoparticles of different sizes and shapes. *Nano Lett* 7(6), 1542–1550.

Chithrani, B.D., Ghazani, A.A., and Chan, W.C., 2006. Determining the size and shape dependence of gold nanoparticles uptake into mammalian cells. *Nano Lett.* 6(4), 662–668.

Cho, J., and Caruso, F., 2005. Investigation of the interaction between ligand stabilized mater Chem. Mater. 17, 4547–4553.

Clement, L., Hurel, C., and Marmier, N., 2013. Toxicity of TiO_2 nanoparticles to cladocerans, algae, rotifers and plants-effects of size and crystalline structure. *Chemosphere* 90, 1083–1090.

Colvin, V.L., 2003. The potential environmental impact of engineered Nanomaterials. *Nat. Biotechnol.* 21, 1166–1170.

Costa, M.V.J.D., and Sharma, P.K., 2016. Effect of copper oxide nanoparticles on growth, morphology, photosynthesis, and antioxidant response in *Oryza sativa*. *Photosynthetica* 54, 110–119.

Cvjetko, P., Milosic, A., domijan, A.M., Vinkovic-Vecek, T., Tolic, S., Peharec-Stefanic, P., et al., 2017. Toxicity of silver ions and differently coated silver nanoparticles in Allium cepa roots. *Ecotoxicol. Environ. Saf.* 137, 18–28.

De Jong, W.H., Hagens, W.I., Krystek, P., Burger, M.C., Sips, A.J., et al., 2008. Particle size-dependent organ distribution of gold nanoparticles after intravenous administration. *Biomaterials* 29, 1912–1919.

Dekkers, S., Krystek, P., Peters, R.J., Lankveld, D.P., Bokkers, B.G., Van Hoeven-Arentez, P.H. et al., 2011. Presence and risks of nanosilica in food products. *Nanotoxicology* 5, 393–405.

Deshmukh, S.P., Patil, S.M., Mullani, S.B., and Delekar, S.D., 2019. Silver nanoparticles as an effective disinfectant: A review. *Mater. Sci. Eng. C-Mater.* 97, 954–965.

Dietz, K.J., and Herth, S., 2011. Plant nanotoxicology. *Trends Plants Sci.* 16, 582–589.

Donaldson, K., Aitken, R., Tran, L., Stone, V., Duffin, R., and Forrest, G., 2006. A carbon nanotube: A review of their properties in relation to pulmonary toxicology and workplace safety. *Toxicol. Sci.* 92, 5–22.

Donaldson, K., and Stone, V., 2002. Current hypotheses on the mechanisms of toxicity of ultrafine particles. *Ann. Ist. Super. Sanita* 39, 405–410.

Elder, A.C.P., Gelein, R., Finkelstein, J.N., Cox, C., and Oberdoster, G., 2000. Pulmonary inflammatory response to inhaled ultrafine particles is modified by age, ozone exposure, and bacterial toxin. *Inhal. Toxicol.* 12, 227–246.

El-Temsah, Y.S., and Joner, E.J., 2012. Ecotoxicological effects on earthworms of fresh and aged nano-sized zero-valent iron (nZVI) in soil. *Chemosphere* 89, 76–82.

European Commission Web Resource. Available online: http://ec.europa.eu/research/industrial_technologies/policy_en.html (accessed on 12 August 2016).

Faust, J.J., Doudrick, K., Yang, Y., Westerhoff, P., and Capco, D.G., 2014. Food grade titanium dioxide distrupts intestinal brush border microvilli in vitro independent of sedimentation. *Cell Biol. Toxicol.* 30, 169–188.

Ferreira, A.J., Cemlyn-Jones, J., and Robalo Cordeiro, C. 2012. Nanoparticles, nanotechnology and pulmonary nanotoxicology. *Rev Port Pneumol.* 19(1), 28–37.

Filon, F.L., Mauro, M., Adami, G., Bovenzi,M., and Crosera, M., 2015. Nanoparticles skin absorption: new aspects for a safety profile evaluation. *Regul. Toxicol. Pharmacol.* 72, 310–322.

Firme, C.P., and Bandaru, P.R., 2010. Toxicity issue in the application of carbon nanotubes to biological system. *Nanomed. Nanotechnol. Biol. Med.* 6, 245–256.

Foster, K.A., Yazdanian, M., and Audus, K.L., 2001. Microparticulate uptake mechanisms of in-vitro cell culture models of the respiratory epithelium. *J Pharma Pharmacol* 53, 57–66.

Franklin, N.M., Rogers, N.J., Apte, S.C., Batley, G.E., Gadd, G.E., and Casey, P.S., 2007. Comparative toxicity of nanoparticulates ZnO, Bulk Zno and $ZnCl_2$ to a freshwater microalga (Pseudokirchneriellasubcapitata): the importance of particle solubility. *Environmental Science and Technology* 41, 8484–8490.

Frohlich, E., and Roblegg, E., 2012. Models for oral uptake of nanoparticles in consumer products. *Toxicology* 291, 10–17.

Frohlich, E.E., and Frohlich, E., 2016. Cytotoxicity of nanoparticles contained in food on intestinal cells and gut microbiota Esther. *Int. J. Mol. Sci.* 17, 509.

Fu, P.P., Xia, Q., Hwang, H.M., Ray, P.C., and YU, H., 2014. Mechanisms if nanotoxicity: generation of reactive oxygen species. *Journal of Food and Drug Analysis* 22, 64–75.

Gottschalk, F., Sonderer, T., Scholz, R.W., and Nowack, B., 2009. Modeled environmental concentrations of engineered Nanomaterials (TiO$_2$, ZnO, Ag, CNT, Fullerences) for different regions. *Environ Sci Technol* 43(24), 9216–9222.

Gottschalk, F., Sun, T.H., and Nowack, B., 2013. Environmental concentrations of engineered Nanomaterials: review of modeling and analytical studies. *Environ Pollut* 181, 287–300.

Handford, C.E., Dean, M., Henchion, M., et al., 2014. Implications of nanotechnology for the agri-food industry: opportunities, benefits and risks. *Trend Food Sci Technol* 40(2), 226–241.

Handy, R.D., Henry, T.B., Scown, T.M., Johnston, B.D., Tyler, and C.R., 2008. Manufactured nanoparticles: their uptake and effects on fish—a mechanistic analysis. *Ecotoxicity* 17, 396–409.

Handy, R.D., Owen, R., Valsami-Jones, E., 2008. The ecotoxicology of nanoparticles and Nanomaterials: current status, knowledge gaps, challenges, and future needs. *Ecotoxicology* 17, 315–325.

Hund-Rinke, K., Simon, M., 2006. Ecotoxic effects of photocatalytic active nanoparticles (TiO$_2$) on algae and daphids. *Environ. Sci. Pollut. Res.* 13, 225–232.

Hansen, S.F., Michelson, E.S., Kamper, A., Borling, P., Stuer-Lauridsen, F., Baun, A., 2008. Categorization framework to aid exposure assessment of Nanomaterials in consumer products. *Ecotoxicology* 17, 438–447.

Hegde, K., Brar, B., S.K., Verma, M., and Surampalli, R.Y., 2016. Current understandings of toxicity, risks and regulations of engineered nanoparticles with respect to environmental microorganisms. *Nanotechnol. Environ. Eng.* 1, 5.

Heinlaan, M., Ivask, A., Blinova, I., Dubourguier, H.C., and Kahru, A., 2008. Toxicity of nanosized and bulk ZnO, CuO and TiO$_2$ to bacteria Vibrio fischeri and crustaceans Daphnia manga and Thamnocephalusplatyurus. *Chemosphere* 71, 1308–1316.

Helaly, M.N., El-Hoseiny, H., El-Sheery, N.I., Rastogi, A., and Kalaji, H.M., 2017. Regulation and physiological role of silicon in alleviating drought stress of mango. *Plant Physiol. Biochem.* 118, 31.

Hoet, P.H., Bruske-Hohlfeid, I., and Salata, O.V., 2004. Nanoparticles-known and unknown health risks. *J Nanobiotechnology* 2, 12–18.

Holsapple, M.P., Farland, W.H., Landry, T,D., Monteiro-Riviere, N.A., Carter, J.M., Walker, N.J., et al. 2005. Research strategies for safety evaluation of nanomaterials, part II: Toxicological and safety evaluation of Nanomaterials, current challenges and data needs. *Toxicol. Sci.* 88, 12–17.

Hoshino, A., Fujioka, K., Oku, T., Nakamura, S., Suga, M., Yamaguchi, Y., et al., 2004. Quantum dots targeted to the assisted organelle in the living cells. *Microbiol. Immunol.* 48, 985–994.

Hossain, Z., Mustafa, G., and Komatsu, S., 2015. Plant responses to nanoparticles stress. *Int. J. Mol. Sci.* 16, 26644–26653.

Hossain, Z., Mustafa, G., Sakata, K., and Komatsu, S. 2016. Insights into the proteomic response of soybean towards Al$_2$O$_3$, ZnO, and Ag nanoparticles stress. *J Hazard Mater.* 304, 291–305.

Hoyt, V.W., and Mason, E., 2008. Nanotechnology: emerging health issues. *J. Chem. Heal. Saf.* 15, 10–15.

Huczko, A., Lange, H., Calko, E., Grubek-Jaworska, H., and Droszcz, P., 2001. Physiological testing of carbon nanotubes: are they asbestos-like? *Fuller. Sci. Technol.* 9, 251–254.

Hussain, S., Hess, K., Gearhart, J., Geiss, K., and Schlager, J., 2005. In vitro toxicity of nanoparticles in BRL 3A rat liver cells. *Toxicol. Vitr.* 19, 975–983.

Ispas, C., Andreescu, D., Patel, A., Goia, Dv., Andreescu, S., et al., 2009. Toxicity and developmental defects of different sizes and shape nickel nanoparticles in zebrafish. *Environmental Science & Technology* 43, 6349–6356.

Jani, P., Halbert, G.W., Langridge, J., Florence, A.T., 1990. Nanoparticles uptake by the rat gastrointestinal mucosa: quantitation and particle size dependency. *J Pharm Pharmacol* 42, 821–826.

Jasim, B., Thomas, R., Mathew, J., Radhakrishnan, E.K., 2016. Plant growth and diosgenin enhancement effect of silver nanoparticles in Fenugreek (*Trigonella foenum-graecum L.*). *Saudi Pharma J.* 25, 443–447.

JieYang, Weidong Cao, and Yukui Rui., 2017. Interaction between nanoparticles and plants: phytotoxicity and defense mechanisms. *Journal of Interactions* 12(1), 158–169.

Jin, H., Heller, D.A., Sharma, R., and Strano, M.S., 2009. Size dependent cellular uptake and expulsion of single walled carbon nanotubes: single particle tracking and a generic uptake model for nanoparticles. *ACS Nano* 3(1), 149–158.

John, F., Reinus Douglas, M.D., and Simon, M.D. 2014. *Gastrointestinal Anatomy and Physiology: The Essentials.* John Wiley & Sons: Oxford, UK, p. 188, ISBN 978-0-470-67484-0.

Joner, E.J., Hartnik, T., and Amundsen, C.E., 2008. *Norwegian Pollution Control Authority Report no. TA 2304/2007.* Bioforsk: Environmental fate and ecotoxicity of engineered nanoparticles, 1–64.

Jovanovic, B., 2015. Critical review of public health regulations of titanium dioxide, a human food additive. *Integr. Environ. Assess Manage.* 11, 10–20.

Junam, Y., and Lead, J.R., 2008. Manufactured nanoparticles: an overview of their chemistry, interaction and potential environmental implications. *Sci. Total Environ.* 400, 396–414.

Kahan, D.M., Braman, D., Slovic, P., Gastil, J., and Cohen, G., 2009. Cultural cognition of the risks and benefits of nanotechnology. *Nat Nanotechnol* 4, 87–90.

Kaida, T., Kobayashi, K., Adachi, M., and Suzuki, F., 2004. Optical characteristics of titanium oxide interference film and the film laminated with oxide and their applications for cosmetics. *J. Cosmet. Sci.* 55, 219–220.

Karakoti, A.S., Hench, L.L., and Seal, S., 2006. The potential toxicity of Nanomaterials—the role of surfaces. *The Journal of the Minerals, Metals & Society.* 58, 77–82.

Kasana, R.C., Panwar, N.R., Kaul, R.K., and Kumar, P., 2017. Biosynthesis and effects of copper nanoparticles on plants. *Environ. Chem. Lett.* 15, 233–240.

Keller, A.A., McFerran, S., Lazareva, A., and Suh, S., 2013. Global life cycle releases of engineered Nanomaterials. *J Nanopart Res* 15(6), 1692.

Klain, S.J., Alvarez, P.J.J., Batley, G.H., Fernades, T.F., Handy, R.D., Lyon, D.Y., et al., 2008. Nanomaterials in the environments: behavior, fate, bioavailability and effects. *Environ. Toxicol. Chem.* 27, 1825–1851.

Larguinho, M., Correia, D., Diniz, M.S., and Baptista, P.V., 2014. Evidence of one-way flow bioaccumulation of gold nanoparticles across two trophic levels. *J. Nanoport. Res.* 16, 1–11.

Larua, C., Laurette, J., Heelin-Boime, N., Khodja, H., Fayard, B., Flank, A.M., et al., 2012. Accumulation, translocation and impact of TiO_2 nanoparticles in wheat (*Triticum aestivum spp.*): influence of diameter and crystal phase. *Sci. Total Environ.* 431, 197–203.

Lee, Y.S., Kim, D.W., Lee, Y.H., Oh, J.H., Yoon, S, Choi, M.S., et al., 2011. Siver nanoparticles induce apoptosis and G2/Marrest via PKC£-dependent signaling in A549 lung cells. *Arcg. Toxicol.* 85, 1529–1540.

Lee, C.W., Mahendra, S., Zodrow, K., Li, D., Tsai, Y.C., Braam, J., and Alvarez, P.J., 2010. Developmental phytotoxicity of metal oxide nanoparticles to Arabodopsis thaliana. *Environ Toxicol Chem.* 29, 669–675.

Li, Z., Hulderman, T., Salmem, R., Chapman, R., Leonard, S.S., Young, S.H., et al., 2007. Cardiovascular effects of pulmonary exposure to single- wall carbon nanotubes. *Environ. Health Perspect.* 115, 377–382.

Lin, W., Huang, Y.W., Zhou, X.D., and Ma, Y., 2006. In vitro toxicity of silica nanoparticles in human lung cancer cells, Toxicol. *Appl. Pharmacol.* 217, 252–259.

Lin, D., and Xing, B., 2008. Root uptake and phytotoxicity of ZnO nanoparticles. *Environ Technol.* 42, 5580–5585.

Liu, L., Wang, X., Yang, X., Fan, W., Wang, X., Wang, N., et al., 2014. Precipitation, characterization, and biotoxicity of nanosized doped ZnO photocatalyst. *International Journal of Photoenergy* 8, doi: 10.1155/2014/475825

Liu, R., and Lal R., 2015. Potentials of engineered nanoparticles as fertilizers for increasing agronomic productions. *Sci. Total Environ.* 514, 131–139.

Lomer, M.C., Hutchinson, C., Volkert, S., Greenfield, Sm., Catterall, A., et al., 2004. Dirtry sources of inorganic microparticles and their and intake in healthy subjects and patients with crohn's disease. *Br J Nutr* 92, 947–955.

Lomer, M.C., Thompson, R.P., Commisso, J., Keen, C.L., and Powell, J.J., 2000. Determination of titanium dioxide in foods using inductively coupled plasma optical emission spectrometry. *Analyst* 125, 2339–2343.

Ma, X., Geisler, Lee J., Deng, Y., and Kolmakov, A., 2010. Interactions between engineered nanoparticles (ENPs) and planta: phytotoxicity, uptake and accumulation. *Sci. Total Environ.* 408, 3053–3061.

Marambio-Jones, C., and Hoek, E.M.V., 2010. A review of the antibacterial effects of silver Nanomaterials and potential implications for human health and the environment. *J. Nanopart. Res.* 12, 1531–1551.

Marano F., Hussain S., Rodrigues- Lima F., Baeza-Saquiban A., and Boland S., 2011. Nanoparticles: molecule targets and cell signaling. *Arch Toxicol* 85(7), 733–741.

Marin, S., Vlasceanu, G.M., Tiplea, R.E., Bucur, I.R., Lemnaru, M., Marin, M.M., and Grumezescu, A.M., 2015. Application and toxicity of silver nanoparticles: a recent review. *Curr. Top. Med. Chem.* 15, 1596–1604.

Martirosyan, A., and Schneider, Y.J., 2014. Engineered Nanomaterials in food: implications for food safety and consumer health. *Int. J. Environ Res Public Health* 11, 5720–5750.

Mazumdar, H., and Ahmad, G.U., 2011. Phytotoxicity effect of silver nanoparticles on *Oryza sativa*. *Int. J. Chem Tech. Res.* 3, 1494–1500.

Mehata, P., C.M., Srivastava, R., Arora, S., and Sharma, A.K., 2016. Impact assessment of silver nanoparticles on plant growth and soil bacterial diversity. *Biotech* 6, 254.

Meyer, K., Rajanahalli, P., Ahamed, M., Rowe, J.J., and Hong, Y., 2011. ZnO nanoparticles induce apoptosis in human dermal fibroblasts via p53 and p38 pathways. *Toxicol. In Vitro* 25, 1721–1726.

Mirzajani, F., Askari, H., Hamzelou, S., Farzaneh, M., and Ghassempour, A., 2013. Effect of silver nanoparticles on *Oryza sativa L.* and its rhizosphere bacteria, Ecotoxicol. *Environ. Saf.* 88, 48–54.

Mineral and Commodity Summaries, 2015. *Titanium and Titanium Dioxide. US Department of the Interior, US Geological Survey*: Reston, VA.

Moghimi, S.M., Hunter, A.C., and Murray, J.C., 2001. Long-circulating and target specific nanoparticles: theory to practice. *Pharmacol Rev* 53, 288–318.

Monteiro-Riviere, N.A., 2010. *Structure and function of skin. In Toxicology of the Skin—Target Organ Series*, Monteiro-Riviere, N.A., Ed. Informa Healthcare: Raleigh, NC, USA, pp. 1–18, ISBN 9781420079173.

Morganti, P., 2010. Use and potential of nanotechnology in cosmetic dermatology. *Clinical Cosmetic Investigation Dermatology*. 3, 5–13.

Muhlfeld, C., Gehr, P., and Rothen-Rutishausar, B., 2008. Translocation and cellular entering mechanisms of nanoparticles in the respiratory tract. *Swiss. Med. Wkly.* 138, 387–391.

Mukhopadhyav, S.S., 2014. Nanotechnology in agriculture: prospects and constraints. *Nanotechnol Sci Appl* 7, 63–71.

Nair, P.M., and Chung, I.M., 2014. Impact of copper oxide nanoparticles exposure on *Arabidopsis thaliana* growth, root system development, root lignifications, and molecular level changes. *Environ. Sci. Pollut. Res. Int.* 21, 12709–12022.

Nabeshi, H., Yoshikawa, T., Matsuyama, K., Nakazato, Y., Arimori, A., Isobe, M., et al., 2012. Amorphous nanosilicans induce consumptive coagulopathy after systemic exposure. *Nanotechnology* 23, 045101.

Nel, A.E., Madler, L., Velegol, D., Xia, T., Hoek, E.M., Somasundaran, P., et al., 2009. Understanding biophysicochemical interaction at the nano-bio interface. *Nat Mater* 8(7), 543–557.

Nguyen, K.C., Seligv, V.L., Massarsky, A., Moon, T.W., Rippstein, P., Tan, J., et al., 2013. Comparison of toxicity of uncoated and coated silver nanoparticles. *J. Phys.* 429, 012025.

Niska, K., Zielinska, E., Radomski, M.W., and Inkielewicz-Stepnial, I., 2018. Metal nanoparticles in dermatology and cosmetology: interaction with human skin cells. *Chem. Biol. Interact*, 1; 295, 38–51.

NNI. Supplement to the President's 2017 Budget-Obama White House, 2017. https://www.nano.gov/sites/default/files/pub_resource/nni_fy17_budget_supplement.pdf (accessed on 20 June 2017).

NNI. National Nanotechnology Initiative Strategic Plan, February 2014. http://nano.gov/sites.default/files/pub_resource/2014_nni_strategic_plan.pdf (accessed on 20 June 2017).

Nohynek, G.J., Lademann, J., Ribaud, C., and Roberts, M.S., 2007. Grey goo on the skin? Nanotechnology, cosmetics and sunscreen safety. *Critical Reviews in Toxicology* 37, 251–277.

Nowack, B., and Bucheli., T.D. 2007. Occurrence and effects of nanoparticles in the environmemts. *Environ. Pollut.* 150, 5–22.

Oberdorster, E., 2004. Manufactured Nanomaterials (fullerenes, C60) induce oxidative stress in the brain of juvenile largemouth bass. *Environ. Health Perspect.* 112, 1058–1062.

Oberdorster, G, Oberdoster, E., and Oberdoster, J., 2005. Nanotoxicology: an emerging discipline evolving from studies of ultrafine particles. *Environ Health Perspect.* 113, 823–839.

Oberdorster G., 2010. Safety assessment for nanotechnology and nanomedicine: concepts of nanotoxicology. *J Intern Med* 267(1), 89–105.

Ozel, R.E., Alkasir, R.S., and Ray K., 2013. Comparative evalution of intenstinal nitric oxide in embryonic zebrafish exposed to metal oxide nanoparticles. *Small* 9, 4250–4261.

Pakrashi, S., Jain, N., Dalai, S., Jayakumar, J., Chandrasekaran, P.T., Raichur, A.M., et al., 2014. In vitro genotoxicity assessment of titanium dioxide nanoparticles by *Allium cepa* root tip assay at high exposure concentrations. *PLoS One* 9, e87789.

Park, Y.H., Bae, H.C., Jang, Y., Jeong, S.H., Lee, H.N., Ryu, W.I., et al., 2013. Effect of the size and surface charge of silica nanoparticles on cutaneous toxicity. *Mol. Cell. Toxicol.* 9, 67–74.

Patra, H.K., Banerjee, S., Chaudhuri, U., Lahiri, P., and Dasgupta, A.K., 2007. Cell selective response to gold nanoparticles. *Nanomedicine* 3, 111–119.

Panariti, A., Miserocchi, G., and Rivolta I., 2012. The effect of nanoparticles uptake on cellular behavior: disrupting or enabling functions? *Nanotechnol Sci Appl* 5, 87.

Panyam, J., Dali, M.M., Sahoo, S.K., Ma, W., Chakrawathi, S., et al., 2003. Polymer degradation and in vitro release of a model protein from poly (D,L-lactide-co-glycolide) nano and microparticles. *Journal of Controlled Released* 92, 173–187.

Perreault, F., Samadani, M., and Dewez, D., 2014. Effect of soluble copper released from copper oxide nanoparticles solubilisation on growth and photosynthetic processes of Lemna gibba L. *Nanotoxicology* 8, 374–382.

Piccinno, F., Gottscalk, F., Seeger, S., and Nowack, B., 2012. Industrial production quantities and uses of ten engineered Nanomaterials in Europe and the world. *J. Nanopart. Res.* 14, 1109.

Pojiak-Blazi M., Jaganjac M, and Zarkoviae N., 2010. *Cell Oxidative Stress: Risk of Metal Nanoparticles*. CRC Press: London, UK; New York, NY, USA.

Poli, G., Leonarduzzi, G., Biasi, F., and Chiarpotto, E., 2004. Oxidative stress and cell signaling. *Curr. Med. Chem.* 11, 1163–1182.

Priester, J.H., Ge, Y., Mielke, R.E., Horst, A.M., Moritz, S.C., Espinosa, K., et al., 2012. Soybean susceptibility to manufactured nanomaterials with evidence for food quality and soil fertility interruption. *Proceed Nat Acad Sci.* 109, E2451–E2456.

Pujalte, I., Dieme, D., Haddad, S., Serventi, A.M., and Bouchard, M., 2017. Toxicokinetics of titanium dioxide (TiO_2) nanoparticles after inhalation in rats. *Toxicol. Lett.* 265, 77–85.

Qiao, H., Liu, W., Gu, H., Wang, D., and Wang, Y., 2015. The transport and deposition of nanoparticles in respiratory system by inhalation. *J. Nanometer.* 2015, 394507.

Qiyu, Y., Liu, Y., Wamg, L., Xu, L., Bai, R., Ji, Y., et al., 2010. Surface chemistry and aspect ratio mediated cellular uptake of Au nanorods. *Biomaterials* 31(30), 7606–7619.

Radomski, A., Jurasz, P., Alonso-Escolano, D., Drews, M., Morandi, M., Malinski, T., and Radomski, M.W., 2005. Nanoparticle-induced platelet aggregation and vascular thrombosis. *Br. J. Pharma.* 146, 882–893.

Rafique, R., Aeshad, M., Khokhar, M.F., Qazi, I.A., Hamza, A., Virk, N., 2014. Growth response of wheat to titanium nanoparticles application. *NUST J. Engin. Sci.* 7, 42–46.

Rancan, F., Gao, Q., Graf, C., Troppens, S., Hadam, S., Hackbarth, S., et al., 2012. Skin penetration and cellular uptake of amorphous silica nanoparticles with variable size, surface functionalization and colloidal stability. *ACS Nano* 6, 6829–6842.

Rani, P.U., Yasur, J., Loke, K.S., and Dutta, D., 2016. Effect of synthetic and biosynthesized silver nanoparticles on growth, physiology and oxidative stress of water hyacinth: Eichhornia crassipes (Mart) Solms. *Acta Physiol Plant* 38, 1–9.

Rastogi, A., Zivcak, M., Sytar, O., Kalaji, H.M., Xiaolan, He., and Mbarki, S., 2017. Impact of metal and metal oxide nanoparticles on plant: a critical review. *Front Chem.* 5, 78.

Reiman, J., Oberle, V., Zuhom, I., and Hoekstra, D., 2004. Size dependent internalization of particles via the pathways of clathrin and caveolae mediated endocytosis. *Biochem J* 377, 159–169.

Rico, C.M., Lee, S.C., Rubenecia, R., Mukherjee, A., Hong, J., Perelta-videa, J.R., et al., 2014. Cerium oxide nanoparticles impact yield and modify nutritional parameters in wheat (*Triticum aestivum* L.). *J. Agri Food Chem.* 62, 9669–9675.

Rim, K.T., Song, S.W., and Kim, H.Y., 2013. Oxidative DNA damage from nanoparticle exposure and its application to workers' health: a literature review. *Saf. Health Work* 4, 177–186.

Runa, S., Hussey, M., and Payne, C.K., 2018. Nanoparticles-cell interactions: relevance for public health. *J. Phys. Chem. B*. 122, 1009–1026.

Salnikov, V., Lukyanenko, Y., Fredrick, C., Lederer, W., and Lukyanenko, V., 2007. Probing the outer mitochondrial membrane in cardiac mitochondria with nanoparticles. *Biophys. J.* 92, 1058–1071.

Schaeubl, N.M., Braydich-stolle, L.K., Schrand, A.M., Miller, J.M., Hutchison, J., et al., 2011. Surface charge of gold nanoparticles mediates mechanism of toxicity. *Nanoscale* 3, 410–420.

Seung Won Shin, In Hyun Song, and Soong Ho Um, 2015. Role of physiological properties in nanoparticles toxicity. *Nanomaterials* 5, 1351–1365.

Siegrist, M., Stampfli, N., Kastenholz, H., et al., 2008. Perceived risks and perceived benefits of different nanotechnology foods and nanotechnology food packaging. *Appetite* 52(2), 282–290.

Siegmann, K., Scherrer, L., and Siegmann, H.C., 1999. Physical and chemical properties of airborne nanoscale particles and how to measure the impact on human health. *J. Mol. Struct.* 458, 191–201.

Sohaebuddin, S.K., Thevenot, P.T., Baker, D., et al., 2010. Nanomaterial cytotoxicity is composition, size and cell type dependent. *Part Fibre Toxicol.* 7(22), 1–17.

Sun, T.Y., Bornhoft, N.A., Hungerbuhler, K., and Nowack, B., 2016. Dynamic probabilistic modeling of environmental emissions of engineered Nanomaterials. *Environ Sci Technol* 50(9), 4701–4711.

Sung, J.H., Ji, J.H., Yoon, J.U., Kim, D.S., Song, M.Y., Jeong, J., et al., 2008. Lung function change in Sprague-Dawley rats after prolonged inhalation exposure to silver nanoparticles. *Inhal. Toxicol.* 20, 567–574.

Technical Specification: Nanotechnologies Terminology and definitions for Nano-Objects Nanoparticle Nanofibre and Nanoplate. ISO/TS 80004-2:2008 Available online: www.iso.org/standard/44278.html (accessed on 12 October 2017).

Teske, S.S., and Detweiler, C.S., 2015. The biomechanisms of metal and metal-oxide nanoparticles interactions with cells. *Int. J. Environ. Res. Public Health* 12, 1112–1134.

Thorek, D.L., and Tsourkas, A., 2008. Size, charge and concentration dependent uptake of iron oxide particles by non-Phagocytic cells. *Biomaterials* 29, 3583–3590.

Trojanowski, R., and Fthenakis, V., 2019. Nanoparticles emissions from residential wood combustion: A critical literature review, characterization and recommendations. *Renew. Sustain. Energy Rev.* 103, 515–528.

Tubana, B.S., Babu, T., and Datnoff, L.E., 2016. A review of silicon in soil and plants and its role in US agriculture: history and future perspectives. *Soil Sci.* 181, 393–411.

U.S. Food and Drug Administration. 1999. Sunscreen drug products for over—the counter human use: final monograph. *Fed Regist.* 64, 27666–27693.

Vance, M.E., Kuiken, T., Vejerano, E.P., Mc Ginnis, S.P., Hochella, H.F., and Hull, D.R., 2015. Nanotechnology in the real word: redeveloping the nanomaterials consumer products inventory. *Beilstein J. Nanotechnology.* 6, 1769–1780.

Vandebriel, R.J., and De Jong, W.H., 2012. A review of mammalian toxicity of ZnO nanoparticles. *Nanotechnol. Sci. Appl.* 5, 61–71.

Vannini, C., Domingo, G., Onelli, E., De Mattia, F., Bruni, I., Marsoni, M., et al., 2014. Phytotoxic and genotoxic effects of silver nanoparticles exposure on germination wheat seedlings. *J Plant Physiol.* 171, 1142–1148.

Wang, S.H., Lee, C.W., Chiou, A., and Wei, P.K., 2010. Size dependent endocytosis of gold nanoparticles studies by three dimension mapping of plasmonic scattering images. *J Nanobiotecchnol* 8(1), 33.

Wang, Y., Aker, W.G., Hwang, H.M., Yedjou, C.G., Yu, H., and Tchounwou P.B., 2011. A study of the mechanism of in vitro Cytotoxicity of metal oxide nanoparticles using catfish primary hepatocytes and human HepG2 cells. *Sci Total Environ.* 409, 4753–4762.

Wang, Z., Li, J., Zhao, J. and Xing B., 2011. Toxicity and internalization of CuO nanoparticles to prokaryotic alga Microcystis aeruginosa as affected by dissolved organic matter. *Environ Sci Technol.* 45, 6032–6034.

Wang, Z.L., 2004. Zinc oxide nanostructures: growth, properties and applications. *J Phys Condens Matter* 16, R829.

Wang, Z., Xu, L., Zhao, J., Wang, X., White, J.C., and Xing B., 2016. CuO nanoparticles interaction with Arabidopsis thaliana: toxicity, parent-progeny transfer, and gene expression. *Environ. Sci. Technol.* 50, 6008–6016.

Weibel, E.R., 1963. *Morphology of the Human Lung.* Springer: New York, NY, USA.

Weibel, E.R., Sapoval, B., and Filoche, M., 2005. Design of peripheral airways for efficient gas exchange. *Respir. Physiol. Neurobiol.* 148, 3–21.

Weir, A., Westerhoff, P., Fabricius, L., Hristovski, K., and Von Goetz, N., 2012. Titanium dioxide nanoparticles in food and personal care products. *Environ. Sci. Technol.* 46, 2242–2250.

Wickett, R.R., and Visscher, M.O., 2006. Structure and function of the epidermal barrier. *AM. J. Infect. Control* 34, 98–110.

Wilson, R.F., 2006. Naotechnology: the challenge of regulating known unknowns. *J. Law Med. Ethics* 34, 704–713.

Winkler, H.C., Suter, M., and Naegeli, H., 2016. Critical review of the safety assessment of nano-structured silica additives in food. *J. Nanobiotechnol.* 14, 44.

Witschi, H.P. 2001. Toxic responses of the respiratory system. In *Casarett and Doull's Toxicity: The Basic Science of Poision*, 6th edn, Klaassen, C.D., Ed. McGraw-Hill: New York, NY, USA, pp. 515–534.

Wu, Y-N., Yang, L-X., Shi, X-Y., Li, I-C., Biazik, J.M., Ratinac, K.R., et al., 2011. The selective growth inhibition of oral cancer by iron core-gold shell nanoparticles through mitochondria- mediated autophagy. *Biomaterials* 32(20), 4565–4573.

Yang, Y.X., Song, Z.M., Cheng, B., Xiang, K., Chen, X.X., et al., 2014. Evaluation of the toxicity of food additive silica nanoparticles on gastrointestinal cells. *J Appl Toxicol* 34, 424–435.

Zhao, M.X., and Zhu, B.J., 2016. The Research and applications of quantum dots as nanocarriers for targeted drug delivery and cancer therapy. *Nanoscale Res. Lett.* 11, 207.

Zhu, H., Han, J., Xiao, J.Q., and Jin, Y., 2008. Uptake, translocation, and accumulation of manufactured iron oxide nanoparticles by pumpkin plants. *J Environ Monitor.* 10, 713–717.

Zhu, M., Nie, G., Meng, H., Xia, T., Nel, A., and Zhao, Y., 2012. Physicochemical properties determine nanomaterial cellular uptake, transport, and fate. *Acc Chem Res* 46(3), 622–631.

Zuvera-Mena, N., Armendari, R., Peralta-Videa, J.R., Gardea-Torresday, J.L., 2016. Effects of silver nanoparticles on radish sprouts: root growth reduction and modifications in nutritional value. *Front. Plant Sci.* 7, 90.

3 Toxicology of Metal Ions, Pesticides, Nanomaterials, and Microplastics

Kamayani Vajpayee, Krupa Kansara, and Ritesh K. Shukla

CONTENTS

3.1 Introduction 35
    3.1.1 Environmental Toxicology 37
3.2 Emerging Environmental Contaminants 37
    3.2.1 Metal Ions..37
        3.2.1.1 Routes of Exposure 38
        3.2.1.2 Toxicity and Mechanism of Action 39
    3.2.2 Pesticides...43
        3.2.2.1 Routes of Exposure 43
        3.2.2.2 Toxicity and Mechanism of Action 44
    3.2.3 Nanomaterials..46
        3.2.3.1 Routes of Exposure 46
        3.2.3.2 Toxicity and Mechanism of Action 47
    3.2.4 Micro- and Nanoplastics 48
        3.2.4.1 Routes of Exposure 49
        3.2.4.2 Toxicity and Mechanism of Action 49
3.3 Conclusion...50
References..51

3.1 INTRODUCTION

Since its inception in the 16th century, toxicology has been perceived as *The Science of Poison*. The term, however, is defined as a multidisciplinary subject dealing with the study of effects stemming from the interaction of physical and chemical agents with living organisms (Costa & Teixeira, 2014). Earlier toxicologists were engaged mostly in the *identification, detection,* and *quantification* of toxic materials. However, today, toxicology has seen a straight shift where toxicologists not only study the toxic effects of several substances on living organisms, but also are engaged in the

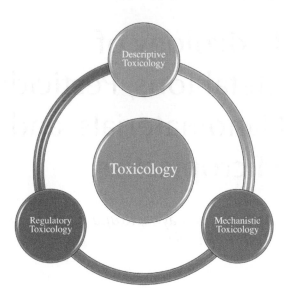

FIGURE 3.1 Major subdivisions of toxicology.

quantification of hazards, estimating the public health aspect of toxic agents, and accessing the novel pharmaceutical drugs (Gerba, 2019).

Given its varied applications in different sectors toxicology can be further divided into the three major subdivisions: *Descriptive Toxicology, Mechanistic Toxicology,* and *Regulatory Toxicology* (Figure 3.1) (Costa & Teixeira, 2014). Additionally, there are other specialized fields like *Forensic Toxicology, Clinical Toxicology, Ecotoxicology, Environmental toxicology,* etc., dealing with the different toxicological aspects of the respective areas. They also work in association with each other to assess and regulate the effect of toxicants (Costa & Teixeira, 2014).

It is a well known fact that the industrial revolution brought about the development of several kinds of industries like the paper industry, glass industry, leather industry, textile industry, etc. These anthropogenic developments besides aiding modernization also contributed by adding several contaminants, namely metal ions (glass, textile, and paper industry, etc.), pesticides, nanomaterials, etc., by directly discharging their chemical waste into the environment. These contaminants, as discussed briefly in the following sections, get incorporated into living systems causing several unrepairable alterations in physiology. Metals ions being readily absorbed through food, air, and water interact with biomolecules causing conformational changes and ultimately causing geno- as well as cytotoxicity (Genchi et al., 2020; Rehman et al., 2019). In a similar manner regular use and discharge of pesticides causes bioaccumulation in the food chain and thus pose a serious threat to living beings (Bhagat et al., 2021; Moebus & Boedeker, 2021). With the increasing awareness towards the toxic effects posed by these anthropogenic actions on biotic and abiotic components of the environment, it becomes necessary to understand the basic behaviour of such contaminants in the environment. Thus, this chapter explores the toxic impact of major environmental contaminants like metal ions, pesticides, micro- and nanoplastics on the biotic system and their fate.

3.1.1 ENVIRONMENTAL TOXICOLOGY

Environmental toxicology as mentioned in the preceding section is a branch of toxicology that deals with the impact of diverse toxicants (contaminants) on the environment. Environmental contaminants are now widely considered as a critical reason for the concern due to the multiple examples of their harmful effects on human health and the environment. Toxic heavy metals notably lead, mercury, arsenic, cadmium, petroleum and polyaromatic hydrocarbons, and various herbicides and pesticides are all commended contaminants. These are commonly referred to as *traditional* or well-known pollutants. However, a large number of synthetic chemicals is also being discharged into the environment each year, either as part of the manufacturing process or as an industrial waste commonly termed as *emerging contaminants* (ECs) (Costa & Teixeira, 2014).

Today, there is a great awareness of the presence of these new ECs in the environment, which has been the subject of a lot of recent research. Studies are now being conducted to evaluate and assess the routes of exposure of these contaminants to animals including humans, their toxic effects, and the underlying mechanism behind the toxic effects. Thus, an attempt has been made in the following sections to briefly evaluate and understand the negative outcome of contaminants like metal ions, pesticides, nanomaterials, and micro-nanoplastics.

3.2 EMERGING ENVIRONMENTAL CONTAMINANTS

For a better understanding of the impact of environmental contaminants, their mode of eliciting toxic reactions and hindering the normal functioning of the cellular physiology, the contaminants have been classified and described as follows.

3.2.1 METAL IONS

Metal toxicity can be induced by both essential metal ions, such as iron and copper, as well as non-essential metals, such as cadmium, lead, and mercury, which are not vital for life but can have deleterious effects when introduced into the human environment, frequently with disastrous ramifications (Nordberg et al., 2007; Egorova & Ananikov, 2017). Most biological systems require approximately 25 elements, including a significant number of metal ions. Humans, however, require 10 key metal ions. Four amongst them, calcium, magnesium, sodium, and potassium, are referred to as "bulk elements" since they account for nearly all (99%) of the metal ion content in the human body. The other five transition metals, namely, zinc, molybdenum, manganese, iron, cobalt, and copper, are considered "trace elements," since they have significantly lower nutritional requirements than the bulk elements but are nonetheless necessary for life (Crichton, 2018). Alternatively, non-essential metals (heavy metals) as the name suggests are not required for the functioning of normal physiology. They are at times defined as dense metals or metalloids that are potentially harmful to life, such as cadmium, lead, mercury, and arsenic.

3.2.1.1 Routes of Exposure

The celebrated dictum of Paracelsus, "What is there that is not poison? All things are poison and nothing is without poison. Solely the dose determines that a thing is not a poison," (Grandjean, 2016) and Claude Bernard's definition of homeostasis (Billman, 2020) significantly demonstrate the principle behind the toxicity of essential metal ions. Every single cell of the body has a precise concentration of these metal ions that is necessary for normal physiological function. However, the toxic effects can be observed when this limit is overreached.

Whilst in the case of heavy metals, there are several ways from which animals, including humans, get exposed to them. Lead-based paint and home dust from the surfaces covered with paints (in air and food) are all sources of low-level environmental exposure to lead (Pb). In addition to this, water from Pb pipes has a major role to play in Pb toxicity. Inhalation and ingestion are the two main pathways for Pb absorption, with inhalation being the more efficient route. Studies suggest the removal of the Pb^{4+}-derived anti-knock compound (tetraethyl lead), which was routinely added to petroleum to increase car engine efficiency, has dramatically reduced blood Pb levels in the urban population. As per the World Health Organisation (2021) humans are also exposed to lead through the regular use of cosmetic products and also from the traditional medicines having lead in them ("Lead poisoning," 2021).

Today, Asia is the leading source of anthropogenic atmospheric mercury (Hg), accounting for more than half of planetary emissions, with substantial Hg contamination in local environments impacted mostly by the chemical industry and gold and mercury mining. According to studies, there has been an order of magnitude increase in Hg in humans, selected Arctic marine mammals, and birds of prey that began in the mid-to-late 19th century and surged in the 20th century. The human contribution to current Hg levels is estimated to be 92 percent (Kumari et al., 2020). The famous Minamata disease and epidemic of methyl mercury poisoning in Iraq (1972) and Japan are the evidence representing human exposure to mercury via food (Kumari et al., 2020).

Cd^{2+} is a ubiquitous environmental pollutant that is primarily absorbed through the respiratory or gastrointestinal systems (QIN et al., 2020). Cigarette smoke and tainted food and beverages are the most common non-industrial sources of exposure. Cd^{2+} is rapidly transferred from soil to plants, and some plant species, such as tobacco, rice, wheat, peanuts, and cocoa, collect substantial levels of Cd^{2+} even in low-Cd^{2+} soils. The highest levels of Cd^{2+} were discovered in topsoil in Europe shortly after the distribution of P_2O_5, implying that soil contamination is caused by the use of rock phosphate fertilizer in intensive arable agriculture (Crichton, 2016). The majority of the Cd^{2+} consumed orally in the instance of Itai–Itai illness came from contaminated rice (Ralston & Raymond, 2010). A representative diagram depicts the possible source of heavy metals, their fate in the human body, cellular system, and in the environment after discharge from the sources (shown in Figure 3.2).

Toxicology of Metal Ions, Pesticides, Nanomaterials, and Microplastics

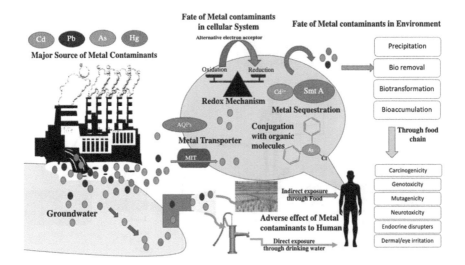

FIGURE 3.2 Major routes of metal ion exposure and their possible fate in the human system.

3.2.1.2 Toxicity and Mechanism of Action

As mentioned in the preceding sections, the metal ions – essential or non-essential – pose a great impact on human health and physiology. Each metal ion shows its characteristic toxic behaviour in a living body. A general toxicity mechanism of heavy metal ions is shown in Figure 3.3. A brief account of the commonly studied metal ions, their toxicity, and the mechanism of action underlying is as follows:

- **Lead (Pb):** Inorganic Pb compounds have been linked to an increased incidence of diseases in a variety of organisms, according to several studies. Pb poisoning has been linked to neurological problems, hypertension, and cognitive deficits, among other things (Patrick L. 2006). According to Chen et al. (2012) exposing zebrafish embryos to modest doses of Pb resulted in embryonic toxicity, behavioral changes, and adult learning/memory deficits.

Another study found that when earthworms were exposed to >2000 mg Pb/kg dw, a high mortality rate was observed. There was a loss of weight, and they could not reproduce (Luo et al., 2014). Gargouri et al. (2013) found that exposing HEK293 kidney cells to lead resulted in decreased cell viability, distortion, cohesion loss, superoxide anions generation, and lipid peroxidation.

Pb slows structural protein synthesis by binding to sulfhydryl protein and decreases the availability and amounts of sulfhydryl antioxidant reserves in the body (Nuran Ercal et al., 2001; Patrick, 2006). When mice are exposed to Pb salts, their calcium homeostasis is compromised (Zhang et al., 2013), and the kidney was found to be the primary organ of damage (Agrawal et al., 2014).

The production of free radicals by Pb causes damage to living systems in two different ways (Nuran Ercal et al., 2001). The first pathway involves the direct formation of reactive oxygen species (ROS) such as O_2, H_2O_2, and hydroperoxides,

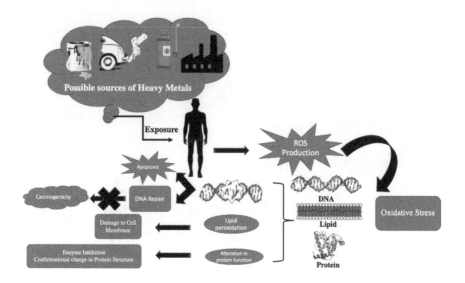

FIGURE 3.3 Mechanisms underlying the cytotoxic and genotoxic behavior of metal ions in the human body.

whereas the second involves the depletion of antioxidants in cells (NuranErcal et al., 2001). The enzymes delta-aminolevulinic acid dehydratase (ALAD) and glutathione reductase (GR) both are susceptible to Pb. Pb exposure has also been linked to the production of reactive nitrogen species (RNS). In vascular endothelial cells, ROS oxidizes nitric oxide to generate peroxynitrite (ONOO), which is highly reactive and can damage DNA and lipids.

Pb binds to protein in erythrocytes. Pb is largely preserved in the bones of animals exposed to it (Hambach et al., 2013). Baranowska-Bosiacka et al. (2009) discovered that mild concentrations of Pb2+ ions moderately inhibited both phosphoribosyl transferases in erythrocytes, which is a crucial mechanism of toxicity. DNA damage, suppression of DNA synthesis or repair, and clastogenicity are all mechanisms linked to Pb's carcinogenic propensity in organisms. Pb damages DNA by producing reactive oxygen species (ROS), which cause DNA oxidation and subsequent damage (Hsu, 2002). Pb can harm mitochondria, trigger apoptosis, and deplete glutathione levels inside cells (Sabath & Robles-Osorio, 2012). It has also been observed to influence gene expression by substituting elements such as Zn in proteins and lowering protein-DNA binding.

Further the studies from Lyu et al. (2017), Singh et al. (2018), and Bellinger et al. (2018) also demonstrate the fatal behavior of lead poisoning.

- **Mercury:** Organic mercury (MeHg) exposure has been associated with mental retardation, cerebral palsy, and dysarthria in children, particularly those exposed in utero (Boucher et al., 2014; Golding et al., 2013). If MeHg is not coupled to thiol ligands, it exists in the human body as a water-soluble compound. It exists as an L-cystein complex after crossing the blood-brain barrier (Guzzi & La

Porta, 2008), and it has been associated with increased antinuclear autoantibody (ANA) and alterations in serum cytokine profile in adults (Nyland et al., 2011). Low-dose mercury has the power to alter a developing embryonal neural system.

The presence of a neurotoxicant such as mercury can stymie the development of behavioral domains such as sensory, motor, and cognitive abilities (Zahir et al., 2005). In a study, prenatal MeHg exposure in offspring was linked to dysplasia of the cerebral and cerebellar cortex, as well as neuronal ectopia (Grandjean et al., 2014). Hg exposure at low concentrations causes oxidative stress, cell cytotoxicity, and an increase in amyloid levels associated with disorders like Parkinson's (Goldman, 2014).

The toxicity of organic mercury, MeHg, stems from its electrophilic nature, which enables it to interact with nucleophiles. Sulfhydryl (–SH) or selenohydryl (–SeH) usually interacts with MeHg. Hg combines with cellular nucleophiles to generate stable complexes (Farina et al., 2011). During interactions, oxidative stress can develop due to the depletion of biomolecule antioxidants, which leads to an increase in ROS production causing DNA damage, denaturation of enzymes, and even apoptosis.

Damage to mitochondria as a result of GSH depletion is one of the key mechanisms related to Hg poisoning. The formation of free radicals is accelerated when mitochondria are damaged. On contact with Hg molecules, sulfhydryl proteins and GSH are depleted, resulting in the production of ROS.

- **Cadmium:** Cadmium is a heavy metal of which the most prevalent form is Cd^{2+}. Because of its low excretion rate, it can be harmful even at low concentrations (Nawrot et al., 2010; Pretto et al., 2014). Workplace exposure to Cd has been associated with malignancies of the lungs, prostate, and kidney in workers working in industries such as metal smelting and battery manufacturing (Nordberg et al., 2005).

Cd toxicity in cellular systems is the result of a biological system's contact with it. Cadmium interferes with sulfhydryl enzymes and other cell ligands, causing oxidative phosphorylation pathways to be disrupted. The resulting complex is re-introduced into circulation. The Cd-metallothione combination is absorbed by the kidney's proximal tubular cells, causing tubular dysfunction and oxidative stress (Patrick, 2003; Sabolic et al., 2006). It affects the respiratory system and promotes the formation of kidney stones (Hambach et al., 2013). Cd exposure to the liver causes hepatocellular damage in humans via the production of metallothionein (Kang et al., 2013). Cd inhibits the activity of numerous enzymes that protect the liver from oxidative stress by upregulating the expression of stress genes. Lipid peroxidation, DNA damage, and protein carbonylation are promoted by the deterioration of hepatic indicators and the increase of oxidative stress (Rashid et al., 2013).

Cd causes testicular necrosis in rats, according to Yang et al. (2003). In their research, rats exposed to Cd had higher levels of MDA and GSH peroxidase (GSH-Px). They discovered that glutathione interacts with intracellular oxygen radicals through the GSH-Px/GSH complex. DNA damage was found to be substantial, and Superoxide dismutase (SOD) activities were shown to be reduced (Yang et al., 2003). Cadmium affects the cellular thiol redox equilibrium. It adds to oxidative stress. Cd indirectly creates ROS and promotes the oxidation of protein, lipid, and DNA in cells, contributing to DNA damage and tumorigenesis (Liu et al., 2009; Wang et al., 2004). Although Cd does not produce free radicals, Waisberg et al. in 2003 reported the formation of nitric oxide, superoxide, and hydroxyl radicals. Cadmium's propensity to substitute iron and copper in many cytoplasmic and membrane proteins is the mechanism behind its indirect creation of free radicals. Through the Fenton reaction, an increase in unbound free Fe and Cu metals contributes to oxidative stress (Wätjen & Beyersmann, 2004). Cu's potential to affect the dissociation of H_2O_2 via the Fenton reaction is enhanced when it is displaced from its binding site. Cadmium also affects cell function and homeostasis by interfering with transport across cell membranes and epithelial (Van Kerkhove et al., 2010).

- **Arsenic:** Arsenic compounds are proven to be carcinogenic to humans and can be found in nature in both organic and inorganic forms. It has been observed that the inorganic species are more toxic and accumulate in exposed organisms (Liao et al., 2004). The trivalent form of As is the most hazardous and frequently interacts with thiol groups in proteins, whereas the pentavalent form has a higher oxidative phosphorylation ability. Hepatocellular carcinoma, melanosis, hyperkeratosis, and parenchymal cell damage have all been linked to chronic low-concentration arsenic exposure (Ahmed et al., 2013; Das et al., 2004; He et al., 2014).

It has been recently discovered that exposing C57BL/6 mice to 10 ppm As for 4 weeks resulted in significant changes in gut microbiota composition. They further suggested that As exposure not only impacts the gut microbiome community in terms of abundance, but it also affects its metabolic profile in terms of function l (Lu et al., 2014). Early-life As exposure was also observed to impair influenza A clearance and aggravate the inflammatory response, resulting in acute and long-term alterations in mouse lung mechanics and airway structure (Ramsey et al., 2013). Chronic exposure can lead to liver and kidney damage, anemia, peripheral neuropathy, mucous membrane irritation, and skin irritation (Rahman et al., 2014).

Arsenic toxicity in animals and humans is closely linked to its metabolism. The mechanism is heavily reliant on the As species involved. Because they have comparable features and structures, As (V) alters the system of organisms by substituting phosphate in essential metabolic reactions. In the human transport system, As(V) can also substitute phosphate in the anion exchange in red blood cells (Hughes et al., 2011). As(III) inhibits a variety of biological enzymes, including pyruvate dehydrogenase, lowering pyruvate to acetyl coenzyme A conversion. Binding to sulfhydryl inhibits enzyme activity. Arsenic decreases glucose absorption into cells, fatty acid oxidation, gluconeogenesis, and acetyl coenzyme A synthesis. As well, it prevents the production of GSH, one of the most effective antioxidants.

In vitro, As(III) combines with thiol-containing compounds including GSH and cystein with dithiols having a stronger affinity than monothiols. As(III) binding to essential thiols prevents vital metabolic reactions, resulting in toxicity. Methylated As(III) also inhibits GSH reductase and thioredoxin reductase, which could be due to

As(III) interacting with key thiol groups, which could cause toxicity. One significant mechanism of action for As toxicity is the generation of reactive oxygen and nitrogen species in the presence of As (Wu et al., 2016).

The formation of reactive species in the context of As is linked to a variety of functions, including DNA repair inhibition, genotoxicity, cell proliferation, and signal transduction (Hughes et al., 2011). H_2O_2, OH radicals, reactive nitrogen species, and peroxyl radicals are some of the reactive species generated. Epigenetic modifications (histone modification, DNA methylation, and noncoding RNA) and induced genomic mutations are two mechanisms currently postulated for malignancies caused by As exposure (Wu et al., 2016).

3.2.2 Pesticides

Any product or mixture of substances intended to kill, repel, or otherwise control a "pest," such as insects, snails, rodents, fungi, bacteria, or weeds, is referred to as a "pesticide" (Bolognesi & Merlo, 2011). The "green revolution" resulted in a large increase in pesticide application, resulting in increased productivity. It is quite difficult to meet a single principle while generating pesticide classification, hence in most circumstances, multiple techniques are preferred. The pesticides are thus, classified into the following categories:

a. Based on assignment: herbicide, pesticide, fungicide, etc.
b. Based on the method of impact: contact, fumigant, and systemic.
c. Based on chemical nature: Organochlorines, organophosphates, carbamates, etc.

3.2.2.1 Routes of Exposure

Pesticides may now be found practically wherever on the planet. Surface runoff, leaching, and/or erosion can all be ways for pesticides originating from human activities to infiltrate water bodies. Meanwhile, pesticide residues can be carried into the atmosphere by drift, evaporation, and wind erosion, resulting in contamination of surface waters, soils, vegetation, and wildlife via precipitation, often at locations far from their source (Dubus, Hollis & Brown, 2000).

Many pesticides circulate in ecosystems and may be acquired by many living animals and may move across food chains due to their cumulative qualities (Wilkinson et al., 2000). Because most creatures in the food web interact with one another, understanding pesticide migration and bioconcentration from dietary exposure are critical for assessing their true environmental consequences (Katagi, 2010). Chemicals can enter surface waters via runoff and erosion (Giddings et al., 2005), but persistent chemicals can also accumulate through other mechanisms such as direct uptake by gills or skin (bioconcentration), uptake of suspended particles (ingestion), and consumption of contaminated food (biomagnification) for aquatic organisms (van der Oost et al., 2003)

Pesticides are absorbed by most terrestrial animals through their skin, respiratory, and gastrointestinal tract surfaces. Different pesticides enter the body through the skin and nasal mucosa (Hodgson, 2010). A few pesticides have been found to generate toxic endpoints in the nasal tissues, and several have been linked to nasal lesions or cancers in experimental animals (Hodgson, 2010). The lung is also a key source of exposure to airborne contaminants that come into contact with blood (Ding & Kaminsky, 2003). Pesticides acquired through all modes of exposure, finally end up in the liver, which is the principal location of pesticide biotransformation for easier elimination through excretion of water-soluble detoxifying products.

3.2.2.2 Toxicity and Mechanism of Action

Pesticides and other associated compounds' toxicity to non-target organisms is still a big concern all over the world. Pesticides can cause a variety of physiological and biochemical changes when they enter the body, searching for mechanisms of toxicity is considerably more difficult than anticipated. Pesticides can have negative consequences by interfering with the hormones or messengers directly impacting the nervous system (e.g., organochlorine pesticides) (Bolognesi & Merlo, 2011), or inducing changes in the activities of certain enzymes directly or indirectly. Due to their autoxidation by molecular oxygen, a significant group of insecticides may directly increase ROS levels in living organisms (Bolognesi & Merlo, 2011). Mostafalou & Abdollahi (2013) have done a lot of work to catalog the molecular pathways of pesticide toxicity systematically. Their research yielded a theoretical interpretation of the causal link between pesticide exposure and human chronic diseases caused by DNA damage (Mostafalou & Abdollahi, 2013). The possible toxicity mechanism of pesticides is shown in Figure 3.4.

The toxicity and the mechanism behind these effects can systematically be studied under the following heads:

- **Pesticides inducing neurotoxicity:** The neurological system is the primary target for organochlorines. Some of them, including variants of the banned pesticide DDT, have been found to cause apoptosis in brain cells by activating mitogen-activated protein kinases (Shinomiya & Shinomiya, 2003). DDT and

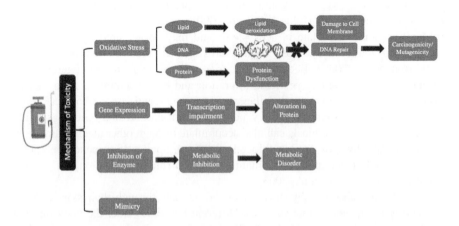

FIGURE 3.4 Flow chart demonstrating the possible toxic mechanisms of pesticides.

pyrethroid toxicity have been linked to the blockage of voltage-gated sodium channels (VGSCs) in the plasmatic membrane of neurons (Silver et al., 2017). Endosulfan's neurotoxic action is most likely due to its ability to inhibit GABAA-gated chloride channels in neurons (Kamijima1 & Casida, 2000).

Because they are irreversible acetylcholinesterase inhibitors, organophosphorus pesticides are also strong neurotoxins (Galloway & Handy, 2003). Non-specific esterases or pseudocholinesterases with a high affinity for butyrylcholine are found in most species. In aquatic species, the activity of AChE has been routinely utilized to identify pesticide exposure to organophosphate or carbamate insecticides (Fulton & Key, 2001). Pesticides containing organophosphates phosphorylate acetylcholinesterase thereby lowering its activity. Exposure to pesticides in high quantities can produce symptoms such as nausea, dizziness, and confusion, as well as hyperactivity, paralysis, respiratory paralysis, and death (John & Shaike, 2015).

The central nervous system has been the focus of many investigations proving the neurotoxicity of 2,4-dichlorophenoxyacetic acid (2,4-D). The generation of free radicals contributes to this toxicity, which results in lower GSH levels and impaired antioxidant enzyme activity, such as superoxide dismutase and catalase (Bukowska, 2003).

- **Reactive oxygen-mediated toxicity:** Many contaminants, including pesticides, have been linked to oxidative stress induction. This induction can happen in a variety of ways: Some chemicals may increase ROS production as a by-product of detoxification pathways; altering the operation of the mitochondrial and endoplasmic reticulum electron transport chains, resulting in ROS overproduction; increasing ROS production by entering autoxidation, the central mechanism for the toxic effects of many environmental toxicants and the toxic effects of numerous environmental toxicants, including pesticides; pesticides can inhibit antioxidant and related enzymes, as well as the formation of antioxidants such as glutathione (van der Oost et al., 2003).

Metals in these pesticides' chemical structures can stimulate ROS generation via the Fenton reaction, which could explain the observed prooxidant activity (Lushchak, 2016). Interestingly, oxidative stress was observed in a variety of animal taxa, including rats, mice, zebrafish, clams, and the sea snail, Hexaplextrunculus, implying that oxidative stress generated by permethrin could be a common mechanism for its toxicity (Wang et al., 2016). Mancozeb, thiram, and disulfiram altered membrane potentials, inhibited ATP-dependent glutamate absorption into synaptic vesicles, and blocked glutamate binding to its receptors, all of which resulted in excitotoxic effects in the brain. It has also been observed that an imbalance between GSH and GSSG, is produced owing to the prooxidant properties of DTC compounds.

- **Carcinogenic and Genotoxic effects of Pesticides:** A few pesticides, such as dithiocarbamates, have been shown to cause cancer. The main concern of these compounds is the development of carcinogenic chemicals like N-nitrosocarbaryl, a carbaryl derivative that is a powerful carcinogen in rats (Bolognesi & Merlo,

2011). Ethylene thiourea, a degradation product of ethylene bisdithiocarbamate fungicides, has been identified as a teratogen, goitrogen, and carcinogen that can interfere with thyroid function and is linked to thyroid cancer in animals.

In vitro experiments of zineb effects on human lymphocytes and CHO cells revealed DNA strand breaks, implying that it may cause cancer if the afflicted cells survived and multiplied. In rat fibroblasts exposed to mancozeb, Calviello et al. (2006) observed DNA single-strand breaks. Many pesticides studied caused a variety of mutations by causing DNA damage. Agrochemical components have a modest genotoxic potential, but occupational exposure to pesticide combinations has been linked to an increase in genotoxic damage in several studies (Calviello et al., 2006).

Genotoxicity is known to be caused by pesticide-induced oxidative stress (Franco et al., 2010). Pesticides have been demonstrated to disrupt cellular redox equilibrium, resulting in increased ROS levels, lipid peroxidation, and antioxidant defense depletion (Lushchak, 2016). Besides these, the studies of Mesnage & Séralini (2018), Bonner & Alavanja (2017), and Rani et al. (2021) also have reported the ill effects of the pesticides, their exposure routes, and fate in living organisms.

3.2.3 Nanomaterials

Nanotechnology has progressed to an advanced level in recent years as a result of major research efforts. It is one of the fastest-growing fields, with applications in nearly every sector, including health care and medicine, cosmetics, material science, food science, and information technology.

Carbon-based nanoparticles (NPs), metal-based nanomaterials, dendrimers, and composites are the four types of nanomaterials present in the human environment. Fullerenes and nanotubes (carbon-based nanomaterials) are employed in films, coatings, and electronics. Metal-based nanomaterials including nanosilver, nanogold, and metal oxides (e.g., titanium dioxide (TiO_2)) are employed in food, cosmetics, and pharmaceuticals. Nanopolymers make up the dendrimers. They're utilized to deliver drugs. Nanoclays, for example, are composites that blend one NP with other nano-sized or bigger particles. They're commonly seen in packaging materials. These nano-sized materials have a high surface-to-volume ratio, resulting in an exponential increase in reactivity, making them appropriate for a wide range of scientific applications. In addition, the mechanical, physical, biological, optical, and chemical properties of the material differ dramatically (López-Serrano et al., 2014).

These unique qualities piqued economic interest while also raising severe concerns about NPs' influence on human health. Several research works have revealed the negative effects of NPs on the environment and organisms including humans (Shukla, Badiye, Vajpayee & Kapoor, 2021). The following sections will hence address the exposure levels and toxic mechanisms behind this novel field.

3.2.3.1 Routes of Exposure

Like several other toxic elements, humans and other organisms get exposed to NPs via ingestion, inhalation, or direct dermal contact. However, human exposure to NPs occurs mostly through ingestion, either directly through food or indirectly by NP

Toxicology of Metal Ions, Pesticides, Nanomaterials, and Microplastics

dissolution from food containers or secondary absorption of breathed particles. They may be transported throughout the body by blood circulation. Probably, their size and surface properties like polarity, hydrophilicity, lipophilicity, and catalytic activity influence their dispersion in the body. Because surface area per unit mass increases as particle size decreases, NPs are expected to have more chemical and biological activity in the organism's body. As a result, smaller NPs may be more harmful than their larger counterparts. According to the theory, the smaller NPs are thought to be taken in by the cells faster than the larger ones. Inhalation is another common way for people to be exposed to NPs in the air. Moreover, human exposure to NPs through skin absorption has also been studied. With pharmacological treatments and cream application, dermal exposure is unavoidable (sunscreens and others). Their epidermal penetration is influenced by several parameters, including the medium of exposure, pH of the medium, temperature, and so on. Blood and macrophages, lymph veins, dendritic cells, and nerve terminals are all found beneath the dermal layer. As a result, particles that are absorbed beneath the various layers of skin are easily transferred through various circulatory systems (Oberdörster et al., 2005; Ganguly et al., 2018).

3.2.3.2 Toxicity and Mechanism of Action

Nanotoxicity is mainly dependent upon the size, shape, surface, and composition of the NPs (shown in Figure 3.5). Several in vitro and in vivo models (*Danio rerio* and *Caenorhabditis elegans*, etc.) are currently being used to elucidate the toxicity posed by these NPs and nanomaterials (NMs) (Chakraborty et al., 2016; Hunt et al., 2013). Several studies have shown the toxic impact of NPs and NMs on the liver, kidney,

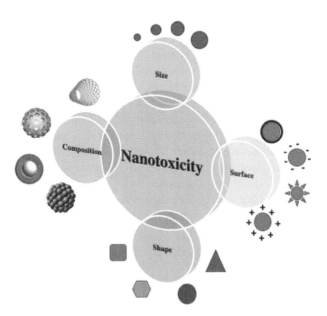

FIGURE 3.5 A schematic demonstration of factors influencing the toxic behavior of nanoparticles.

lungs, skin, cardiovascular system, and immune system of the organism (Sahu & Hayes, 2017). But the major routes by which they elicit the toxic response still remain unanswered.

The role of metal ions in the toxicity of metallic NPs is still being debated. According to the findings, metal ions may be the cause for the toxicity of their corresponding metallic NPs. In a study on zinc nanoparticles (ZnO NPs) researchers suggested that the metal NPs can affect the cell's "metal hemostasis," causing changes in all of the cell's vital activities (Lebrun et al., 2014).

For a wide range of substances, oxidative stress is a significant mechanism of toxicity. The damage to mitochondrial membranes caused by cellular oxidative stress is an early sign of cellular stress. Damage to the mitochondrial membrane causes mitochondrial malfunction, which is a crucial stage in cell injury and death. The toxicity of nanomaterials appears to be influenced by oxidative stress. NPs of different sizes and chemical compositions can preferentially localize in mitochondria, causing oxidative stress and cellular damage. In cultured human embryonic kidney (HEK293) cells, Wang et al. (2009) studied the mechanism of cytotoxicity of SiO2 NPs. They noticed a dose-dependent reduction in cell viability, a rise in intracellular ROS, and a drop in GSH content; they reported the cytotoxicity of SiO_2 NPs in HEK293 cells was linked to oxidative stress. These findings suggested that oxidative stress may play a role in NP toxicity. This may not be true for all NPs, though. In HepG2 and $CaCo_2$ cells treated to 20 nm nanosilver, Sahu et al. (2016) found no direct involvement of cellular oxidative stress.

Inflammation, which is mediated by the generation of inflammatory mediators like cytokines, appears to play a key role in nanomaterial toxicity. Dendritic cells exposed to TiO_2 NPs produced more pro-inflammatory cytokines, as reported by Schanen et al. (2009). In human monocytes, Ainslie et al. (2009) reported the generation of inflammatory cytokines and ROS produced by titanium, silicon oxide, and polycaprolactone NPs. In comparison to the positive control, lipopolysaccharide, these NPs caused a considerably lower inflammatory response. Titanium NPs were found to be more inflammatory than silicon oxide and polycaprolactone NPs.

Silver NPs were found to have pro-inflammatory effects in primary rat brain microvessel endothelial cells in vitro, according to Trickler et al. (2010). These NPs caused cytotoxicity and a pro-inflammatory response that was size- and duration-dependent. In comparison to the larger 80 nm silver particles, the smaller 25 nm silver particles caused much more harmful effects at lower concentrations and shorter exposure times, according to their research.

Thus, both oxidative stress and inflammation may play a role in nanomaterial toxicity. However, the precise chemical processes of nanotoxicity are still unknown.

3.2.4 Micro- and Nanoplastics

Microplastics are plastic materials with a microsize dimension (typically ranging from 1 m to 5 mm) (Verla et al., 2019).Their prevalence in the environment is steadily increasing, especially in the ocean, as evidenced by an increase in the frequency and quantity of plastic consumed by seabirds. They have also recently been quantified in considerable quantities in the soil ecosystem and plant tissues. The

Toxicology of Metal Ions, Pesticides, Nanomaterials, and Microplastics   49

fact that microplastics are ubiquitous and bioavailable for injection by sea animals, soil, and plants growing on microplastic polluted soils is one of the key possible hazards associated with them. They have been found in seabirds, fish, sandhoppers, sea turtles, crustaceans, and mussels, soil invertebrates like collembolan, oligochaeta (e.g., earthworms), and isopods, and plants like wheat plants. Their presence has also been found in table salts, drinkable water, and human excreta (Verla et al., 2019).

3.2.4.1 Routes of Exposure

Human exposure to microplastics can be direct (through intake of microplastic-contaminated water, soil, or salt) or indirect (by consumption of microplastic-contaminated seafood and plants). Further, they can be inhaled via airborne microplastics. Seafood consumption may pose the biggest risk, as it is a key source of protein and the sea is a hotspot for plastic trash contamination (Verla et al., 2019). Fibrous microplastics that do not match the criteria for airborne fibers (i.e., length greater than 5 mm, diameter less than 3 mm, and aspect (length-to-diameter) ratio greater than 3–1) can enter the body through the mouth, nose, or skin (Verla et al., 2019).

3.2.4.2 Toxicity and Mechanism of Action

Particle toxicity occurs when a critical mass of microplastics is localized, triggering an immunological response from the body that the body can not control. A possible toxicity mechanism of micro- and nanoplastics is shown in Figure 3.6.

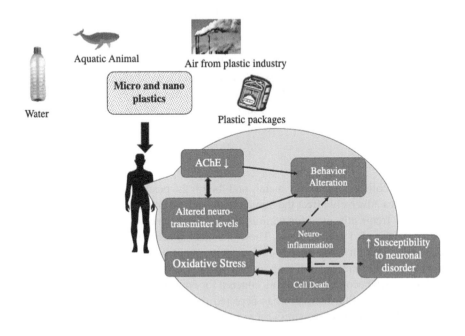

FIGURE 3.6 An illustrative diagram showing exposure routes and the toxic mechanism of microplastics.

A slew of research has looked at the impact of pure microplastics/nanoplastics (MPs/NPs) in mammalian models in the last three years (largely mouse). Ingested MPs/NPs were discovered in the stomach liver and kidney of mice. Reduced mucus secretion, gut barrier failure intestinal inflammation, and gut microbiota dysbiosis are all pathological alterations in the gut. Inflammation and cholesterol accumulation or changes in lipid profiles, as well as changes in lipid metabolism indicators, are all known liver diseases (Yong et al., 2020; Wang et al., 2020).

Disorders in energy metabolism and bile acid metabolism are two other metabolic issues identified by omics-type analysis. A study using mice fed PS MPs, on the other hand, found neither histologically identifiable lesions nor strong inflammatory reactions. There were no significant behavioral alterations or abnormalities in rats fed NPs in neurobehavioral research.

However, a few pieces of research have examined the effect of pure MPs/NPs on human cells in culture. Despite demonstrating some degree of cellular uptake, a number of these investigations revealed no or minor evidence of cellular toxicity except at very high MP/NP concentrations. When PET NPs produced by laser ablation were tested on the human gut adenocarcinoma epithelial line Caco-2, the researchers discovered a proclivity for NP uptake and bridging of a Caco-2 cells-based intestinal barrier model. In a variety of human cell lines, a few other investigations have documented some degree of cellular toxicity or pathogenic impact. Prietl and colleagues demonstrated that 20 nm PS NPs are readily absorbed by human monocytic cells and are cytotoxic. Larger NPs (100 and 1000 nm) induced the production of cytokines like IL-6 and IL-8 from monocytes and macrophages, as well as a detectable degree of respiratory burst in monocytes. In T98G and HeLa cells, Schirinzi and colleagues observed low but detectable levels of ROS generation and cytotoxicity induced by MPs (Yong et al., 2020; Smith et al., 2018).

Nevertheless, the trials with pure MPs/NPs on human cells described thus far did not show severe cytotoxic or cytostatic effects, but they did show the possibility of low-to-moderate unfavorable effects depending on the cell type, MP/NP sizes, and degree of cellular absorption. ROS generation and pro-inflammatory reactions were also seen.

3.3 CONCLUSION

It is well established that heavy metals and their ions, pesticides, nanomaterials, and microplastics adversely affect the human body and the environment. These chemical species are frequently used for various applications, hence, their discharge into the environment is also very frequent. These chemical species are constantly increasing due to their low biocompatibility in the environment which enhances their persistence in the environment for a long time. Therefore, these chemical species alone or combined have become an emerging concern for environmental toxicity studies.

Available scientific literature has depicted that these chemical species generate ROS that leads to oxidative stress on exposure to the cellular system which further leads to disrupt the cellular membrane, protein dysfunction, and DNA damage. As the metabolic ability of these chemical species are very poor they accumulate in the human body on exposure and are responsible for carcinogenicity, mutagenicity, and sometimes disordering physiological systems (neuronal or endocrine disruption). Discharging these chemical species into the main water body of the environment is a major concern that directly or indirectly affects human health and also disturbs the aquatic ecosystem significantly.

Though the toxicological studies of these chemical species have been conducted on an individual basis, they may not depict the accurate toxicological data in a real environmental condition where these chemical species are combined and show additive, potentiation, synergistic preventive, or inhibitory effect.

Therefore, it is the need of the hour that toxicological studies be carried out in a manner where the study should elucidate the combined effects of these chemical species on human health and the environment which not only provide accurate information about their adverse effects but may also help to take preventive measures in this direction.

REFERENCES

Agrawal, S., Flora, G., Bhatnagar, P., & Flora, S. J. (2014). Comparative oxidative stress, metallothionein induction and organ toxicity following chronic exposure to arsenic, lead and mercury in rats. *Cellular and Molecular Biology (Noisy-le-Grand, France)*, 60(2), 13–21.

Ahmed, M., Habibullah-Al-Mamun, M., Parvin, E., Akter, M., & Khan, M. (2013). Arsenic induced toxicity and histopathological changes in gill and liver tissue of freshwater fish, tilapia (*Oreochromis mossambicus*). *Experimental and Toxicologic Pathology*, 65(6), 903–909. https://doi.org/10.1016/j.etp.2013.01.003

Ainslie, K., Tao, S., Popat, K., Daniels, H., Hardev, V., Grimes, C., & Desai, T. (2009). In vitro inflammatory response of nanostructured titania, silicon oxide, and polycaprolactone. *Journal of Biomedical Materials Research Part A*, 91A(3), 647–655. https://doi.org/10.1002/jbm.a.32262

B Billman G. E. (2020). Homeostasis: The Underappreciated and Far Too Often Ignored Central Organizing Principle of Physiology. *Frontiers in Physiology*, 11, 200. https://doi.org/10.3389/fphys.2020.00200

B Bolognesi, C., & Merlo, F. (2011). Pesticides: Human Health Effects. *Encyclopedia Of Environmental Health*, 9(8), 438–453. https://doi.org/10.1016/b978-0-444-52272-6.00592-4

Baranowska-Bosiacka, I., Dziedziejko, V., Safranow, K., Gutowska, I., Marchlewicz, M., & Dołęgowska, B. et al. (2009). Inhibition of erythrocyte phosphoribosyltransferases (APRT and HPRT) by Pb^{2+}: a potential mechanism of lead toxicity. *Toxicology*, 259(1–2), 77–83. https://doi.org/10.1016/j.tox.2009.02.005

Bellinger, D., Malin, A., & Wright, R. (2018). The Neurodevelopmental Toxicity of Lead: History, Epidemiology, and Public Health Implications. Linking Environmental Exposure To Neurodevelopmental Disorders, 2, 1-26. https://doi.org/10.1016/bs.ant.2018.03.009

Bhagat, J., Nishimura, N., & Shimada, Y. (2021). Toxicological interactions of microplastics/nanoplastics and environmental contaminants: current knowledge and future perspectives. *Journal of Hazardous Materials*, 405, 123913. https://doi.org/10.1016/j.jhazmat.2020.123913

Bonner, M., & Alavanja, M. (2017). Pesticides, human health, and food security. *Food and Energy Security*, 6(3), 89–93. https://doi.org/10.1002/fes3.112

Boucher, O., Muckle, G., Jacobson, J., Carter, R., Kaplan-Estrin, M., & Ayotte, P. et al. (2014). Domain-specific effects of prenatal exposure to PCBs, mercury, and lead on infant cognition: results from the environmental contaminants and child development study in Nunavik. *Environmental Health Perspectives*, 122(3), 310–316. https://doi.org/10.1289/ehp.1206323

Bukowska, B. (2003). Effects of 2,4-D and its metabolite 2,4-dichlorophenol on antioxidant enzymes and level of glutathione in human erythrocytes. *Comparative Biochemistry and Physiology Part C: Toxicology & Pharmacology*, 135(4), 435–441. https://doi.org/10.1016/s1532-0456(03)00151-0

C Crichton, R. (2016). Metal Toxicity – An Introduction. In R. Crichton, R. Ward & R. Hider, Metal Chelation in Medicine (1st ed., pp. 1–23). The Royal Society of Chemistry's. Retrieved 9 September 2022, from https://pubs.rsc.org/en/content/ebook/978-1-78262-064-8.

Calviello, G., Piccioni, E., Boninsegna, A., Tedesco, B., Maggiano, N., Serini, S., Wolf, F. I., & Palozza, P. (2006). DNA damage and apoptosis induction by the pesticide Mancozeb in rat cells: involvement of the oxidative mechanism. *Toxicology and Applied Pharmacology*, 211(2), 87–96. https://doi.org/10.1016/j.taap.2005.06.001

Chakraborty, C., Sharma, A. R., Sharma, G., & Lee, S. S. (2016). Zebrafish: A complete animal model to enumerate the nanoparticle toxicity. *Journal of Nanobiotechnology*, 14(1), 65. https://doi.org/10.1186/s12951-016-0217-6

Chen, J., Chen, Y., Liu, W., Bai, C., Liu, X., & Liu, K. et al. (2012). Developmental lead acetate exposure induces embryonic toxicity and memory deficit in adult zebrafish. *Neurotoxicology and Teratology*, 34(6), 581–586. doi: 10.1016/j.ntt.2012.09.001

Costa, S., & Teixeira, J. (2014). Toxicology. *Encyclopedia Of Toxicology*, 718–720. https://doi.org/10.1016/b978-0-12-386454-3.00440-1

Crichton, R. (2019). Biological inorganic chemistry (3rd ed., pp. 1–16). Academic Press, an imprint of Elsevier.

Das, H., Mitra, A., Sengupta, P., Hossain, A., Islam, F., & Rabbani, G. (2004). Arsenic concentrations in rice, vegetables, and fish in Bangladesh: a preliminary study. *Environment International*, 30(3), 383–387. https://doi.org/10.1016/j.envint.2003.09.005

Ding, X., & Kaminsky, L. (2003). Human extrahepatic cytochromes P450: function in xenobiotic metabolism and tissue-selective chemical toxicity in the respiratory and gastrointestinal tracts. *Annual Review of Pharmacology and Toxicology*, 43(1), 149–173. https://doi.org/10.1146/annurev.pharmtox.43.100901.140251

Dubus, I., Hollis, J., & Brown, C. (2000). Pesticides in rainfall in Europe. *Environmental Pollution*, 110(2), 331–344. doi: 10.1016/s0269-7491(99)00295-x

Egorova, K., & Ananikov, V. (2017). Toxicity of metal compounds: knowledge and myths. *Organometallics*, 36(21),4071–4090. https://doi.org/10.1021/acs.organomet.7b00605

Farina, M., Rocha, J., & Aschner, M. (2011). Mechanisms of methylmercury-induced neurotoxicity: evidence from experimental studies. *Life Sciences*, 89(15–16), 555–563. https://doi.org/10.1016/j.lfs.2011.05.019

Franco, R., Li, S., Rodriguez-Rocha, H., Burns, M., & Panayiotidis, M. (2010). Molecular mechanisms of pesticide-induced neurotoxicity: relevance to Parkinson's disease. *Chemico-Biological Interactions*, 188(2), 289–300. https://doi.org/10.1016/j.cbi.2010.06.003

Fulton, M. H., & Key, P. B. (2001). Acetylcholinesterase inhibition in estuarine fish and invertebrates as an indicator of organophosphorus insecticide exposure and effects. *Environmental Toxicology and Chemistry*, 20(1), 37–45. https://doi.org/10.1897/1551-5028(2001)020<0037:aiiefa>2.0.co;2

Galloway, T., & Handy, R. (2003). Immunotoxicity of organophosphorous pesticides. *Ecotoxicology (London, England)*, 12(1–4), 345–363. https://doi.org/10.1023/a:1022579416322

Ganguly, P., Breen, A., & Pillai, S. (2018). Toxicity of nanomaterials: exposure, pathways, assessment, and recent advances. *ACS Biomaterials Science & Engineering*, 4(7), 2237–2275. https://doi.org/10.1021/acsbiomaterials.8b00068

Gargouri, M., Magné, C., Dauvergne, X., Ksouri, R., El Feki, A., Metges, M., & Talarmin, H. (2013). Cytoprotective and antioxidant effects of the edible halophyte Sarcocornia perennis L. (swampfire) against lead-induced toxicity in renal cells. *Ecotoxicology and Environmental Safety*, 95, 44–51. doi: 10.1016/j.ecoenv.2013.05.011

Genchi, G., Carocci, A., Lauria, G., Sinicropi, M., & Catalano, A. (2020). Nickel: human health and environmental toxicology. *International Journal of Environmental Research and Public Health*, 17(3), 679. https://doi.org/10.3390/ijerph17030679

Gerba, C. (2019). *Environmental and Pollution Science* (3rd ed., pp. 511–540). Academic Press.

Giddings, J. M., Anderson, T. A., Hall, L. W., Jr, Kendall, R. J., Richards, R. P., Solomon, K. R., & Williams, W. M. (2005). *A Probabilistic Aquatic Ecological Risk Assessment of Atrazine in North American Surface Waters*. Pensacola, FL: SETAC Press.

Golding, J., Steer, C. D., Hibbeln, J. R., Emmett, P. M., Lowery, T., & Jones, R. (2013). Dietary predictors of maternal prenatal blood mercury levels in the ALSPAC birth cohort study. *Environmental Health Perspectives*, 121(10), 1214–1218. https://doi.org/10.1289/ehp.1206115

Goldman SM (2014) Environmental toxins and Parkinson's disease. *Annu Rev Pharmacol Toxicol*, 54, 141–164.

Grandjean, P. (2016). Paracelsus revisited: the dose concept in a complex world. *Basic & Clinical Pharmacology & Toxicology*, 119(2), 126–132. doi: 10.1111/bcpt.12622

Grandjean, P., Weihe, P., Debes, F., Choi, A., & Budtz-Jørgensen, E. (2014). Neurotoxicity from prenatal and postnatal exposure to methylmercury. *Neurotoxicology and Teratology*, 43, 39–44. https://doi.org/10.1016/j.ntt.2014.03.004

Guzzi, G., & La Porta, C. (2008). Molecular mechanisms triggered by mercury. *Toxicology*, 244(1), 1–12. https://doi.org/10.1016/j.tox.2007.11.002

Hambach, R., Lison, D., D'Haese, P., Weyler, J., De Graef, E., & De Schryver, A. et al. (2013). Co-exposure to lead increases the renal response to low levels of cadmium in metallurgy workers. *Toxicology Letters*, 222(2), 233–238. https://doi.org/10.1016/j.toxlet.2013.06.218

He, J., Wang, M., Jiang, Y., Chen, Q., Xu, S., & Xu, Q. et al. (2014). Chronic arsenic exposure and angiogenesis in human bronchial epithelial cells via the ROS/miR-199a-5p/HIF-1 α /COX-2 pathway. *Environmental Health Perspectives*, 122(3), 255–261. https://doi.org/10.1289/ehp.1307545

Hodgson, E. (2010). Metabolism of pesticides. In: Krieger R (ed), *Hayes' Handbook of Pesticide Toxicology* (3rd ed.) (pp. 893–921). New York: Academic Press.

Hsu, P. (2002). Antioxidant nutrients and lead toxicity. *Toxicology*, 180(1), 33–44. https://doi.org/10.1016/s0300-483x(02)00380-3

Hughes, M., Beck, B., Chen, Y., Lewis, A., & Thomas, D. (2011). Arsenic exposure and toxicology: a historical perspective. *Toxicological Sciences*, 123(2), 305–332. https://doi.org/10.1093/toxsci/kfr184

Hunt, P., Marquis, B., Tyner, K., Conklin, S., Olejnik, N., Nelson, B., & Sprando, R. (2013). Nanosilver suppresses growth and induces oxidative damage to DNA inCaenorhabditis elegans. *Journal of Applied Toxicology*, 33(10), 1131–1142. https://doi.org/10.1002/jat.2872

John, E., & Shaike, J. (2015). Chlorpyrifos: pollution and remediation. *Environmental Chemistry Letters*, 13(3), 269–291. https://doi.org/10.1007/s10311-015-0513-7

Kamijima, M., & Casida, J. E. (2000). Regional modification of [3H] ethynylbicycloorthobenzoate binding in mouse brain GABAA receptor by endosulfan, fipronil, and avermectin B1a. *Toxicol Appl Pharmacol.*, 163, 188–194.

Kang, M., Cho, S., Lim, Y., Seo, J., & Hong, Y. (2013). Effects of environmental cadmium exposure on liver function in adults. *Occupational and Environmental Medicine*, 70(4), 268–273. https://doi.org/10.1136/oemed-2012-101063

Katagi T. (2010). Bioconcentration, bioaccumulation, and metabolism of pesticides in aquatic organisms. *Reviews of environmental contamination and toxicology*, 204, 1–132. https://doi.org/10.1007/978-1-4419-1440-8_1

Kumari, S., Amit, Jamwal, R., Mishra, N., & Singh, D. (2020). Recent developments in environmental mercury bioremediation and its toxicity: a review. *Environmental Nanotechnology, Monitoring & Management*, 13, 100283. https://doi.org/10.1016/j.enmm.2020.100283

Lebrun, V., Tron, A., Scarpantonio, L., Lebrun, C., Ravanat, J.-L., Latour, J.-M. et al. (2014). Efficient oxidation and destabilization of Zn(Cys)4Zinc fingers by singlet oxygen. *Angew. Chem. Int. Ed.* 53(35), 9365–9368. 10.1002/anie.201405333

Liao, C., Tsai, J., Ling, M., Liang, H., Chou, Y., & Yang, P. (2004). Organ-specific toxicokinetics and dose? Response of arsenic in tilapia oreochromis mossambicus. *Archives of Environmental Contamination and Toxicology*, 47(4), 502–510. https://doi.org/10.1007/s00244-004-3105-2

Liu, J., Qu, W., & Kadiiska, M. (2009). Role of oxidative stress in cadmium toxicity and carcinogenesis. *Toxicology and Applied Pharmacology*, 238(3), 209–214. https://doi.org/10.1016/j.taap.2009.01.029

López-Serrano, A., Olivas, R. M., Landaluze, J. S., & Cámara, C. (2014). Nanoparticles: a global vision. Characterization, separation, and quantification methods. Potential environmental and health impact. *Anal. Methods* 6(1), 38–56. 10.1039/c3ay40517f

Lu, K., Abo, R., Schlieper, K., Graffam, M., Levine, S., & Wishnok, J. et al. (2014). Arsenic exposure perturbs the gut microbiome and its metabolic profile in mice: an integrated metagenomics and metabolomics analysis. *Environmental Health Perspectives*, 122(3), 284–291. https://doi.org/10.1289/ehp.1307429

Luo, W., Verweij, R., & van Gestel, C. (2014). Determining the bioavailability and toxicity of lead contamination to earthworms requires using a combination of physicochemical and biological methods. *Environmental Pollution*, 185, 1–9. doi: 10.1016/j.envpol.2013.10.017

Lushchak, V. (2016). Contaminant-induced oxidative stress in fish: a mechanistic approach. *Fish Physiology and Biochemistry*, 42(2), 711–747. https://doi.org/10.1007/s10695-015-0171-5

Lyu, M., Yun, J., Chen, P., Hao, M., & Wang, L. (2017). Addressing toxicity of lead: progress and applications of low-toxic metal halide perovskites and their derivatives. *Advanced Energy Materials*, 7(15), 1602512. https://doi.org/10.1002/aenm.201602512

Mesnage, R., & Séralini, G. (2018). Editorial: toxicity of pesticides on health and environment. *Frontiers in Public Health*, 6. https://doi.org/10.3389/fpubh.2018.00268

Moebus, S., & Boedeker, W. (2021). Case fatality as an indicator for the human toxicity of pesticides – a systematic scoping review on the availability and variability of severity indicators of pesticide poisoning. *International Journal of Environmental Research and Public Health*, 18(16), 8307. https://doi.org/10.3390/ijerph18168307

Mostafalou, S., & Abdollahi, M. (2013). Pesticides and human chronic diseases: evidences, mechanisms, and perspectives. *Toxicology and Applied Pharmacology*, 268(2), 157–177. https://doi.org/10.1016/j.taap.2013.01.025

Nawrot, T., Staessen, J., Roels, H., Munters, E., Cuypers, A., & Richart, T. et al. (2010). Cadmium exposure in the population: from health risks to strategies of prevention. *Biometals*, 23(5), 769–782. https://doi.org/10.1007/s10534-010-9343-z

Nordberg, G., Jin, T., Hong, F., Zhang, A., Buchet, J., & Bernard, A. (2005). Biomarkers of cadmium and arsenic interactions. *Toxicology and Applied Pharmacology*, 206(2), 191–197. https://doi.org/10.1016/j.taap.2004.11.028

Nordberg, G., Nordberg, M., Fowler, B., & Friberg, L. (2007). *Handbook on the Toxicology of Metals* (3rd ed.). Burlington: Elsevier Science.

Nuran Ercal, B., Hande Gurer-Orhan, B., & Nukhet Aykin-Burns, B. (2001). Toxic metals and oxidative atress part I: mechanisms involved in metal induced oxidative damage. *Current Topics in Medicinal Chemistry*, 1(6), 529–539. https://doi.org/10.2174/1568026013394831

Nyland, J., Fillion, M., Barbosa, F., Shirley, D., Chine, C., & Lemire, M. et al. (2011). Biomarkers of methylmercury exposure immunotoxicity among fish consumers in amazonian Brazil. *Environmental Health Perspectives*, 119(12), 1733–1738. https://doi.org/10.1289/ehp.1103741

Oberdörster, G., Maynard, A., Donaldson, K., Castranova, V., Fitzpatrick, J., & Ausman, K. et al. (2005). Principles for characterizing the potential human health effects from exposure to nanomaterials: elements of a screening strategy. *Particle And Fibre Toxicology*, 2(1), 1–35. https://doi.org/10.1186/1743-8977-2-8

Patrick, L. (2003). Toxic metals and antioxidants: part II. The role of antioxidants in arsenic and cadmium toxicity. *Alternative Medicine Review: A Journal of Clinical Therapeutic*, 8(2), 106–128.

Patrick, L. (2006). Lead toxicity part II: the role of free radical damage and the use of antioxidants in the pathology and treatment of lead toxicity. *Alternative Medicine Review: A Journal of Clinical Therapeutic*, 11(2), 114–127.

Pretto, A., Loro, V. L., Morsch, V. M., Moraes, B. S., Menezes, C., Santi, A., & Toni, C. (2014). Alterations in carbohydrate and protein metabolism in silver catfish (Rhamdiaquelen) exposed to cadmium. *Ecotox Environ Safe* 100, 188–192.

QIN, S., LIU, H., NIE, Z., RENGEL, Z., GAO, W., LI, C., & ZHAO, P. (2020). Toxicity of cadmium and its competition with mineral nutrients for uptake by plants: a review. *Pedosphere*, 30(2), 168–180. https://doi.org/10.1016/s1002-0160(20)60002-9

Rahman, M. K., Choudhary, M. I., Arif, M., & Morshed, M. M. (2014). Dopamine-β-hydroxylase activity and levels of its cofactors and other biochemical parameters in the serum of arsenicosis patients of Bangladesh. *International Journal of Biomedical Science*, 10(1), 52–60.

Ralston, N., & Raymond, L. (2010). Dietary selenium's protective effects against methylmercury toxicity. *Toxicology*, 278(1), 112–123. doi: 10.1016/j.tox.2010.06.004

Ramsey, K., Foong, R., Sly, P., Larcombe, A., & Zosky, G. (2013). Early life arsenic exposure and acute and long-term responses to Influenza A infection in mice. *Environmental Health Perspectives*, 121(10), 1187–1193. https://doi.org/10.1289/ehp.1306748

Rani, L., Thapa, K., Kanojia, N., Sharma, N., Singh, S., & Grewal, A. et al. (2021). An extensive review on the consequences of chemical pesticides on human health and environment. *Journal of Cleaner Production*, 283, 124657. https://doi.org/10.1016/j.jclepro.2020.124657

Rashid, K., Sinha, K., & Sil, P. (2013). An update on oxidative stress-mediated organ pathophysiology. *Food and Chemical Toxicology*, 62, 584–600. https://doi.org/10.1016/j.fct.2013.09.026

Rehman, M., Liu, L., Wang, Q., Saleem, M., Bashir, S., Ullah, S., & Peng, D. (2019). Copper environmental toxicology, recent advances, and future outlook: a review. *Environmental*

Science and Pollution Research, 26(18), 18003–18016. https://doi.org/10.1007/s11 356-019-05073-6

Sabath, E., & Robles-Osorio, M. L. (2012). Medio ambiente y riñón: nefrotoxicidad por metalespesados [Renal health and the environment: heavy metal nephrotoxicity]. Nefrologia 32, 279–286.

Sabolic, I., Herakkramberger, C., Antolovic, R., Breton, S., & Brown, D. (2006). Loss of basolateral invaginations in proximal tubules of cadmium-intoxicated rats is independent of microtubules and clathrin. Toxicology, 218(2–3), 149–163. https://doi.org/10.1016/j.tox.2005.10.009

Sahu, S., & Hayes, A. (2017). Toxicity of nanomaterials found in human environment. Toxicology Research and Application, 1, 239784731772635. https://doi.org/10.1177/2397847317726352

Sahu, S., Roy, S., Zheng, J., & Ihrie, J. (2016). Contribution of ionic silver to genotoxic potential of nanosilver in human liver HepG2 and colon $CaCo_2$ cells evaluated by the cytokinesis-block micronucleus assay. Journal of Applied Toxicology, 36(4), 532–542. https://doi.org/10.1002/jat.3279

Schanen, B., Karakoti, A., Seal, S., Drake III, D., Warren, W., & Self, W. (2009). Exposure to titanium dioxide nanomaterials provokes inflammation of an in vitro human immune construct. ACS Nano, 3(9), 2523–2532. https://doi.org/10.1021/nn900403h

Shinomiya, N., & Shinomiya, M. (2003). Dichlorodiphenyltrichloroethane suppresses neurite outgrowth and induces apoptosis in PC_{12} pheochromocytoma cells. Toxicology Letters, 137(3), 175–183. https://doi.org/10.1016/s0378-4274(02)00401-0

Shukla, R., Badiye, A., Vajpayee, K., & Kapoor, N. (2021). Genotoxic Potential of Nanoparticles: Structural and Functional Modifications in DNA. Frontiers In Genetics, 12(728250), 1–16. https://doi.org/10.3389/fgene.2021.728250

Silver, K., Dong, K., & Zhorov, B. S. (2017). Molecular Mechanism of Action and Selectivity of Sodium Ch annel Blocker Insecticides. Current Medicinal Chemistry, 24(27), 2912–2924. https://doi.org/10.2174/0929867323666161216143844

Singh, N., Kumar, A., Gupta, V., & Sharma, B. (2018). Biochemical and molecular bases of lead-induced toxicity in mammalian systems and possible mitigations. Chemical Research in Toxicology, 31(10), 1009–1021. https://doi.org/10.1021/acs.chemrestox.8b00193

Smith, M., Love, D., Rochman, C., & Neff, R. (2018). Microplastics in seafood and the implications for human health. Current Environmental Health Reports, 5(3), 375–386. https://doi.org/10.1007/s40572-018-0206-z

Trickler, W., Lantz, S., Murdock, R., Schrand, A., Robinson, B., & Newport, G. et al. (2010). Silver nanoparticle induced blood-brain barrier inflammation and increased permeability in primary rat brain microvessel endothelial cells. Toxicological Sciences, 118(1), 160–170. https://doi.org/10.1093/toxsci/kfq244

van der Oost, R., Beyer, J., & Vermeulen, N. (2003). Fish bioaccumulation and biomarkers in environmental risk assessment: a review. Environmental Toxicology and Pharmacology, 13(2), 57–149. https://doi.org/10.1016/s1382-6689(02)00126-6

Van Kerkhove, E., Pennemans, V., & Swennen, Q. (2010). Cadmium and transport of ions and substances across cell membranes and epithelia. Biometals, 23(5), 823–855. https://doi.org/10.1007/s10534-010-9357-6

Verla, A., Enyoh, C., Verla, E., & Nwarnorh, K. (2019). Microplastic–toxic chemical interaction: a review study on quantified levels, mechanism and implication. SN Applied Sciences, 1(11). https://doi.org/10.1007/s42452-019-1352-0

Wätjen, W., & Beyersmann, D. (2004). Cadmium-induced apoptosis in C6 glioma cells: influence of oxidative stress. Biometals : an international journal on the role of metal ions in biology, biochemistry, and medicine, 17(1), 65–78. https://doi.org/10.1023/a:1024405119018

Waisberg, M., Joseph, P., Hale, B., & Beyersmann, D. (2003). Molecular and cellular mechanisms of cadmium carcinogenesis. *Toxicology*, 192(2–3), 95–117. https://doi.org/10.1016/s0300-483x(03)00305-6

Wang, F., Gao, F., Lan, M., Yuan, H., Huang, Y., & Liu, J. (2009). Oxidative stress contributes to silica nanoparticle-induced cytotoxicity in human embryonic kidney cells. *Toxicology In Vitro*, 23(5), 808–815. https://doi.org/10.1016/j.tiv.2009.04.009

Wang, W., Ge, J., & Yu, X. (2020). Bioavailability and toxicity of microplastics to fish species: a review. *Ecotoxicology and Environmental Safety*, 189, 109913. https://doi.org/10.1016/j.ecoenv.2019.109913

Wang, X., Martínez, M., Dai, M., Chen, D., Ares, I., & Romero, A. et al. (2016). Permethrin-induced oxidative stress and toxicity and metabolism. a review. *Environmental Research*, 149, 86–104. https://doi.org/10.1016/j.envres.2016.05.003

Wang, Y., Fang, J., Leonard, S., & Krishna Rao, K. (2004). Cadmium inhibits the electron transfer chain and induces Reactive Oxygen Species. *Free Radical Biology and Medicine*, 36(11), 1434–1443. https://doi.org/10.1016/j.freeradbiomed.2004.03.010

Who.int. (2021). *Lead Poisoning*. Retrieved 1 November 2021, from www.who.int/news-room/fact-sheets/detail/lead-poisoning-and-health.

Wilkinson, C., Christoph, G., Julien, E., Kelley, J., Kronenberg, J., McCarthy, J., & Reiss, R. (2000). Assessing the risks of exposures to multiple chemicals with a common mechanism of toxicity: how to cumulate? *Regulatory Toxicology and Pharmacology*, 31(1), 30–43. doi: 10.1006/rtph.1999.1361

Wu, X., Cobbina, S., Mao, G., Xu, H., Zhang, Z., & Yang, L. (2016). A review of toxicity and mechanisms of individual and mixtures of heavy metals in the environment. *Environmental Science and Pollution Research*, 23(9), 8244–8259. https://doi.org/10.1007/s11356-016-6333-x

Yang, J. (2003). Cadmium-induced damage to primary cultures of rat Leydig cells. *Reproductive Toxicology*, 17(5), 553–560. https://doi.org/10.1016/s0890-6238(03)00100-x

Yong, C., Valiyaveettil, S., & Tang, B. (2020). Toxicity of microplastics and nanoplastics in mammalian systems. *International Journal of Environmental Research and Public Health*, 17(5), 1509. https://doi.org/10.3390/ijerph17051509

Zahir, F., Rizwi, S., Haq, S., & Khan, R. (2005). Low dose mercury toxicity and human health. *Environmental Toxicology and Pharmacology*, 20(2), 351–360. https://doi.org/10.1016/j.etap.2005.03.007

Zhang, J., Cao, H., Zhang, Y., Zhang, Y., Ma, J., & Wang, J. et al. (2013). Nephroprotective effect of calcium channel blockers against toxicity of lead exposure in mice. *Toxicology Letters*, 218(3), 273–280. https://doi.org/10.1016/j.toxlet.2013.02.005

4 Natural Organic Matter
A Ubiquitous Adsorbent in Aquatic Systems to Probe Nanoparticles Behavior and Effects in Ecosystem

Abdulkhaik Mansuri and Ashutosh Kumar

CONTENTS

4.1 Introduction	59
4.2 Release of Engineered Nanomaterials in Environmental Matrices	61
4.3 Environmental Impact of Engineered Nanomaterials	62
4.3.1 Different Routes of Nanomaterial Exposure on the Aquatic Organisms	62
4.3.2 Fate of Engineered Nanomaterials in the Environment	63
4.3.3 Interactions Between the Natural Organic Matter and Engineered Nanomaterial	64
4.3.4 Effect of Natural Organic Matter in the Toxicity of the Nanoparticles to the Organisms and Food Chain	65
4.4 Trophic Transfer and Biomagnification Potential of Engineered Nanomaterial	67
4.5 Impact of Engineered Nanomaterials on the Producers of the Ecosystem	68
4.6 Conclusion	68
Acknowledgments	74
References	74

4.1 INTRODUCTION

The global nanotechnology market was valued at $1,055.1 million in 2018, and is projected to reach $2,231.4 million by 2025, growing at a CAGR of 10.5% from 2019 to 2025.[1] The major advantages of nanomaterial (NM) use is its ultrasmall size and large surface area-to-volume ratio that improves the quality of the product. NMs are between 1 and 100 nm, with at least one dimension. The most explored sectors of nanotechnology uses are health care and fitness, sports, biomedicine, textiles, cosmetics, paints, agriculture, and wastewater treatment (Chen et al., 2014; Gogos et al., 2012; Joseph & Morrison, 2006; Kahru & Dubourguier, 2010; Kansara et al., 2018; Patel et al., 2018; Robinson & Morrison, 2009; Scott & Chen, 2013). The number of nanoproducts in the consumer product market has exceeded 1800,

DOI: 10.1201/9781003244349-4

and numerous unregistered products are also available in the consumer market (Foss Hansen et al., 2016; Vance et al., 2015). This increased production and heavy usage of NPs has increased their unintended release in aquatic systems.

Silicon dioxide (SiO_2), titanium dioxide (TiO_2), and silver NPs are some of the most commonly used NPs in consumer products and industries that have been found in natural waters even after the wastewater treatment processes. Although the predicted concentrations of these NPs in aquatic systems are much lower than the concentrations known to affect aquatic organisms. Importantly, the concentration will change with the proliferation of NPs in different consumer products in the future. Once these particles are released into the natural environment, they may interact with the aquatic colloids, such as natural organic matter (NOM), humic substances, and salt ions, which affect the behaviour, transportation, transformation, and bioavailability of NPs (Kansara et al., 2021a; Kansara et al., 2021b). NOM are classified into three major classes: (1) rigid biopolymers, such as polysaccharides and peptidoglycans produced by phytoplankton or bacteria; (2) fulvic compounds, mostly from terrestrial sources, originating from the decomposition products of plants; and (3) flexible biopolymers, composed of aquagenic refractory organic matter from recombination of microbial degradation products. The final concentration of NOM in aquatic systems depends on the biogeochemical conditions and climate but typically ranges from 0.1 to 10 mg/mL. NOM is considered as a pool of substances that contains numerous functional groups such as thiols, phenolic-OH, quinone, aldehydes, carboxyls, and methoxyls. Additionally, NOM chelates the metals and absorbs organic toxicants, which play important roles in the cycle and transport of inorganic, organic, and ions. NOM is usually adsorbed on the surface of the NMs by different electrostatic interactions, ligand exchange, hydrogen bonding and hydrophobic interactions, and cation binding, which affect their dispersity and bioavailability. Apart from NOM, salt ions, protein content, and the presence of molecular clusters enable nucleation leading to agglomeration/aggregation, thereby modulating the bioavailability of engineered nanomaterials (ENMs) in the environment. Also, biomolecules such as proteins or polymers present in the ecosystem form a layer over the ENMs, named "corona," which plays an important role in their biological fate. It has also been shown that it is not only the ENMs alone but the corona too that governs the properties of the "particle-plus-corona" compound in the biological system (Kumar & Dhawan, 2019; Patel & Kumar, 2019).

Great progress has been made in the last few years in understanding the bioavailability, toxicity, and food chain impact of ENMs. Most of the studies in trophic transfer assessment have shown the movement of ENM up to two trophic levels with a few studies up to three trophic levels (Chae & An, 2016; Dalai et al., 2014; Ghafari et al., 2008; Holbrook et al., 2008; Judy et al., 2011; Judy et al., 2012; Kim et al., 2016; Kim et al., 2013; Kumar et al., 2011b; Majumdar et al., 2016; Mielke et al., 2013; Mortimer et al., 2016; Pakrashi et al., 2014; Werlin et al., 2011; Zhao et al., 2013; Zhu et al., 2010). The bioaccumulation and biomagnification of these ENMs can also alter the reproductive potential, bacterivory, swimming behaviour, and growth rate in some aquatic animals such as *Tetrahymena*, Daphnia, and Zebrafish (Bouldin et al., 2008; Kansara et al., 2019; Mielke et al., 2013; Werlin et al., 2011; Zhu et al., 2010). The spread of ENMs in the ecosystem hierarchy from the individual organism to population and community will pose a risk to ecosystem services such as carrying capacity, biogeochemical cycling, nutrient recycling, and energy transformation (Holden et al., 2013).

4.2 RELEASE OF ENGINEERED NANOMATERIALS IN ENVIRONMENTAL MATRICES

The release of ENMs in the environment has increased exponentially due to the discovery of new applications in different areas from electronics to biotechnology (Figure 4.1). It has been reported that the NM applications in cosmetics, paints, pigments, and coatings are the major sources of their direct release in the environment. While wastewater treatment plants and aquatic environment are the sinks of ENM disposal after completion of the lifecycle of nano-based products, it has also been estimated that the 189,200 and 69,200 metric tons of NPs are released per year globally into landfills and water, respectively (Keller & Lazareva, 2014).

NMs enter the environment at various stages of their lifecycle: during production, transport, research and development, fabrication, consumer use, waste treatment plant, and disposal (Bernhardt et al., 2010; Gottschalk et al., 2009; Holden et al., 2013; Keller & Lazareva, 2014; Lee et al., 2010; Walser et al., 2012). As the ENMs enter

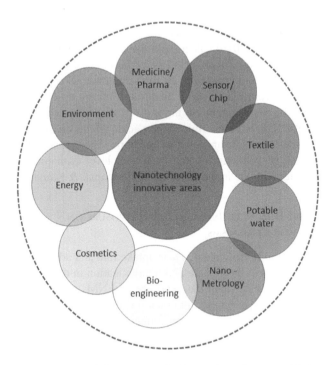

FIGURE 4.1 The present and future areas for innovation of nano-enabled products in the world market.

the environment they pass through various routes and interact with the biological, chemical, and physical environmental entities, which may prevent their behavior and transport in the environmental matrices.

At present, the predicted concentrations of ENMs in the natural water bodies are very low; even the three most used NPs (i.e., nTiO$_2$, nZnO, and nAg) are present at a level of ng/L. As we all know the risk of any new chemical entity is a function of its hazard potential and exposure. The toxicity of various types of NMs has been reported in a range of models like bacteria, paramecium, daphnia, tetrahymena, fishes, organ-specific cell lines, mice, and plants (Bernhardt et al., 2010; Caceres-Velez et al., 2018; Cupi et al., 2015; Deng et al., 2016; Gupta et al., 2019; Jung et al., 2018; Lee et al., 2016; Shang et al., 2017; Wang et al., 2015; Wirth et al., 2012, Senapati et al., 2021, Kumar et al., 2011a). However, the toxic/lethal concentrations of NMs in different model organisms are several folds higher than their detected concentration in the environment. It can also not be ruled out that the increase in the number and extent of applications of NMs will raise concentrations very rapidly due to improper disposal and/or lack of management practices.

4.3 ENVIRONMENTAL IMPACT OF ENGINEERED NANOMATERIALS

The studies on environmental toxicity of ENMs have been conducted using different aquatic model organisms such as bacteria (*E. coli* and *Pseudomonas aeruginosa*), ciliated protozoan (*Tetrahymena* spp.), crustacean (*Daphnia* spp.), and fish (*Danio rerio*). Bondarenko et al. (2016) categorized the hazard ranking of the seven most characterized ENMs as Ag>ZnO>CuO>TiO2>MWCNTs>SiO2>Au, based on the eco-toxicity assays in model organisms. At the same time, as per Foss Hansen et al. (2016) nanoproducts from the category health and fitness pose the highest risk to human health and the environment. So far, most of the studies carried out under the risk assessment of ENMs have determined their acute effects on the model organisms after direct exposure through an aqueous medium. The impact of ENMs at the organisms, population, community, and ecosystem level is majorly dependent on their exposure routes and fate in the environment (Senapati & Kumar, 2018).

4.3.1 DIFFERENT ROUTES OF NANOMATERIAL EXPOSURE ON THE AQUATIC ORGANISMS

ENMs are a special kind of pollutant. Their uptake rate and adverse effects particularly depend on the routes of exposure and internalization in the cell and organisms. In the aquatic environment, when the exposure of ENMs to the organisms occurs particularly through the aqueous medium then it is defined as bioconcentration (Hou et al., 2013). The passive uptake mechanism may lie in the process (Thomann et al., 1992; van der Oost et al., 2003). While the exposure of ENMs through all possible routes including contaminated food is defined as bioaccumulation (Dalai et al., 2014). The increase in the concentration of pollutants over time is also considered as a major factor the bioaccumulation. The transfer of ENMs from prey to predator with more than one time higher magnitude is defined as biomagnification. The eco-toxicological

Natural Organic Matter

studies on the established pollutants have suggested biomagnification of any pollutant as the most dangerous to human health and the environment.

4.3.2 FATE OF ENGINEERED NANOMATERIALS IN THE ENVIRONMENT

ENMs, upon entry into the aquatic environment, interact with biotic and abiotic components (Kansara et al., 2021). The biological entities (bacteria, protozoans, algae, plants, and their exudates and biomacromolecules) interact with ENMs and alter their fate by the transformation of their original properties (Figure 4.2). Earlier, it was shown that bacterial cells are ubiquitously present and have a high ratio of surface area to their volume; thus, the cells interact with and absorb high levels of ENMs (Holden et al., 2014). Additionally, the presence of exopolymeric substances (EPS) such as lipopolysaccharides (LPS) and siderophores (in *Pseudomonas aeruginosa*) on the outer membrane of bacterial cells promotes the adsorption of ENMs from the aquatic environment (Chojnacka, 2010; Jucker et al., 1998; Zhang et al., 2011). It has also been observed that the long-chain LPS in *E. coli* cells adheres more strongly to $nTiO_2$ than short-chain LPS (Li & Logan, 2004). Apart from bacteria, ciliated protozoans such as *Tetrahymena* have also been reported to secrete mucus from mucous membranes under stress conditions, which affects the fate of ENMs in the medium (Ghafari et al., 2008). Hetero-agglomeration and co-sedimentation of metal oxide NP with algal cells have also been observed (Ma et al., 2015).

Similarly, the interactions of ENMs with abiotic factors such as clay particles and humic acid (HA) can influence the hetero-agglomeration and sedimentation in the aquatic environmental matrices (Zhou et al., 2012). It has also been observed that the

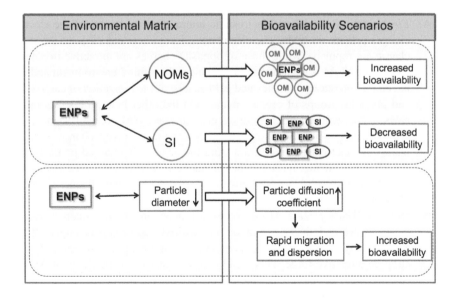

FIGURE 4.2 The bioavailability scenarios of the different ENPs in an environmental Matrix.

biomacromolecules (carbohydrates and proteins), inorganic acids (e.g., humic acid, fulvic acid, hematite, lignite, etc.), and other inorganic substances (e.g., sulphates, sulphides, phosphates, nitrates, etc.) by adsorption or absorption, agglomeration or aggregation, sedimentation, speciation, ionization, capping, and biouptake affects the fate of ENMs (Glenn & Klaine, 2013; Hernandez-Viezcas et al., 2013; Kumar et al., 2011; Lowry et al., 2010; Sun et al., 2009) (Figure 4.2). The change in the fate of ENMs may cause positive or negative effects on ENM stability. The humic and fulvic acids (FA) are known to increase the dispersion of ENMs in the environment, while the clay particles, algae, and exo-polysaccharides destabilize ENMs in the natural environment (Domingos et al., 2009; Lowry et al., 2010; Ma et al., 2015; Zhou et al., 2012).

Such transformations on ENMs behavior in the environment alter their bioavailability and toxicity to the producers, consumers, and decomposers of the aquatic ecosystem (Gupta et al., 2016). The lipophilic coating of ENMs in environment can also influence the bioaccumulation and biomagnification potential in the food chain (Hoet et al., 2004; Holbrook et al., 2008).

4.3.3 Interactions Between the Natural Organic Matter and Engineered Nanomaterial

NOM and ENMs both have extreme levels and complexity, hence their interaction mechanism has also been attributed mainly to the intrinsic nature of the NMs, organic matter, surrounding environmental conditions (pH, ionic strength, temperature), and biotic factors. The formulation of corona on the surface of NMs is built by several forces such as Van der Waals forces and electrostatic and electrosteric forces. These forces determine the structure of the protein corona that form at these interfaces.

A. *Electrostatic interaction:* Electrostatic interactions occur due to the electric charges present on NMs. The electrostatic forces between NMs of the same charge are repulsive while those of opposite charges are attractive in nature. Generally, NOM is negatively charged under the natural environment and the net charge increases with elevated pH because of the ionization of carboxylic and phenolic groups of organic matter. On the other hand, the solvation of NMs or the addition of capping agents can make NM charged. Hence, the NMs that carry a positive charge can react with the negatively charged organic matter by electrostatic interaction. And the negatively charged particles can absorb organic matter through other interactions, which is reflected by the fact that all-natural colloids are negatively charged despite their distinct composition and surface natures (Kumar & Dhawan, 2019).

B. *Van der Waals forces:* These forces originate due to the fluctuations of electrons, which induces a small dipole (positive and negative charges) in the NPs, thereby inducing a dipole moment in the atoms of the adjacent NPs creating an attractive force. This could preferentially occur between NOM and carbon NMs (Kumar & Dhawan, 2019).

C. *Ligand exchange:* This is assumed to be the most dominant mechanism accounting for NOM and metal oxide NP interaction. The acidic functional group (-COOH and phenolic – OH) of organic matter and hydroxyls on the surface of metal NPs were responsible for the reaction. Generally, in the ligand exchange three steps are involved: (A) protonation of surface hydroxyl of NMs allowing them to more exchangeable; (B) the outer sphere complexation of the protonated hydroxyl group with –COOH groups or –OH of organic matter; and (C) ligand exchange.

D. *Electrosteric forces:* The electrosteric forces are based on the steric repulsion between molecules adsorbed on NMs, which is necessary to prevent their aggregation, agglomeration, and coagulation in the solutions. Electrosteric repulsion occurs due to the attachment of polymers or surfactants on the surface of NMs. In the presence of NOM the stability of silver (Ag), titanium dioxide (TiO_2), aluminum oxide (Al_2O_3), zinc oxide (ZnO), iron (Fe) NPs, fullerenes, and carbon nanotubes (CNTs) increases as protein corona around NPs can be stabilized by electrostatic repulsion, which is achieved by providing a steric barrier (Patel & Kumar, 2019).

E. *Solvation and solphobic forces:* Solvation is defined as the process of surrounding solute with solvent. The water molecules present in the medium bind the hydrophilic NMs with sufficient energy to form steric layers on their surfaces, making the NMs difficult to adhere to each other. Therefore, the solvation forces increase particle stability through hydrophilic repulsion. "Solvophobic" forces enable the NMs to aggregate when the attractive interaction due to hydrogen bonding or other polar interactions between solvent particles becomes stronger than the interaction between solvent and solute. Solvation forces play a role in interactions between SiO_2 NPs in water.

4.3.4 Effect of Natural Organic Matter in the Toxicity of the Nanoparticles to the Organisms and Food Chain

NOM is a major factor that alters the fate and effects of NMs by hindering their bioavailability to the biological system being administered. Metal oxide NPs such as Cu_2O, Ag, ZnO, and TiO_2 have been reported to induce the eco-toxicity in the aquatic model against several model organisms from bacterium to a higher-order vertebrate (Kumar et al., 2011a). The NPs exhibit toxicity to organisms by different mechanisms, which are direct mechanical injury leading to membrane permeability, changing respiration of the cells, internalization and permeabilization, DNA damages, ROS generation, etc. (Vallabani et al., 2019).

The bioavailability of the engineered NPs is highly affected by the presence of NOM in the surroundings. NOM such as Humic Acid, Fulvic Acid, Suwannee River Fulvic Acid, and Pony Lake Fulvic Acid contribute to the varied toxicity by adhering to the surface of the active NPs by the different types of force (i.e., electrostatic interactions, van der Waals forces, etc.). This can lead to the surface stabilization of the NPs by generating a stabilizing layer around the NPs. This increases the overall stability of the NPs and can render them to remain viable for longer durations, facilitating movement and exposure to organisms in the aquatic system it is subjected to. It has also been found that the bacterial assemblage faces more amplified biotoxicity by AgNPs after its stabilization with terrestrial humic acid and river humic acid. The

AgNPs stabilized in river humic acid showed higher antibacterial activity than terrestrial humic acid, because of its higher hydrophobicity (Dasari & Hwang, 2010). In the case of CuO NPs in the presence of Suwannee River Fulvic Acid, there was increased dispersion of the surface stabilized NPs which increased the toxicity of the particles towards the prokaryotic algae *Microcystis aeruginosa* due to increased ROS generation and cellular uptake (Wang et al., 2011). Higher cellular uptake directly correlates to increased internal permeation and fatal damage, which can be caused by the NPs. In other experiments conducted of the interaction of NPs with NOM showed higher biotoxicity of the NPs; TiO_2 in the presence of HA increased toxicity in *Danio rerio* under both dark and light conditions due to high internalization (Yang et al., 2013). *Pseudokirchneriella subcapitata* subjected to ZnO and TiO_2 in the presence of Suwannee River Fulvic Acid did not show any effect of NOM on the toxicity of the NPs (Neale et al., 2015).

On the other hand, numerous experiments have shown that the toxicity of the NPs tends to decrease after reaction with the NOM. The NOM usually develops a coat around the NPs and prevents direct contact of bare NPs with the cells, which alleviates its toxicity due to membrane damage in absence of the NOM. Studies performed on *Bacillus subtilis, Pseudomonas putida,* and *Pseudomonas aeruginosa* with ZnO and different types of NOM showed that tannic acid plays an important role by developing a stable complex with the Zn^{2+} ions (dissociated from the ZnO NPs) and reducing the overall toxicity of the particles (Li et al., 2010). In another work, *Anaebena sp.* exposed to ZnO NPs in the presence of humic acid showed Zn^{2+} toxicity was abrogated by the presence of humic acid substances and alleviated any sort of direct injury from the NPs (Tang et al., 2015). The toxicity is usually caused by the dissolution of free metal ions of the NPs rather than the bound forms with NOM (Senapati & Kumar, 2018).

Pony Lake Fluvic Acid, another NOM that is highly rich in sulphur and nitrogen, showed a significant reduction in NP toxicity. The increased concentration of sulphur and nitrogen interacts quickly with the metal NPs and reduces the overall bioavailability to the organisms, such as *Shewanella oneidensis* subjected to AgNPs (Gunsolus et al., 2015) and *Caenorhabditis elegans* with citrate-coated AgNPs (Yang et al., 2014). AgNPs when administered to *Daphnia magna* in the presence of Poney Lake Fluvic Acid showed a drastic 70% decrease in toxicity of the NPs (Jung et al., 2018).

In general, the NOM interacts with the NPs by different forces and makes a coat around the NPs, providing it an overall charge that in turn prevents its interaction with the organisms by developing a steric repulsion amongst the particle and the organism. The algae *Chlorella sp.* showed decreased toxicity to the humic acid bound TiO_2 due to increased electrostatic repulsion, thus leading to reduced ROS generation and limiting of the toxicity of the particles (Lin et al., 2012). Many times the presence of external factors leads to agglomeration of NPs, which drastically reduces its bioavailability to the organisms and could directly relate to the decrease in the toxic effect of the NPs. Japanese medaka embryos showed alleviated toxic levels of Ag and aged Ag upon the presence of humic acid due to reduced bioavailability (Kim et al., 2013). Similarly, reduced bioavailability was observed in the case of *Danio rerio* that had been administered AgNPs in the presence of humic acid resulting in decreased toxicity (Caceres-Velez et al., 2018). All the different experiments conducted over the long-time span clearly show that it's very difficult to ascertain how NPs would interact in the aquatic system in the presence of NOM. There is yet a lack of data to confirm and predict a trendline of the behaviour of the NPs in the system.

4.4 TROPHIC TRANSFER AND BIOMAGNIFICATION POTENTIAL OF ENGINEERED NANOMATERIAL

The unique properties of ENMs such as small size, high reactivity, long-term persistence, poor water solubility, and slow degradability can lead to their bioaccumulation and biomagnification in the environmental food chain (Hoet et al., 2004; Holbrook et al., 2008). It was well-established in earlier studies on potential environmental pollutants such as dichlorodiphenyltrichloroethane (DDT), lead, asbestos, and mercury that if the substance is highly accumulative then it can pose more threats to human health and environment (Holbrook et al., 2008; Werlin et al., 2011). Studies have shown the transfer of ENMs from first to second trophic level with increasing concentration, but reaching the third trophic level the concentration of NMs decreases several-fold.

In two trophic food chain, Werlin et al. (2011) showed the trophic transfer of CdSe quantum dots (QDs) from *Pseudomonas aeruginosa* to *Tetrahymena thermophile* with five times higher concentration (BMF ~5). Chen et al. (2015) showed the biomagnification of TiO_2 NPs from *Scenedesmus obliquus* to *Daphnia magna* with a biomagnification factor of 7.8 to 2.2. (Judy et al., 2011; Unrine et al., 2012). An increase in the concentration of Au NPs up in the food chain from *Lycopersicon esculentum* to *Manduca sexta* with a factor of 6.2–11.6 was observed. Holbrook et al. (2008) showed the trophic transfer quantum dots from protozoans to rotifer, with a biomagnification factor of 0.62 to 0.29. Zhu et al. investigated the trophic transfer of TiO_2 NPs from *Daphnia magna* to *Danio rerio* with a biomagnification factor of 0.1. (Mortimer et al., 2016) have shown the trophic transfer of carbon nanotubes without biomagnification in a microbial food chain comprised of *Pseudomonas aeruginosa* as prey and *Tetrahymena thermophile* as predator organisms.

In three trophic level food chain, Majumdar et al. (2016) demonstrated the trophic transfer of CeO_2 NPs with five times higher concentration in a food chain from *Phaseolus vulgaris* to *Epilachna varivestis* and then to *Podisus maculiventris*. (Lee & An, (2015) provided evidence on the trophic transfer of QDs up to three trophic levels among from *Astasia longa* to *Moina macrocopa* to *Danio rerio* with a biomagnification factor of 0.1.

Most reported studies regarding biomagnification were performed under laboratory-controlled conditions without the influence of any environmental factors, which can change the fate of ENMs. Hence, the observations made from these studies raise several questions such as 1) Will the obtained data be extrapolated to a realistic environment? 2) Will it be similar to the other organisms of the same genera? Most of the available data on biomagnification has been performed in aquatic organisms. Therefore, more work is needed in terrestrial organisms.

4.5 IMPACT OF ENGINEERED NANOMATERIALS ON THE PRODUCERS OF THE ECOSYSTEM

The available literature on the effects of ENMs in plants shows that ENMs may accumulate in agriculture crops and/or increase concentrations of the constituted metal of NMs in the fruits or grains of the particular plant (Gardea-Torresdey et al., 2014). The accumulation of ENMs can cause harmful effects on the yield and productivity of plants, alter the nutritional value of food crops, and also transfer them across the trophic levels. The algal cells are also the primary producers of the aquatic environment; the accumulation of ENMs on the algal cells can transfer the ENMs across the trophic levels of a food chain (Zhang et al., 2012). The most available literature on the bioaccumulation and trophic transfer studies on the producers to primary consumers is listed in Table 4.1. The studies on the phytotoxicity of the most commonly used ENMs (Ag, ZnO, TiO_2, Cu/CuO, and carbon-based) are summarised in Table 4.2. It was observed that ENMs can pose multiple adverse effects on plants such as alteration of physiological processes (Darlington et al., 2009; El-Temsah & Joner, 2012; Kumari et al., 2009; Lee et al., 2008; Lin & Xing, 2008; Stampoulis et al., 2009; Yan et al., 2013), gene expression (Ghosh et al., 2015; Ghosh et al., 2011; Khodakovskaya et al., 2012; Khodakovskaya et al., 2011), DNA damage (Katti et al., 2015), and increased formation of ROS (Begum & Fugetsu, 2012; Hernandez-Viezcas et al., 2011; Liu et al., 2013). These phytotoxicological studies were focused on certain parameters such as rate of seed germination, root and shoot development, yield measurement, biomass growth, DNA damage, cell wall damage, transpiration potential, cell cycle, leaf growth, flowering delay, and germination index. No studies have looked at the mechanisms of toxicity of ENMs in plants. Thus we need to explore this in the future to better understand the safety of nanotechnology in plants. Additionally, no relevant standard toxicity assays have been developed to study the effect of photoperiod on the toxicity of photocatalytic NPs. The selection of experimental duration is also a matter of concern while conducting phytotoxicity assays of ENMs.

4.6 CONCLUSION

The chapter discussed the current status of the release of ENMs in the aquatic environment and their accumulation in the different trophic level organisms of the food chain. In present scenarios, the concentrations of engineered NMs are very low in the natural environment, but the unique physicochemical properties of engineered NMs have the potential to enhance bioavailability, which can enable their biomagnification and consequent toxicity in aquatic organisms. The available data have shown that ENMs can transfer from one trophic level to another in the terrestrial, aquatic, and estuarine food chain. But the potential trophic transfer depends on the type of NPs. For example, quantum dots, gold, and cerium oxide NPs transfer to another trophic level with increasing concentration (biomagnification factor >1) in contrast to zinc oxide and titanium dioxide NPs. The biomagnification of the contaminants can be very harmful to ecosystems as we learned from the past in the case of mercury and DDT. Therefore, it is an urgent demand to test the biomagnification of ENMs in

Natural Organic Matter

TABLE 4.1
Accumulation of Engineered Nanomaterials in the Producers and Consumers of the Environmental Food Chain

Nanoparticle	Test Organism	Mode of Exposure	Concentration	Test Duration	BAF/BCF/BMF or TTF or TMF/BSAF/Percent	Type of Food Chain	References
Producers: Plants-Primary consumers							
Tannate coated Au NPs	Tobacco hornworm (*Manduca sexta*) caterpillars	Tomato leaf tissue	100 mg/L Au	7 d	BAF = 0.16	Terrestrial	(Judy et al., 2012)
CeO$_2$ & ZnO NPs	cucumber plants	Soil	400, & 800 mg/kg	40 d	1.27 mg of Ce and 110 mg Zn per kg dry weight	Terrestrial	(L. Zhao et al., 2013)
Al$_2$O$_3$ NPs	*Ceriodaphnia dubia*	Algae (*Chlorella ellipsoids*)	120 μg/mL	72 h	BMF = 0.19	Aquatic	(Pakrashi et al., 2014)
AuNPs (5, 10, & 15 nm)	*Manduca sexta* (tobacco hornworm)	*Nicotiana tabacum* L. cv Xanthi	100 mg Au/L	41 h	BMF = 6.2, 11.6 & 9.6 for 5, 10, & 15 nm	Terrestrial	(Judy et al., 2011)
Au Nanorods (CTAB coated)	Estuarine mesocosm (sea water, sediment, sea grass, microbes, biofilms, snails, clams,shrimp and fish)	Water	7.08 × 10^8 particles/mL	12 d	BCF (L kg^{-1} DW) Biofilm = 15300, Sea grass = 8.21, Shrimp = 115, Fish = 474, Snail = 167, Clam = 22 800	Estuarine	(Ferry et al., 2009)
ZnO NPs	Soybean (*Glycine max*)	Root	500 mg/L	49 days	BCF Leaf = 0.69, Stem = 0.25, Root = 0.24, Nodule = 0.07	Terrestrial	(Hernandez-Viezcas et al., 2013)
ZnO NPs	Peanut (*Arachis hypogaea* L.)	Leaves	133 mg/L	110 d	Zn bioaccumulation in leaf = 42%	Terrestrial	(Prasad et al., 2012)
TiO$_2$ NPs	Tomato	Root	50, 100, 1000, 2500, 5000 mg/L	6-week old plant to maturity	% Accumulation Stem = 11.0–11.4%, Leaves = 6.1–8.4%	Terrestrial	(Song et al., 2013)

(*continued*)

TABLE 4.1 (Continued)
Accumulation of Engineered Nanomaterials in the Producers and Consumers of the Environmental Food Chain

Nanoparticle	Test Organism	Mode of Exposure	Concentration	Test Duration	BAF/BCF/BMF or TTF or TMF/BSAF/Percent	Type of Food Chain	References
CeO$_2$ NPs	Soybean	root	1000 mg/L	49 d	BCF Root = 0.21, Nodule = 0.011, Stem = 0.0001, Leaf = 3 × 10^{-7}	Terrestrial	(Song et al., 2013)
ZnO NPs	Isopods (*Porcellio scaber*)	Diet (NPs contaminated leaves)	6.2 and 2.5 g ZnO per kg dry leaf	28 d	BAF (kg dry leaf per kg dry biomass) = 0.1 and 0.2	Terrestrial	(Pipan-Tkalec et al., 2010)
Consumers: Primary consumers (invertebrates)-Top consumers (vertebrates)							
TiO$_2$ NPs	*Ceriodaphnia dubia*	diet (*Scenedesmus obliquus*) and Water	1, 2, 4,8, 16,32 & 64 µg/mL	21 d	BAF = 214.38 (L/Kg) BMF = 0.218	Aquatic	(Dalai et al., 2014)
TiO$_2$ nanoparticle aggregates	*Daphnia magna*	Water and Diet (*Pseudokirchneriella subcapitata*)	1 mg/L	72 h	BCF = 118062.84 L/Kg BAF = 1232.28 L/Kg	Aquatic	(Zhu, Chang, et al., 2010)
Nanocrystalline fullerenes as C60	*Daphnia magna*	Water	0.2 mg/L	24 h	BCF (L/kg DW) = 15 000 and 437 500 for mother and baby daphnia	Aquatic	(Tao et al., 2009)
AgNP-citrate, AgNP-PVP	Polychaete (*Nereis virens*)	Sediment	0.75, 7.5 and 75 mg/kg dry sediment weight	28 d	Ag accumulation = 32–44%, relative to the sediments	Estuarine Sediment	(Wang et al., 2013)
CdSe core – ZnS shell Carboxylated and Biotinylated QDs	Rotifer (*Brachionus calyciflorus*)	Ciliated protozoan (*Tetrahymena pyriformis*)	Cd^{2+}/g QDs Carboxylated = 985.4 ng Biotinylated = 624.3 ng	48 h	BMF (dry weight) = Carboxylated QDs-0.62 Biotinylated QDs-0.29	Aquatic	(Holbrook et al., 2008)
TiO$_2$ nanoparticle	*Tetrahymena thermophila*	Bacteria (*Pseudomonas aeruginosa*)	0.10 mg/mL	16 h	BMF = 0.16	Aquatic	(Mielke et al., 2013)

Natural Organic Matter

TiO$_2$ nanoparticle	Zebrafish (Danio rerio)	Daphnia magna	1 mg/L	21 d	BMF = 0.009	Aquatic	(Zhu, Wang, et al., 2010)
AgNP (Coated with carbonate)	Daphnia magna	Water	0.02–0.5 mg/L	48 h	BCF = 46000 104 (L kg^{-1} dry Weight)	Aquatic	(Zhao & Wang, 2011)
C60	Daphnia magna	Water and Diet	30 mg/L	48 h	BCF (L kg^{-1} DW) (30 mg/L) = 38000	Aquatic	(Oberdörster et al., 2006)
C60	Daphnia magna	Water	0.5–2 mg/L	24 h uptake & 2 d depuration	BCF (L/kg DW) (0.5 mg/L) = 95000 (2 mg/L) = 25000	Aquatic	(Tervonen et al., 2010)
MWCNT	Daphnia magna	Water	0.04–0.4 mg/L	48 h uptake & 2 d depuration	BCF (L/kg Dry biomass) = 350 000–440 000	Aquatic	(Petersen et al., 2009)
CdSe QDs (Citrate coated)	Ciliated protozoa (Tetrahymena thermophila)	QDs loaded bacteria (Pseudomonas aeruginosa)	75 mg/L	24 h	BMF (DW) ¼ 5.4	Aquatic	(Werlin et al., 2011)
SWCNT and MWCNT	Oligochaetes (Lumbriculus variegates)	River sediment	0.03 and 0.37 mg CNT/kg dry-sediment	28 d exposure	BSAF (kg OC kg^{-1} lipid) SWCNT = 0.28 MWCNT = 0.40	Sediment	(Petersen et al., 2008)
MWCNT	Oligochaetes (Lumbriculus variegates)	River sediment	0.37 mg CNT/g dry sediment	28 d exposure	BSAF (kg dry sediment per kg dry biomass) = 0.67	Sediment	(Petersen et al., 2010)
TiO$_2$ NPs	Polychaets (Arenicola marina)	Marine Sediment	1–3 g TiO$_2$ per kg sediment	10 d	BSAF (kg OC per kg Lipid) = 0.156–0.196	Sediment	(Galloway et al., 2010)
TiO$_2$ NPs	Carp (Cyprinus carpio)	Water	10 mg/L	25 d	BCF (L kg^{-1} whole DW) = 495	Aquatic vertebrate	(Sun et al., 2007)
TiO$_2$ NPs	Carp (Cyprinus carpio)	Water	10 mg/L	20 d	BCF (L kg^{-1} whole body DW) = 325	Aquatic vertebrate	(Zhang et al., 2007)
TiO$_2$ NPs	Carp (Cyprinus carpio)	Water	10 mg/L	20 d	BCF (L kg^{-1} whole body DW) = 617	Aquatic vertebrate	(Sun et al., 2009)

(continued)

TABLE 4.1 (Continued)
Accumulation of Engineered Nanomaterials in the Producers and Consumers of the Environmental Food Chain

Nanoparticle	Test Organism	Mode of Exposure	Concentration	Test Duration	BAF/BCF/BMF or TTF or TMF/BSAF/Percent	Type of Food Chain	References
TiO$_2$ NPs	Zebrafish (*Danio rerio*)	Water	0.1–1 mg/L	14 d	BCF (L kg^{-1} DW) = 25 & 181	Aquatic vertebrate	(Zhu, Wang, et al., 2010)
CdSe-ZnS QD (coated with Poly (acrylic acid)-octylamine)	Zebrafish (*Danio rerio*)	QD loaded Daphnids	Adult fish = 10 QD/ Daphnia Juvenile fish = 1 mL of QD/ 3000 Artemia	21 d	BMF = 0.04 and 0.004 for adult and juvenile zebrafish	Aquatic	(Lewinski et al., 2011)
TiO$_2$ NPs	Earthworm (*Eisenia fetida*)	artificial soil	0.1–5 g/kg dry soil	7 d	BSAF (kg dry soil per kg wet biomass) = 0.01–0.02	Terrestrial	(Hu et al., 2010)
ZnO NPs	Earthworm (*Eisenia fetida*)	artificial soil	0.1–5 g/kg dry soil	7 d	BSAF (kg dry soil per kg wet biomass) = 0.01–0.37	Terrestrial	(Hu et al., 2010)

XAS: X-ray absorption spectroscopy, **ICP-MS**: inductively coupled plasma mass spectrometry, **HR-ICP-MS**: High resolution inductively coupled plasma mass spectrometry, **ICP-AES**: inductively coupled plasma atomic emission spectroscopy, **ICP-OES**: inductively coupled plasma–optical emission spectrometry, **AAS**: Atomic Absorption Spectrophotometer, **LSC**: Liquid scintillation counting.

TABLE 4.2
The Direct Toxicological Impact of Ag, ZnO, TiO$_2$, Cu/CuO and Carbon Based Engineered Nanomaterials in Plants

NP type	Concentration	Plant Species	Toxicity	References
Ag	10 mg/L	Barley, flax	Decreases seed germination and reduces shoot length	(El-Temsah & Joner, 2012)
Ag	100, 500, 1000 mg/L	zucchini	Reduced transpiration, biomass and root growth	(Stampoulis et al., 2009)
Ag	25–100 mg/L	*Allium cepa*	Effect mitosis cell cycle with altered metaphase. The fragmentation and disruptions in the cell wall Chromosomal aberrations	(Kumari et al., 2009)
Cu	200–800 mg/L	Mungbean and Wheat	Effect the normal growth of seedling and shoot	(Lee et al., 2008)
Cu	1000 mg/L	zucchini	Reduce the biomass production, root growth	(Darlington et al., 2009)
ZnO	1000 mg/L	ryegrass	Reduced biomass production, contracted root tips, damage in the epidermis and root cap, vacuolated and collapsed cortical cells	(Lin & Xing, 2008)
ZnO	2000 mg/L	corn, reddish, rape, ryegrass, lettuce, cucumber	Reduced seed germination and root growth	(Lin & Xing, 2007)
ZnO	1000 mg/L	zucchini	Reduced biomass production	(Stampoulis et al., 2009)
ZnO	2000–4000 mg/L	soybean	Decreased root growth	(Lopez-Moreno et al., 2010)
SWCNT	400 mg/L	Rice	Flowering delay, decreased in the yield	(Lin et al., 2009)
SWCNT	104, 315, 1750 mg/L	tomato	Reduction in the root growth	(Canas et al., 2008)
MWCNT	1000 mg/L	zucchini	Reduced biomass production	(Stampoulis et al., 2009)
MWCNT	2000 mg/L	lettuce	Reduced root length	(Lin & Xing, 2007)
MWCNT	20, 40, 80 mg/L	rice	Cell shrinkage, plasma membrane detachment from cell wall, apoptosis	(Tan et al., 2009)
MWCNT	50–200 mg/L	*Lycopersicon esculentum*	Increased the expression of stress related gene	(Khodakovskaya et al., 2011)
MWCNT	0.1 mg/L	*Nicotiana tabacum*	Increased the expression of genes related to, cell-wall assembly, cell growth, regulation of cell cycle progression and synthesis of aquaporins.	(Khodakovskaya et al., 2012)

(continued)

TABLE 4.2 (Continued)
The Direct Toxicological Impact of Ag, ZnO, TiO$_2$, Cu/CuO and Carbon Based Engineered Nanomaterials in Plants

NP type	Concentration	Plant Species	Toxicity	References
TiO$_2$	300, 1000 mg/L	maize	Repressed transpiration, hydraulic conductivity and leaf growth	(Asli & Neumann, 2009)
TiO$_2$	2–10 mM mg/L	Allium cepa, Nicotiana tabacum, Zea mays, Vicia narbonensis, Arabidopsis thaliana	DNA damage was observed using comet assay, Chromosomal aberrations, Micronuclei test was positive, Altered mitotic activity	(Ruffini Castiglione et al., 2011)
ZnO	500–4000 mg/L	Proposis juliflora-velutina	Increases in the catalase activity in whole plant, Decrease in ascorbate peroxidase activity in roots, while increases in leaves and stem.	(Hernandez-Viezcas et al., 2011)

organisms that have not yet been investigated for such effects and are the most relevant to ecosystem services.

ACKNOWLEDGMENTS

Financial assistance by The Gujarat Institute for Chemical Technology (GICT) for the Establishment of a Facility for environmental risk assessment of chemicals and nanomaterials is acknowledged.

NOTE

1 www.primescholars.com/articles/market-analysis-for-nanomaterials-2020-97209.html.

REFERENCES

Asli, S., & Neumann, P. M. (2009). Colloidal suspensions of clay or titanium dioxide nanoparticles can inhibit leaf growth and transpiration via physical effects on root water transport. *Plant Cell Environ*, *32*(5), 577–584. https://doi.org/10.1111/j.1365-3040.2009.01952.x

Begum, P., & Fugetsu, B. (2012). Phytotoxicity of multi-walled carbon nanotubes on red spinach (Amaranthus tricolor L) and the role of ascorbic acid as an antioxidant. *J Hazard Mater*, *243*, 212–222. https://doi.org/10.1016/j.jhazmat.2012.10.025

Bernhardt, E. S., Colman, B. P., Hochella, M. F., Jr., Cardinale, B. J., Nisbet, R. M., Richardson, C. J., & Yin, L. (2010). An ecological perspective on nanomaterial impacts in the environment. *J Environ Qual*, *39*(6), 1954–1965. https://doi.org/10.2134/jeq2009.0479

Bondarenko, O. M., Heinlaan, M., Sihtmae, M., Ivask, A., Kurvet, I., Joonas, E., Jemec, A., Mannerstrom, M., Heinonen, T., Rekulapelly, R., Singh, S., Zou, J., Pyykko, I., Drobne, D., & Kahru, A. (2016). Multilaboratory evaluation of 15 bioassays for (eco)toxicity screening and hazard ranking of engineered nanomaterials: FP7 project NANOVALID. *Nanotoxicology, 10*(9), 1229–1242. https://doi.org/10.1080/17435390.2016.1196251

Bouldin, J. L., Ingle, T. M., Sengupta, A., Alexander, R., Hannigan, R. E., & Buchanan, R. A. (2008). Aqueous toxicity and food chain transfer of Quantum DOTs in freshwater algae and Ceriodaphnia dubia. *Environ Toxicol Chem, 27*(9), 1958–1963. https://doi.org/10.1897/07-637.1

Caceres-Velez, P. R., Fascineli, M. L., Sousa, M. H., Grisolia, C. K., Yate, L., de Souza, P. E. N., Estrela-Lopis, I., Moya, S., & Azevedo, R. B. (2018). Humic acid attenuation of silver nanoparticle toxicity by ion complexation and the formation of a Ag(3+) coating. *J Hazard Mater, 353*, 173–181. https://doi.org/10.1016/j.jhazmat.2018.04.019

Canas, J. E., Long, M., Nations, S., Vadan, R., Dai, L., Luo, M., Ambikapathi, R., Lee, E. H., & Olszyk, D. (2008). Effects of functionalized and nonfunctionalized single-walled carbon nanotubes on root elongation of select crop species. *Environ Toxicol Chem, 27*(9), 1922–1931. https://doi.org/10.1897/08-117.1

Chae, Y., & An, Y. J. (2016). Toxicity and transfer of polyvinylpyrrolidone-coated silver nanowires in an aquatic food chain consisting of algae, water fleas, and zebrafish. *Aquat Toxicol, 173*, 94–104. https://doi.org/10.1016/j.aquatox.2016.01.011

Chen, H., Seiber, J. N., & Hotze, M. (2014). ACS Select on nanotechnology in food and agriculture: a perspective on implications and applications. *J Agric Food Chem, 62*(6), 1209–1212. https://doi.org/10.1021/jf5002588

Chen, J., Li, H., Han, X., & Wei, X. (2015). Transmission and accumulation of nano-TiO_2 in a 2-step food chain (Scenedesmus obliquus to Daphnia magna). *Bull Environ Contam Toxicol, 95*(2), 145–149. https://doi.org/10.1007/s00128-015-1580-y

Chojnacka, K. (2010). Biosorption and bioaccumulation – the prospects for practical applications. *Environ Int, 36*(3), 299–307. https://doi.org/10.1016/j.envint.2009.12.001

Cupi, D., Hartmann, N. B., & Baun, A. (2015). The influence of natural organic matter and aging on suspension stability in guideline toxicity testing of silver, zinc oxide, and titanium dioxide nanoparticles with Daphnia magna. *Environ Toxicol Chem, 34*(3), 497–506. https://doi.org/10.1002/etc.2855

Dalai, S., Iswarya, V., Bhuvaneshwari, M., Pakrashi, S., Chandrasekaran, N., & Mukherjee, A. (2014). Different modes of TiO_2 uptake by Ceriodaphnia dubia: relevance to toxicity and bioaccumulation. *Aquat Toxicol, 152*, 139–146. https://doi.org/10.1016/j.aquatox.2014.04.002

Darlington, T. K., Neigh, A. M., Spencer, M. T., Nguyen, O. T., & Oldenburg, S. J. (2009). Nanoparticle characteristics affecting environmental fate and transport through soil. *Environ Toxicol Chem, 28*(6), 1191–1199. https://doi.org/10.1897/08-341.1

Dasari, T. P., & Hwang, H. M. (2010). The effect of humic acids on the cytotoxicity of silver nanoparticles to a natural aquatic bacterial assemblage. *Sci Total Environ, 408*(23), 5817–5823. https://doi.org/10.1016/j.scitotenv.2010.08.030

Deng, C.-H., Gong, J.-L., Zeng, G.-M., Jiang, Y., Zhang, C., Liu, H.-Y., & Huan, S.-Y. (2016). Graphene–CdS nanocomposite inactivation performance toward Escherichia coli in the presence of humic acid under visible light irradiation. *Chemical Engineering Journal, 284*, 41–53. https://doi.org/https://doi.org/10.1016/j.cej.2015.08.106

Domingos, R. F., Tufenkji, N., & Wilkinson, K. J. (2009). Aggregation of titanium dioxide nanoparticles: role of a fulvic acid. *Environ Sci Technol, 43*(5), 1282–1286. https://doi.org/10.1021/es8023594

El-Temsah, Y. S., & Joner, E. J. (2012). Impact of Fe and Ag nanoparticles on seed germination and differences in bioavailability during exposure in aqueous suspension and soil. *Environ Toxicol, 27*(1), 42–49. https://doi.org/10.1002/tox.20610

Ferry, J. L., Craig, P., Hexel, C., Sisco, P., Frey, R., Pennington, P. L., Fulton, M. H., Scott, I. G., Decho, A. W., Kashiwada, S., Murphy, C. J., & Shaw, T. J. (2009). Transfer of gold nanoparticles from the water column to the estuarine food web. *Nat Nanotechnol, 4*(7), 441–444. https://doi.org/10.1038/nnano.2009.157

Foss Hansen, S., Heggelund, L. R., Revilla Besora, P., Mackevica, A., Boldrin, A., & Baun, A. (2016). Nanoproducts – what is actually available to European consumers? [10.1039/C5EN00182J]. *Environmental Science: Nano, 3*(1), 169–180. https://doi.org/10.1039/C5EN00182J

Galloway, T., Lewis, C., Dolciotti, I., Johnston, B. D., Moger, J., & Regoli, F. (2010). Sublethal toxicity of nano-titanium dioxide and carbon nanotubes in a sediment dwelling marine polychaete. *Environ Pollut, 158*(5), 1748–1755. https://doi.org/10.1016/j.envpol.2009.11.013

Gardea-Torresdey, J. L., Rico, C. M., & White, J. C. (2014). Trophic transfer, transformation, and impact of engineered nanomaterials in terrestrial environments. *Environ Sci Technol, 48*(5), 2526–2540. https://doi.org/10.1021/es4050665

Ghafari, P., St-Denis, C. H., Power, M. E., Jin, X., Tsou, V., Mandal, H. S., Bols, N. C., & Tang, X. S. (2008). Impact of carbon nanotubes on the ingestion and digestion of bacteria by ciliated protozoa. *Nat Nanotechnol, 3*(6), 347–351. https://doi.org/10.1038/nnano.2008.109

Ghosh, M., Bhadra, S., Adegoke, A., Bandyopadhyay, M., & Mukherjee, A. (2015). MWCNT uptake in Allium cepa root cells induces cytotoxic and genotoxic responses and results in DNA hyper-methylation. *Mutat Res, 774*, 49–58. https://doi.org/10.1016/j.mrfmmm.2015.03.004

Ghosh, M., Chakraborty, A., Bandyopadhyay, M., & Mukherjee, A. (2011). Multi-walled carbon nanotubes (MWCNT): induction of DNA damage in plant and mammalian cells. *J Hazard Mater, 197*, 327–336. https://doi.org/10.1016/j.jhazmat.2011.09.090

Glenn, J. B., & Klaine, S. J. (2013). Abiotic and biotic factors that influence the bioavailability of gold nanoparticles to aquatic macrophytes. *Environ Sci Technol, 47*(18), 10223–10230. https://doi.org/10.1021/es4020508

Gogos, A., Knauer, K., & Bucheli, T. D. (2012). Nanomaterials in plant protection and fertilization: current state, foreseen applications, and research priorities. *J Agric Food Chem, 60*(39), 9781–9792. https://doi.org/10.1021/jf302154y

Gottschalk, F., Sonderer, T., Scholz, R. W., & Nowack, B. (2009). Modeled environmental concentrations of engineered nanomaterials (TiO(2), ZnO, Ag, CNT, Fullerenes) for different regions. *Environ Sci Technol, 43*(24), 9216–9222. https://doi.org/10.1021/es9015553

Gunsolus, I. L., Mousavi, M. P., Hussein, K., Buhlmann, P., & Haynes, C. L. (2015). Effects of humic and fulvic acids on silver nanoparticle stability, dissolution, and toxicity. *Environ Sci Technol, 49*(13), 8078–8086. https://doi.org/10.1021/acs.est.5b01496

Gupta, G. S., Dhawan, A., & Shanker, R. (2016). Montmorillonite clay alters toxicity of silver nanoparticles in zebrafish (Danio rerio) eleutheroembryo. *Chemosphere, 163*, 242–251. https://doi.org/10.1016/j.chemosphere.2016.08.032

Gupta, G. S., Kansara, K., Shah, H., Rathod, R., Valecha, D., Gogisetty, S., Joshi, P., & Kumar, A. (2019). Impact of humic acid on the fate and toxicity of titanium dioxide nanoparticles in Tetrahymena pyriformis and zebrafish embryos. *Nanoscale Advances, 1*(1), 219–227.

Hernandez-Viezcas, J. A., Castillo-Michel, H., Andrews, J. C., Cotte, M., Rico, C., Peralta-Videa, J. R., Ge, Y., Priester, J. H., Holden, P. A., & Gardea-Torresdey, J. L. (2013).

In situ synchrotron X-ray fluorescence mapping and speciation of CeO(2) and ZnO nanoparticles in soil cultivated soybean (Glycine max). *ACS Nano, 7*(2), 1415–1423. https://doi.org/10.1021/nn305196q

Hernandez-Viezcas, J. A., Castillo-Michel, H., Servin, A. D., Peralta-Videa, J. R., & Gardea-Torresdey, J. L. (2011). Spectroscopic verification of zinc absorption and distribution in the desert plant Prosopis juliflora-velutina (velvet mesquite) treated with ZnO nanoparticles. *Chem Eng J, 170*(1–3), 346–352. https://doi.org/10.1016/j.cej.2010.12.021

Hoet, P. H., Nemmar, A., & Nemery, B. (2004). Health impact of nanomaterials? *Nat Biotechnol, 22*(1), 19. https://doi.org/10.1038/nbt0104-19

Holbrook, R. D., Murphy, K. E., Morrow, J. B., & Cole, K. D. (2008). Trophic transfer of nanoparticles in a simplified invertebrate food web. *Nat Nanotechnol, 3*(6), 352–355. https://doi.org/10.1038/nnano.2008.110

Holden, P. A., Nisbet, R. M., Lenihan, H. S., Miller, R. J., Cherr, G. N., Schimel, J. P., & Gardea-Torresdey, J. L. (2013). Ecological nanotoxicology: integrating nanomaterial hazard considerations across the subcellular, population, community, and ecosystems levels. *Acc Chem Res, 46*(3), 813–822. https://doi.org/10.1021/ar300069t

Holden, P. A., Schimel, J. P., & Godwin, H. A. (2014). Five reasons to use bacteria when assessing manufactured nanomaterial environmental hazards and fates. *Curr Opin Biotechnol, 27*, 73–78. https://doi.org/10.1016/j.copbio.2013.11.008

Hou, W. C., Westerhoff, P. and Posner, J. D. (2013). Biological Accumulation of Engineered Nanomaterials: A Review of Current Knowledge. *Environ Sci Process Impacts, 15*, 103–122. https://doi.org/10.1039/C2EM30686G

Hu, C. W., Li, M., Cui, Y. B., Li, D. S., Chen, J., & Yang, L. Y. (2010). Toxicological effects of TiO$_2$ and ZnO nanoparticles in soil on earthworm Eisenia fetida. *Soil Biology and Biochemistry, 42*(4), 586–591. https://doi.org/https://doi.org/10.1016/j.soilbio.2009.12.007

Joseph, T., & Morrison, M. (2006). *Nanotechnology in Agriculture and Food*. European Nanotechnology Gateway.

Jucker, B. A., Zehnder, A. J. B., & Harms, H. (1998). Quantification of polymer interactions in bacterial adhesion. *Environmental Science & Technology, 32*(19), 2909–2915. https://doi.org/10.1021/es980211s

Judy, J. D., Unrine, J. M., & Bertsch, P. M. (2011). Evidence for biomagnification of gold nanoparticles within a terrestrial food chain. *Environ Sci Technol, 45*(2), 776–781. https://doi.org/10.1021/es103031a

Judy, J. D., Unrine, J. M., Rao, W., & Bertsch, P. M. (2012). Bioaccumulation of gold nanomaterials by Manduca sexta through dietary uptake of surface contaminated plant tissue. *Environ Sci Technol, 46*(22), 12672–12678. https://doi.org/10.1021/es303333w

Jung, Y., Metreveli, G., Park, C. B., Baik, S., & Schaumann, G. E. (2018). Implications of pony lake fulvic acid for the aggregation and dissolution of oppositely charged surface-coated silver nanoparticles and their ecotoxicological effects on *Daphnia magna*. *Environ Sci Technol, 52*(2), 436–445. https://doi.org/10.1021/acs.est.7b04635

Kahru, A., & Dubourguier, H. C. (2010). From ecotoxicology to nanoecotoxicology. *Toxicology, 269*(2–3), 105–119. https://doi.org/10.1016/j.tox.2009.08.016

Kansara, K., Bolan, S., Radhakrishnan, D., Palanisami, T., Al-Muhtaseb, A. H., Bolan, N., Vinu, A., Kumar, A., & Karakoti, A. (2021 a). A critical review on the role of abiotic factors on the transformation, environmental identity and toxicity of engineered nanomaterials in aquatic environment. *Environ Pollut, 296*, 118726. https://doi.org/10.1016/j.envpol.2021.118726

Kansara, K., Paruthi, A., Misra, S. K., Karakoti, A. S., & Kumar, A. (2019). Montmorillonite clay and humic acid modulate the behavior of copper oxide nanoparticles in aqueous environment and induces developmental defects in zebrafish embryo. *Environ Pollut*, *255*(Pt 2), 113313. https://doi.org/10.1016/j.envpol.2019.113313

Kansara, K., Patel, P., Shukla, R. K., Pandya, A., Shanker, R., Kumar, A., & Dhawan, A. (2018). Synthesis of biocompatible iron oxide nanoparticles as a drug delivery vehicle. *Int J Nanomedicine*, *13*(T-NANO 2014 Abstracts), 79–82. https://doi.org/10.2147/IJN.S124708

Kansara, K., Sathish, C. I., Vinu, A., Kumar, A., & Karakoti, A. S. (2021 b). Assessment of the impact of abiotic factors on the stability of engineered nanomaterials in fish embryo media. *Emergent Materials*, *4*(5), 1339–1350. https://doi.org/10.1007/s42247-021-00224-3

Katti, D. R., Sharma, A., Pradhan, S. M., & Katti, K. S. (2015). Carbon nanotube proximity influences rice DNA. *Chemical Physics*, *455*, 17–22. https://doi.org/https://doi.org/10.1016/j.chemphys.2015.03.015

Keller, A. A., & Lazareva, A. (2014). Predicted releases of engineered nanomaterials: from global to regional to local. *Environmental Science & Technology Letters*, *1*(1), 65–70. https://doi.org/10.1021/ez400106t

Khodakovskaya, M. V., de Silva, K., Biris, A. S., Dervishi, E., & Villagarcia, H. (2012). Carbon nanotubes induce growth enhancement of tobacco cells. *ACS Nano*, *6*(3), 2128–2135. https://doi.org/10.1021/nn204643g

Khodakovskaya, M. V., de Silva, K., Nedosekin, D. A., Dervishi, E., Biris, A. S., Shashkov, E. V., Galanzha, E. I., & Zharov, V. P. (2011). Complex genetic, photothermal, and photoacoustic analysis of nanoparticle-plant interactions. *Proc Natl Acad Sci USA*, *108*(3), 1028–1033. https://doi.org/10.1073/pnas.1008856108

Kim, J. I., Park, H. G., Chang, K. H., Nam, D. H., & Yeo, M. K. (2016). Trophic transfer of nano-TiO$_2$ in a paddy microcosm: a comparison of single-dose versus sequential multi-dose exposures. *Environ Pollut*, *212*, 316–324. https://doi.org/10.1016/j.envpol.2016.01.076

Kim, J. Y., Kim, K. T., Lee, B. G., Lim, B. J., & Kim, S. D. (2013). Developmental toxicity of Japanese medaka embryos by silver nanoparticles and released ions in the presence of humic acid. *Ecotoxicol Environ Saf*, *92*, 57–63. https://doi.org/10.1016/j.ecoenv.2013.02.004

Kumar, A., & Dhawan, A. (2019). *Nanoparticle–Protein Corona: Biophysics to Biology* (Vol. 40). Royal Society of Chemistry.

Kumar, A., Dhawan, A., & Shanker, R. (2011). The need for novel approaches in ecotoxicity of engineered nanomaterials. *J Biomed Nanotechnol*, *7*(1), 79–80. https://doi.org/10.1166/jbn.2011.1211

Kumar, A., Pandey, A. K., Singh, S. S., Shanker, R., & Dhawan, A. (2011a). Cellular response to metal oxide nanoparticles in bacteria. *J Biomed Nanotechnol*, *7*(1), 102–103. https://doi.org/10.1166/jbn.2011.1222

Kumar, A., Pandey, A. K., Singh, S. S., Shanker, R., & Dhawan, A. (2011b). Engineered ZnO and TiO(2) nanoparticles induce oxidative stress and DNA damage leading to reduced viability of Escherichia coli. *Free Radic Biol Med*, *51*(10), 1872–1881. https://doi.org/10.1016/j.freeradbiomed.2011.08.025

Kumari, M., Mukherjee, A., & Chandrasekaran, N. (2009). Genotoxicity of silver nanoparticles in Allium cepa. *Sci Total Environ*, *407*(19), 5243–5246. https://doi.org/10.1016/j.scitotenv.2009.06.024

Lee, J., Mahendra, S., & Alvarez, P. J. (2010). Nanomaterials in the construction industry: a review of their applications and environmental health and safety considerations. *ACS Nano*, *4*(7), 3580–3590. https://doi.org/10.1021/nn100866w

Lee, S. W., Park, S. Y., Kim, Y., Im, H., & Choi, J. (2016). Effect of sulfidation and dissolved organic matters on toxicity of silver nanoparticles in sediment dwelling organism, Chironomus riparius. *Sci Total Environ, 553*, 565–573. https://doi.org/10.1016/j.scitotenv.2016.02.064

Lee, W. M., & An, Y. J. (2015). Evidence of three-level trophic transfer of quantum dots in an aquatic food chain by using bioimaging. *Nanotoxicology, 9*(4), 407–412. https://doi.org/10.3109/17435390.2014.948517

Lee, W. M., An, Y. J., Yoon, H., & Kweon, H. S. (2008). Toxicity and bioavailability of copper nanoparticles to the terrestrial plants mung bean (Phaseolus radiatus) and wheat (Triticum aestivum): plant agar test for water-insoluble nanoparticles. *Environ Toxicol Chem, 27*(9), 1915–1921. https://doi.org/10.1897/07-481.1

Lewinski, N. A., Zhu, H., Ouyang, C. R., Conner, G. P., Wagner, D. S., Colvin, V. L., & Drezek, R. A. (2011). Trophic transfer of amphiphilic polymer coated CdSe/ZnS quantum dots to Danio rerio. *Nanoscale, 3*(8), 3080–3083. https://doi.org/10.1039/c1nr10319a

Li, B., & Logan, B. E. (2004). Bacterial adhesion to glass and metal-oxide surfaces. *Colloids Surf B Biointerfaces, 36*(2), 81–90. https://doi.org/10.1016/j.colsurfb.2004.05.006

Li, M., Pokhrel, S., Jin, X., Mädler, L., Damoiseaux, R., & Hoek, E. M. (2010). Stability, bioavailability, and bacterial toxicity of ZnO and iron-doped ZnO nanoparticles in aquatic media. *Environmental science & technology, 45*(2), 755–761. https://doi.org/10.1021/es102266g

Lin, D., & Xing, B. (2007). Phytotoxicity of nanoparticles: inhibition of seed germination and root growth. *Environ Pollut, 150*(2), 243–250. https://doi.org/10.1016/j.envpol.2007.01.016

Lin, D., & Xing, B. (2008). Root uptake and phytotoxicity of ZnO nanoparticles. *Environ Sci Technol, 42*(15), 5580–5585. https://doi.org/10.1021/es800422x

Lin, D., Ji, J., Long, Z., Yang, K., & Wu, F. (2012). The influence of dissolved and surface-bound humic acid on the toxicity of TiO(2) nanoparticles to Chlorella sp. *Water Res, 46*(14), 4477–4487. https://doi.org/10.1016/j.watres.2012.05.035Lin, S., Reppert, J., Hu, Q., Hudson, J. S., Reid, M. L., Ratnikova, T. A., Rao, A. M., Luo, H., & Ke, P. C. (2009). Uptake, translocation, and transmission of carbon nanomaterials in rice plants. *Small, 5*(10), 1128–1132. https://doi.org/10.1002/smll.200801556

Liu, Q., Zhang, X., Zhao, Y., Lin, J., Shu, C., Wang, C., & Fang, X. (2013). Fullerene-induced increase of glycosyl residue on living plant cell wall. *Environ Sci Technol, 47*(13), 7490–7498. https://doi.org/10.1021/es4010224

Lopez-Moreno, M. L., de la Rosa, G., Hernandez-Viezcas, J. A., Peralta-Videa, J. R., & Gardea-Torresdey, J. L. (2010). X-ray absorption spectroscopy (XAS) corroboration of the uptake and storage of CeO(2) nanoparticles and assessment of their differential toxicity in four edible plant species. *J Agric Food Chem, 58*(6), 3689–3693. https://doi.org/10.1021/jf904472e

Lowry, G. V., Hotze, E. M., Bernhardt, E. S., Dionysiou, D. D., Pedersen, J. A., Wiesner, M. R., & Xing, B. (2010). Environmental occurrences, behavior, fate, and ecological effects of nanomaterials: an introduction to the special series. *J Environ Qual, 39*(6), 1867–1874. https://doi.org/10.2134/jeq2010.0297

Ma, S., Zhou, K., Yang, K., & Lin, D. (2015). Heteroagglomeration of oxide nanoparticles with algal cells: effects of particle type, ionic strength and pH. *Environ Sci Technol, 49*(2), 932–939. https://doi.org/10.1021/es504730k

Majumdar, S., Trujillo-Reyes, J., Hernandez-Viezcas, J. A., White, J. C., Peralta-Videa, J. R., & Gardea-Torresdey, J. L. (2016). Cerium biomagnification in a terrestrial food chain: influence of particle size and growth stage. *Environ Sci Technol, 50*(13), 6782–6792. https://doi.org/10.1021/acs.est.5b04784

Mielke, R. E., Priester, J. H., Werlin, R. A., Gelb, J., Horst, A. M., Orias, E., & Holden, P. A. (2013). Differential growth of and nanoscale TiO(2) accumulation in Tetrahymena thermophila by direct feeding versus trophic transfer from Pseudomonas aeruginosa. *Appl Environ Microbiol, 79*(18), 5616–5624. https://doi.org/10.1128/AEM.01680-13

Mortimer, M., Petersen, E. J., Buchholz, B. A., Orias, E., & Holden, P. A. (2016). Bioaccumulation of multiwall carbon nanotubes in tetrahymena thermophila by direct feeding or trophic transfer. *Environ Sci Technol, 50*(16), 8876–8885. https://doi.org/10.1021/acs.est.6b01916

Neale, P. A., Jämting, Å. K., O'Malley, E., Herrmann, J., & Escher, B. I. (2015). Behaviour of titanium dioxide and zinc oxide nanoparticles in the presence of wastewater-derived organic matter and implications for algal toxicity. *Environmental Science: Nano, 2*(1), 86–93.

Oberdörster, E., Zhu, S., Blickley, T. M., McClellan-Green, P., & Haasch, M. L. (2006). Ecotoxicology of carbon-based engineered nanoparticles: effects of fullerene (C60) on aquatic organisms. *Carbon, 44*(6), 1112–1120. https://doi.org/https://doi.org/10.1016/j.carbon.2005.11.008

Pakrashi, S., Dalai, S., Chandrasekaran, N., & Mukherjee, A. (2014). Trophic transfer potential of aluminium oxide nanoparticles using representative primary producer (Chlorella ellipsoides) and a primary consumer (Ceriodaphnia dubia). *Aquat Toxicol, 152*, 74–81. https://doi.org/10.1016/j.aquatox.2014.03.024

Patel, P., & Kumar, A. (2019). Factors affecting a nanoparticle's protein corona formation. In Ashutosh Kumar and Alok Dhawan (Eds.) *Nanoparticle–Protein Corona* (pp. 61–79). The Royal Society of Chemistry.

Patel, P., Kansara, K., Singh, R., Shukla, R. K., Singh, S., Dhawan, A., & Kumar, A. (2018). Cellular internalization and antioxidant activity of cerium oxide nanoparticles in human monocytic leukemia cells. *Int J Nanomedicine, 13*(T-NANO 2014 Abstracts), 39–41. https://doi.org/10.2147/IJN.S124996

Petersen, E. J., Akkanen, J., Kukkonen, J. V., & Weber, W. J., Jr. (2009). Biological uptake and depuration of carbon nanotubes by Daphnia magna. *Environ Sci Technol, 43*(8), 2969–2975. https://doi.org/10.1021/es8029363

Petersen, E. J., Huang, Q., & Weber, W. J. (2008). Ecological uptake and depuration of carbon nanotubes by Lumbriculus variegatus. *Environ Health Perspect, 116*(4), 496–500. https://doi.org/10.1289/ehp.10883

Petersen, E. J., Huang, Q., & Weber, W. J., Jr. (2010). Relevance of octanol-water distribution measurements to the potential ecological uptake of multi-walled carbon nanotubes. *Environ Toxicol Chem, 29*(5), 1106–1112. https://doi.org/10.1002/etc.149

Pipan-Tkalec, Z., Drobne, D., Jemec, A., Romih, T., Zidar, P., & Bele, M. (2010). Zinc bioaccumulation in a terrestrial invertebrate fed a diet treated with particulate ZnO or $ZnCl_2$ solution. *Toxicology, 269*(2–3), 198–203. https://doi.org/10.1016/j.tox.2009.08.004

Prasad, T., Sudhakar, P., Sreenivasulu, Y., Latha, P., Munaswamy, V., Reddy, K. R., Sreeprasad, T., Sajanlal, P., & Pradeep, T. (2012). Effect of nanoscale zinc oxide particles on the germination, growth and yield of peanut. *Journal of Plant Nutrition, 35*(6), 905–927.

Robinson, D., & Morrison, M. (2009). *Nanotechnology Developments for the Agrifood Sector – Report of the ObservatoryNANO*. Institute of Nanotechnology, UK.

Roco, M.C., Mirkin, C.A., Hersam, M.C. (2011). *Nanotechnology Research Directions for Societal Needs in 2020*; Science Policy Reports Series. Springer: New York.

Ruffini Castiglione, M., Giorgetti, L., Geri, C., & Cremonini, R. (2011). The effects of nano-TiO_2 on seed germination, development and mitosis of root tip cells of Vicia narbonensis L. and Zea mays L. *Journal of Nanoparticle Research, 13*(6), 2443–2449. https://doi.org/10.1007/s11051-010-0135-8

Scott, N., & Chen, H. (2013). Nanoscale science and engineering for agriculture and food systems. *Industrial Biotechnology*, *9*(1), 17–18.

Senapati, V. A., & Kumar, A. (2018). Nanoparticles and the aquatic environment: application, impact and fate. In Alok Dhawan, Ashutosh Kumar, Sanjay Singh and Rishi Sanker (Eds.) *Nanobiotechnology* (pp. 299–322). CRC Press.

Senapati, V. A., & Kumar, A. (2018). ZnO nanoparticles dissolution, penetration and toxicity in human epidermal cells. Influence of pH. *Environmental Chemistry Letters*, *16*(3), 1129–1135. https://doi.org/10.1007/s10311-018-0736-5

Shang, E., Li, Y., Niu, J., Zhou, Y., Wang, T., & Crittenden, J. C. (2017). Relative importance of humic and fulvic acid on ROS generation, dissolution, and toxicity of sulfide nanoparticles. *Water Res*, *124*, 595–604. https://doi.org/10.1016/j.watres.2017.08.001

Song, U., Jun, H., Waldman, B., Roh, J., Kim, Y., Yi, J., & Lee, E. J. (2013). Functional analyses of nanoparticle toxicity: a comparative study of the effects of TiO_2 and Ag on tomatoes (Lycopersicon esculentum). *Ecotoxicol Environ Saf*, *93*, 60–67. https://doi.org/10.1016/j.ecoenv.2013.03.033

Stampoulis, D., Sinha, S. K., & White, J. C. (2009). Assay-dependent phytotoxicity of nanoparticles to plants. *Environ Sci Technol*, *43*(24), 9473–9479. https://doi.org/10.1021/es901695c

Sun, H., Zhang, X., Niu, Q., Chen, Y., & Crittenden, J. C. (2007). Enhanced accumulation of arsenate in carp in the presence of titanium dioxide nanoparticles. *Water, Air, and Soil Pollution*, *178*(1), 245–254.

Sun, H., Zhang, X., Zhang, Z., Chen, Y., & Crittenden, J. C. (2009). Influence of titanium dioxide nanoparticles on speciation and bioavailability of arsenite. *Environ Pollut*, *157*(4), 1165–1170. https://doi.org/10.1016/j.envpol.2008.08.022

Tan, X.-m., Lin, C., & Fugetsu, B. (2009). Studies on toxicity of multi-walled carbon nanotubes on suspension rice cells. *Carbon*, *47*(15), 3479–3487.

Tang, Y., Li, S., Lu, Y., Li, Q., & Yu, S. (2015). The influence of humic acid on the toxicity of nano-ZnO and Zn^{2+} to the Anabaena sp. *Environ Toxicol*, *30*(8), 895–903. https://doi.org/10.1002/tox.21964

Tao, X., Fortner, J. D., Zhang, B., He, Y., Chen, Y., & Hughes, J. B. (2009). Effects of aqueous stable fullerene nanocrystals (nC60) on Daphnia magna: evaluation of sub-lethal reproductive responses and accumulation. *Chemosphere*, *77*(11), 1482–1487. https://doi.org/10.1016/j.chemosphere.2009.10.027

Tervonen, K., Waissi, G., Petersen, E. J., Akkanen, J., & Kukkonen, J. V. (2010). Analysis of fullerene-C60 and kinetic measurements for its accumulation and depuration in Daphnia magna. *Environ Toxicol Chem*, *29*(5), 1072–1078. https://doi.org/10.1002/etc.124

Thomann, R. V., Connolly, J. P., & Parkerton, T. F. (1992). An equilibrium model of organic chemical accumulation in aquatic food webs with sediment interaction. *Environmental Toxicology and Chemistry: An International Journal*, *11*(5), 615–629.

Unrine, J. M., Shoults-Wilson, W. A., Zhurbich, O., Bertsch, P. M., & Tsyusko, O. V. (2012). Trophic transfer of Au nanoparticles from soil along a simulated terrestrial food chain. *Environ Sci Technol*, *46*(17), 9753–9760. https://doi.org/10.1021/es3025325

Vallabani, N. V. S., Sengupta, S., Shukla, R. K., & Kumar, A. (2019). ZnO nanoparticles-associated mitochondrial stress-induced apoptosis and G2/M arrest in HaCaT cells: a mechanistic approach. *Mutagenesis*, *34*(3), 265–277. https://doi.org/10.1093/mutage/gez017

van der Oost, R., Beyer, J., & Vermeulen, N. P. (2003). Fish bioaccumulation and biomarkers in environmental risk assessment: a review. *Environ Toxicol Pharmacol*, *13*(2), 57–149. https://doi.org/10.1016/s1382-6689(02)00126-6

Vance, M. E., Kuiken, T., Vejerano, E. P., McGinnis, S. P., Hochella, M. F., Jr., Rejeski, D., & Hull, M. S. (2015). Nanotechnology in the real world: Redeveloping the nanomaterial consumer products inventory. *Beilstein J Nanotechnol, 6*, 1769–1780. https://doi.org/10.3762/bjnano.6.181

Walser, T., Limbach, L. K., Brogioli, R., Erismann, E., Flamigni, L., Hattendorf, B., Juchli, M., Krumeich, F., Ludwig, C., Prikopsky, K., Rossier, M., Saner, D., Sigg, A., Hellweg, S., Gunther, D., & Stark, W. J. (2012). Persistence of engineered nanoparticles in a municipal solid-waste incineration plant. *Nat Nanotechnol, 7*(8), 520–524. https://doi.org/10.1038/nnano.2012.64

Wang, Y., Miao, A. J., Luo, J., Wei, Z. B., Zhu, J. J., & Yang, L. Y. (2013). Bioaccumulation of CdTe quantum dots in a freshwater alga Ochromonas danica: a kinetics study. *Environ Sci Technol, 47*(18), 10601–10610. https://doi.org/10.1021/es4017188

Wang, Z., Li, J., Zhao, J., & Xing, B. (2011). Toxicity and internalization of CuO nanoparticles to prokaryotic alga Microcystis aeruginosa as affected by dissolved organic matter. *Environ Sci Technol, 45*(14), 6032–6040. https://doi.org/10.1021/es2010573

Wang, Z., Quik, J. T., Song, L., Van Den Brandhof, E. J., Wouterse, M., & Peijnenburg, W. J. (2015). Humic substances alleviate the aquatic toxicity of polyvinylpyrrolidone-coated silver nanoparticles to organisms of different trophic levels. *Environ Toxicol Chem, 34*(6), 1239–1245. https://doi.org/10.1002/etc.2936

Werlin, R., Priester, J. H., Mielke, R. E., Kramer, S., Jackson, S., Stoimenov, P. K., Stucky, G. D., Cherr, G. N., Orias, E., & Holden, P. A. (2011). Biomagnification of cadmium selenide quantum dots in a simple experimental microbial food chain. *Nat Nanotechnol, 6*(1), 65–71. https://doi.org/10.1038/nnano.2010.251

Wirth, S. M., Lowry, G. V., & Tilton, R. D. (2012). Natural organic matter alters biofilm tolerance to silver nanoparticles and dissolved silver. *Environ Sci Technol, 46*(22), 12687–12696. https://doi.org/10.1021/es301521p

Yan, S., Zhao, L., Li, H., Zhang, Q., Tan, J., Huang, M., He, S., & Li, L. (2013). Single-walled carbon nanotubes selectively influence maize root tissue development accompanied by the change in the related gene expression. *J Hazard Mater, 246–247*, 110–118. https://doi.org/10.1016/j.jhazmat.2012.12.013

Yang, S. P., Bar-Ilan, O., Peterson, R. E., Heideman, W., Hamers, R. J., & Pedersen, J. A. (2013). Influence of humic acid on titanium dioxide nanoparticle toxicity to developing zebrafish. *Environ Sci Technol, 47*(9), 4718–4725. https://doi.org/10.1021/es3047334

Yang, X., Jiang, C., Hsu-Kim, H., Badireddy, A. R., Dykstra, M., Wiesner, M., Hinton, D. E., & Meyer, J. N. (2014). Silver nanoparticle behavior, uptake, and toxicity in Caenorhabditis elegans: effects of natural organic matter. *Environ Sci Technol, 48*(6), 3486–3495. https://doi.org/10.1021/es404444n

Zhang, P., Ma, Y., Zhang, Z., He, X., Guo, Z., Tai, R., Ding, Y., Zhao, Y., & Chai, Z. (2012). Comparative toxicity of nanoparticulate/bulk Yb(2)O(3) and YbCl(3) to cucumber (Cucumis sativus). *Environ Sci Technol, 46*(3), 1834–1841. https://doi.org/10.1021/es2027295

Zhang, W., Rittmann, B., & Chen, Y. (2011). Size effects on adsorption of hematite nanoparticles on E.coli cells. *Environ Sci Technol, 45*(6), 2172–2178. https://doi.org/10.1021/es103376y

Zhang, X., Sun, H., Zhang, Z., Niu, Q., Chen, Y., & Crittenden, J. C. (2007). Enhanced bioaccumulation of cadmium in carp in the presence of titanium dioxide nanoparticles. *Chemosphere, 67*(1), 160–166. https://doi.org/10.1016/j.chemosphere.2006.09.003

Zhao, C. M., & Wang, W. X. (2011). Comparison of acute and chronic toxicity of silver nanoparticles and silver nitrate to Daphnia magna. *Environ Toxicol Chem, 30*(4), 885–892. https://doi.org/10.1002/etc.451

Zhao, L., Sun, Y., Hernandez-Viezcas, J. A., Servin, A. D., Hong, J., Niu, G., Peralta-Videa, J. R., Duarte-Gardea, M., & Gardea-Torresdey, J. L. (2013). Influence of CeO_2 and ZnO nanoparticles on cucumber physiological markers and bioaccumulation of Ce and Zn: a life cycle study. *J Agric Food Chem*, *61*(49), 11945–11951. https://doi.org/10.1021/jf404328e

Zhou, D., Abdel-Fattah, A. I., & Keller, A. A. (2012). Clay particles destabilize engineered nanoparticles in aqueous environments. *Environ Sci Technol*, *46*(14), 7520–7526. https://doi.org/10.1021/es3004427

Zhu, X., Chang, Y., & Chen, Y. (2010). Toxicity and bioaccumulation of TiO_2 nanoparticle aggregates in Daphnia magna. *Chemosphere*, *78*(3), 209–215. https://doi.org/10.1016/j.chemosphere.2009.11.013

Zhu, X., Wang, J., Zhang, X., Chang, Y., & Chen, Y. (2010). Trophic transfer of TiO(2) nanoparticles from Daphnia to zebrafish in a simplified freshwater food chain. *Chemosphere*, *79*(9), 928–933. https://doi.org/10.1016/j.chemosphere.2010.03.022

5 Analysis of Emerging Contaminants in Water Samples
Advances and Challenges

Sivaperumal P., Nikhil P. K., Smith C., Shaikh S., Rajnish G., Rupal T., and Tejal G. M.

CONTENTS

5.1 Introduction 85
5.2 Environmental Chemistry of Emerging Micro-pollutants 87
5.3 Sampling, Extraction, and Quantification 89
    5.3.1 Sample Collection 89
    5.3.2 Sample Extraction 89
    5.3.3 Sample Preparation Techniques 90
    5.3.4 Water Analysis 92
    5.3.5 Optimization of GC-MS/MS 94
    5.3.6 Optimization of LC-HR/MS 105
5.4 Conclusions 106
Acknowledgments 109
References 111

5.1 INTRODUCTION

Human activity such as personal care, healthcare, and industrial operations entails the use of chemicals that produces or generate waste (Tremblay *et al.*, 2011). Whilst most contaminants degrade with the passage of time, there are certain breeds of contaminants that are exclusive due to their resistance to degradation. These therefore tend to accumulate in the environment thereby eventually becoming a hazard to living organisms (Tremblay *et al.*, 2011). This resistance to degradation calls for legislative intervention because of their potential to cause slow but sure elevation in concentrations in our water bodies (Petrovic *et al.*, 2003). Emerging pollutants, usually categorised as synthetic (man-made compounds) and natural chemicals, include products that are intensively used in large quantities on a day to day basis (Petrovic´ *et al.*, 2003, Valsecchia *et al.*, 2015). The last few decades have seen reporting of extensive studies that have proven the presence of these emerging pollutants (parent compound) and their metabolites in groundwater, wastewater, drinking water, etc., to name a few (Janna, 2011).

However, the most common playground of emerging contaminants are the industrial and domestic wastewater treatment plants (WWTPs), which receive effluents from human activities as well as that of the industrial discharges (Giulia, 2009; Zuloaga et al., 2012).

Introduction of newer and sensitive analytical equipment with passing time is a boon for detection of chemicals in complex sample matrices. Whilst on the one hand they have increased specificity for estimations at nano-levels, these have in turn also drawn attention to newer compounds that were until now outside the ambit of monitoring and regulation. These chemicals of emerging concern (CECs) or emerging contaminants as per their current nomenclature were hitherto undetected and were not considered as being a potential risk (Zwiener and Frimmel, 2004) until long. However, it is also imperative to note that the term emerging contaminants could be a bit misleading since these contaminants are not necessarily new in origin (Söderström, 2009). CECs by current nomenclature encompass a diverse group of compounds, which include algal and cyanobacterial toxins, plasticizers, brominated and organophosphate flame retardants, pharmaceuticals, hormones and endocrine-disrupting compounds (EDCs), personal care products (PCPs), drugs of abuse as well as resultant metabolites, polar pesticides and their by-products/end-products, disinfection by-products, organometallics, nanomaterials, per-fluorinated compounds (PFCs), surfactants and their metabolites (Petrovic et al., 2010), etc. The current pace of development has led to extensive and widespread usage of products containing CECs. This has, in turn, led to an extensive environmental distribution of CECs with potential eco-toxicological effects despite exceedingly low concentrations attracting increased and sustained interest amongst researchers, regulatory authorities, and the public at large courtesy the efficacy of newer analytical techniques (Zwiener et al., 2004; Söderström, 2009).

Studies have indicated that CECs present in fertilizers have a strong possibility of getting into agricultural produce (food). These would in turn have the potential to have a negative impact on food quality as well as be detrimental to human health (Farré and Barcelo, 2013) over a period of time. Stormwater runoff has also been identified as another possible source of emerging pollutants since it is often a reservoir of flame retardants, plasticizers, pesticides, and the like (Janna, 2011). The environment today is therefore a man-made reservoir of a host of emerging contaminants (Metcalfe, 2014; Shafrir and Avisar, 2012; Rivera-Utrilla et al., 2013). And, it is pollutants such as pharmaceuticals, EDCs, and PCPs that are of great interest amongst these breeds of emerging contaminants.

This chapter intends to delve into recent advances in estimating emerging pollutants and detail sample preparation methodologies for environmental samples. Advantages and disadvantages of these sample preparation techniques are reviewed in addition to examining the future perspectives with respect to newer sample preparation methods, each specific to different types of emerging contaminants. The very fact that there is a myriad of emerging pollutants makes detailing of CECs even more exhaustive. This chapter focuses chiefly on emerging contaminants such as pharmaceuticals, EDCs, and PCPs present in the environment.

Emerging contaminants present in pharmaceuticals and personal care products as well as the many endocrine disruptors are ever-present in environmental matrices such as soil, water, and sediment, among others. These being capable of causing unwanted imbalances in living organisms, determination of these pollutants requires separation techniques capable of high efficiency and selectivity coupled with higher sensitivities since the direct analysis of these pollutants in complex matrices is often a tedious task. It is for this reason that rapid, inexpensive, efficient, and environmentally friendly sample preparation techniques have been developed as a precursor to chromatographic quantification.

Emerging contaminants such as polychlorinated biphenyls (PCB), polybrominated biphenyls (PBB), dichlorodiphenyltrichloroethane (DDT), dioxins, phthalates, alkylphenols, bisphenol A (BPA), etc. (Campbell *et al.*, 2006; Diamanti-Kandarakis *et al.*, 2009; Gore *et al.*, 2015) are commonly present compounds used in industrial solvents or lubricants, pesticides, plasticizers, etc., to name a few (Diamanti-Kandarakis *et al.*, 2009; Nohynek *et al.*, 2013).

Post introduction into the environment through run-offs from farmlands and effluents from wastewater treatment plants these contaminants usually bio-accumulate in groundwater, river, and lakes (Auriol *et al.*, 2006) and thereafter in aquatic organisms as a consequence of their stability and persistence. Their levels of presence in sediments and sludge often being in traces necessitates the requirement of an efficient pre-treatment of these environmental samples prior to detection and estimation procedures (Gorga *et al.*, 2014; Azzouz and Ballesteros, 2016; Cavaliere *et al.*, 2016).

5.2 ENVIRONMENTAL CHEMISTRY OF EMERGING MICRO-POLLUTANTS

In recent years, advances in instrumentation have yielded significant progress in the detection of CECs in environmental matrices. Continuous industrial development has propitiated the generation of a wide variety of chemicals, the application of which has proliferated in day-to-day activities. These substances, or CECs as they are now referred to, are of a huge concern to society as a result of their persistence and stability (Manzetti and van der Spoel, 2015). Ubiquitous in nature their toxicological effects are however not yet fully known (Stasinakis, 2012). Research has established that prolonged exposure to EDCs, including alkylphenols, phthalates, parabens, or BPA, provokes abnormal growth as well as disorders of the reproductive system in wildlife and humans (Masuo and Ishido, 2011; Jiménez-Díaz *et al.*, 2015; Azzouz and Ballesteros, 2016). Polybrominated diphenyl ethers (PBDE), as well as PCPs, have been demonstrated to cause neurotoxicity as well as alter normal endocrine functions (McDonald, 2002; Brausch and Rand, 2011; Liu and Wong, 2013; Kaw and Kannan, 2017).

Modernization coupled with urban expansion has led to an exponential increase in the generation of global wastes (Manzetti and van der Spoel, 2015). With a rise in the breed of pollutants, WWTPs are today improperly equipped to remove these emerging contaminants present in household and industrial wastes. Present mostly at trace levels (µg/L or lower) their insufficient removal leads to the discharge of an aqueous effluent containing micro-organisms in addition to the prevailing organic and inorganic substances (Christodoulou and Stamatelatou, 2016), which in turn further pollutes the environment it mixes with (Healy et al., 2017). There are several routes for disposal of sewage sludge. However, most of these result in deposition of treated or untreated sludge into the environment thereby elevating potential eco-toxicological risks (Zuloaga et al., 2012; Manzetti and van der Spoel, 2015). Sludge deposition on agricultural land can cause groundwater contamination from leaching or land runoffs that in turn could further lead to hazardous environmental consequences or degradation of human health (Healy et al., 2017).

Adverse health effects such as an increased risk of cancer or those involving the nervous system, liver, kidney, and eye issues among several other diseases (Zwiener and Frimmel, 2004; Alavanja et al., 2004; Espada, 2001; Yu et al., 2016; Bouchard et al., 2011; Handal et al., 2007; Gorchev and Ozolins, 1984, Jurewicz and Hanke, 2008) have been reported from prolonged exposure to above-maximum contaminant levels of organic pollutants such as pesticides, volatile organic compounds (VOCs), polycyclic aromatic hydrocarbons (PAHs), phenolic compounds, etc. (US EPA, 2010). Considering these implications, accurate monitoring and analysis of liquid waste samples forms an important mandate for environmental agencies (Koester and Clement, 1993).

Extraction and isolation of compounds from sample matrices is a necessary precursor to instrumental analysis and determination of CECs. In addition to clean-up, ideal sample preparation technique should aid in enrichment of compounds to improve sensitivity. Literature review on emerging pollutants analysis in environmental samples has revealed that traditional and modern sample preparation techniques have recently obtained greater importance. This is chiefly because direct analysis in complex matrices and samples wherein these pollutants are present in minute quantities is difficult. And this in turn, necessitates an extensive review of a number of sample preparation techniques for their application for extraction as well as pre-concentration. The strength and weakness of each technique are also dependent on the type of procedure adopted, its accuracy and sensitivity, repeatability, simplicity, and cost involved.

There are available today a host of well-established sample preparation techniques for determination of organic contaminants in water samples. These include liquid-liquid extraction (LLE) (Barceló, 1993; Mahara, et al., 1998; Suffet and Faust, 1972; Hennion and Pichon, 1994), solid-phase extraction (SPE) (Hennion et al., 1994; Erger and Schmidt, 2014; Pihlström et al., 1997), and solid-phase micro-extraction (SPME) (Arthur and Pawlyszin, 1990; Souza-Silva et al., 2015). Whilst LLE is a simple technique approved by the US Environmental Protection Agency (Barceló, 1993) and that is commonly also used for water samples, SPE as a technique is most extensively used due to its attractive advantages such as flexibility, simplicity, high selectivity, automation and rapidity, inclusion of higher enrichment factors as well as absence of emulsions and the ease for use of different sorbent materials; over traditional techniques such as LLE. SPE has also been reported as using smaller volumes of potentially toxic solvents as compared to LLE, though a significant amount of disposables is still required. It is for reasons such as these, that alternative techniques such as dispersive solid phase (micro) extraction (DSPE/DSPME) are gaining prominence. This technique has been demonstrated to display advantages (over traditional

SPE) such as rapidity without the need for conditioning, simplicity, superior enrichment, and cost-effectiveness as a result of reduced sample size, sorbent amount, solvents, and resultant waste. However, the main disadvantage of DSPE/DSPME is the inability to change solvents between extraction and pre-concentration steps.

On the contrary, determination of emerging pollutants in solid samples (soils, sediments, and food matrices) has not received much attention. There is therefore a demand for development of sample preparation techniques with improved selectivity, sensitivity, and reduced matrix effects specific to complex matrices in solid samples. Development and improvement of sample preparation techniques are directly related to a spike in the use of chromatographic techniques hyphenated to mass spectrometry and routinely also used for identification and quantification of a wide spectrum of emerging pollutants. However, that no guidelines or legislative intervention are currently available to regulate their presence (Petrovic et al., 2003; Sanchez-Prado et al., 2010; Patiño et al., 2015 Zavala et al., 2015; Bletsou et al., 2015) makes it worrisome.

5.3 SAMPLING, EXTRACTION, AND QUANTIFICATION

5.3.1 SAMPLE COLLECTION

An oft neglected, but the most important part of analysis is sample collection. A well-defined protocol for sample collection plays an important role in accurate reporting.

Sample collection protocols differ from contaminant to contaminant (i.e., agricultural waste run-offs, industrial effluents, wastes from households, percolations, and run-offs from improperly disposed hazardous wastes, etc.) and the same should be studied and adhered to prior to embarking on sample collection. A figurative detailing of the source of emerging contaminants is depicted in Figure 5.1.

The first and primary step in sample collection is selection of the target site to be sampled. It is imperative that this site is representative of the contaminated water sample that is the focus of the study design. That the type of sample collected has a non-biased spread of contaminants is also an important factor to be kept in mind during sample collection. Multiple samples may either be collected from the identified site to ensure absence of bias. Care should also be taken to collect a sufficient amount of the water sample in inert sample containers such that it is adequate for the study as well as replicate analysis if the same may be needed. Post collection of the water sample, it may be properly coded with all relevant information attributed to it and stored in insulated boxes prior to transfer to the laboratory for further analysis.

5.3.2 SAMPLE EXTRACTION

Extraction and clean-up of collected samples post sampling constitutes a cardinal step towards separation and determination of analytes of interest. Post collection of water samples, this step chiefly involves removal of interferences and concentrating the analytes/contaminants to detectable levels. The three main reasons for it to be properly carried out are:

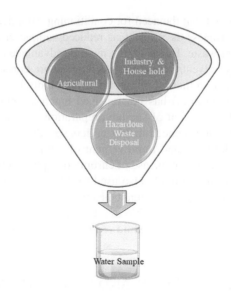

FIGURE 5.1 Source of emerging contaminants in water samples.

1. Removal of interferences that would otherwise affect determination of analytes;
2. To concentrate analytes to detectable concentrations; and,
3. To aid in performing solvent switching to desired solvent conditions for detection.

Sample preparation and clean-up necessitate pH adjustment to optimize extraction efficiency. Studies have suggested adjusting pH to acidic, basic, or neutral depending on the type of the analyte (Wille *et al.*, 2010a). Studies have also hinted at the use of chelating agents such as di- or tetrasodium ethylenediaminetetraacetate (Na$_2$EDTA/ Na$_4$EDTA), quenching agents like ascorbic acid, and/or other preservatives prior to initiating extraction (Ferrer *et al.*, 2010). Techniques commonly used for extraction of the emerging contaminants constitute LLE, SPE, SPME, and dSPE. Post extraction the analytes are then subjected to detection using an array of advanced instrumental techniques widely available today. A pictorial description of steps involved in collection of samples. extraction and detection techniques employed is given in Figure 5.2.

5.3.3 Sample Preparation Techniques

A direct analysis of emerging pollutants in complex matrices is difficult because their concentrations are usually low, in addition to their association with the sample matrix (Farré *et al.*, 2010). This usually makes it difficult to predict their route of transport as well as their fate in the environment. Lack of information about them is another impending factor (Petrovic´ *et al.*, 2003). It is for these reasons that the step of sample preparation needs to be properly designed prior to execution (Petrovic´ *et al.*, 2003).

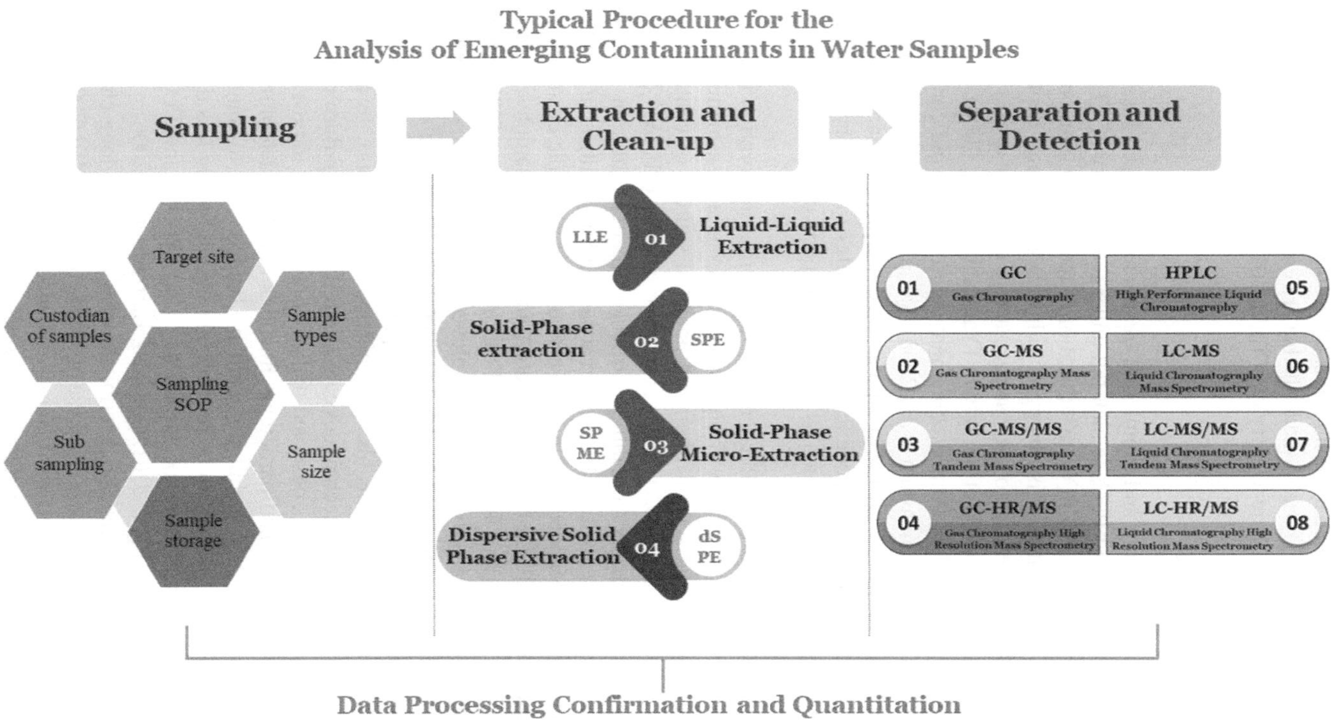

Analysis of Emerging Contaminants in Water Samples 91

FIGURE 5.2 Typical procedure for sampling, extraction, and detection of emerging contaminants.

Sample preparation as a procedure may include filtration, pH adjustment, extraction, and clean-up as well as some other pre-concentration procedures to ensure that analytes are found at suitable concentrations (Padrón et al., 2014). The current trend in sample preparation is the use of methods that are environmentally friendly, low in cost, automated, and simplistic to carry out (Yoon, 2010).

5.3.4 Water Analysis

Several aspects of sampling and sample pre-treatment have been assessed as being essential for the analysis of PFCs, which fall under the category of CECs. Sampling depth plays a crucial role in PFC quantification. Studies have demonstrated that a decrease in PFC concentration is observed with increasing water depth (Yamashita et al., 2008). This can be due to their surface-active character and in ocean waters due to the global ocean-circulation system (Yamashita et al., 2008). Furthermore, it is also important to select sample equipment carefully to avoid contamination as well as PFC adsorption (Yamashita et al., 2008). Sampling practices advocate rinsing sample collection bottles with (semi-) polar solvents prior to sampling (Petrovic et al., 2010). Old practices of acidification of samples to conserve them are now discouraged to prevent volatilization as well as adsorption onto the sample container (Petrovic et al., 2010).

The surface-active nature of PFCs makes them prone to sorption onto the filter material, which is why filtration of aqueous samples is also a matter of concern. It has been reported that several filters are potential sources of contamination for perfluorooctanoic acid (PFOA) and perfluorononanoate (PFNA), hence except in aqueous samples visibly containing particulate matter, filtration as a sample-preparation step may be avoided. In a study carried out by Schultz et al., 2006 centrifugation has been recommended as an alternative step for sample clean-up. Usually, low concentrations of PFCs (pg/L–ng/L) are found in water samples, necessitating their pre-concentration and isolation. Application of LLE and/or SPE, followed by solvent evaporation has been commonly employed and its use has even been reported in the third inter-laboratory study carried out on PFCs (Van Leeuwen et al., 2011). That SPE as a technique was predominantly applied in nearly 80% of water samples (Van Leeuwen et al., 2011) was also documented in this report. In a comparative study for sample preparation techniques carried out by González-Barreiro et al., 2006 between LLE and SPE, it was observed that when using LLE, the overall PFC concentration of the aqueous and particulate fractions could be determined, but since filtration was avoided, this technique could be limited only to long-chain PFCs. Apparently, SPE was observed to be best suited to PFCs having less than 10 carbon atoms, including even perfluorooctanesulfonic acid (PFOS) and PFOA, the most important contaminants (EPA, 2012; González-Barreiro et al., 2006).

An SPE approach detailing the determination of short and long-chain PFCs has been described as well, post the publication of which, the ISO 25101/2006 method was established. This method used weak anion-exchange (Oasis WAX) SPE cartridges. Similarly, use of C18 and certain oasis hydrophilic-lipophilic balance (HLB) cartridges for the purpose of PFC enrichment has also been oft-reported (González-Barreiro et al., 2006; Wille et al., 2010b). An alternative and direct determination of

PFCs in aqueous samples has also been reported (Schultz et al., 2006) wherein, PFCs were quantified using large volume injection without any sample pre-treatment. Limit of quantitation (LOQs) obtained thereby was in the 1 ng/L range.

LC coupled to quadrupole mass spectrometry (MS) in the negative-ionization mode is an instrumental method preferred for the determination of ionic PFCs in environmental matrices, including both perfluorosulfonic acid (PFSA) and perfluorinated carboxylic acid (PFCA). Post LC separation, triple quadrupole mass spectrometry (QqQ-MS) has been reported to provide excellent quantification for PFCs even at extremely low concentrations. Contrastingly, it has been reported that multi-stage tandem mass spectrometry (MSn) when used along with triple quadrupole (QqQ) or ion-trap mass spectrometry (IT-MS) for detection of PFSAs, majorly including PFOS, efficiency was drastically reduced. These substances were observed to exhibit extremely high stability at extreme conditions and it was due to this that, the use of QqQ-MS or IT-MS resulted in PFOS fragments of [SO_3] and [FSO_3], with m/z ions of 80 and 99, respectively. Unfortunately, interferences that co-elute along with PFOS were observed to have similar retention times and also one such similar transition.

Research also has mentioned another analytical approach for PFC determination. This powerful technique involves use of high-resolution full-scan accurate-mass measurements wherein quadrupole time-of-flight mass spectrometry (QToF-MS) was established as an optimal detector for PFC quantification as a result of its capability to combine high selectivity with high sensitivity. There is also mention in the literature about the use of liquid chromatography time-of-flight mass spectrometry (LC-ToF-MS) for the determination of 14 PFCs in liquid samples of sewage, surface water, and seawater having LOQs in the range of 2–200 ng/L (Wille et al., 2010b). Scanty reference also mentions using high resolution mass spectrometry (HRMS) via an Orbitrap instrument for PFC determination. However, there is the absence of sufficient literature to establish its applicability. Ionic PFCAs and PFSAs have also been analyzed using Gas Chromatography (GC). However, there is a need for this analysis to be preceded with a derivatization with a mixture of N, N-dicyclohexylcarbodiimide or iso-propanol and 2,4-difluoroaniline as per requirements (Langlois et al., 2007). Non-ionic PFCs, such as fluorotelomer alcohols and fluoroalkyl sulfonamides, were reported to be directly amenable to GC due to their higher volatility.

Langlois et al., 2007 reported a high resolution gas chromatography (HRGC) method for ionic and non-ionic PFCs wherein separation of PFOS isomers was carried out using high-resolution characteristics of a capillary GC-column that was coated with 5%-phenyl 95%-methylpolysiloxane. However, fine-tuning the method is pertinent to overcome certain characteristic difficulties during PFC analysis.

One of the major problems in the analysis of PFCs is background contamination of analytical blanks, especially for low-ng/L concentration range analysis (Taniyasu et al., 2005). It is due to these reasons that use of Teflon materials is avoided throughout extraction procedures. Prior to their analysis, it is therefore recommended to replace with clean and dry glass materials, the internal parts of high performance liquid chromatography (HPLC) systems, which are usually made of Polytetrafluoroethylene (PTFE) along with stainless steel and polyether ether ketone (PEEK). Another commonly used precaution is to place between the pump and the injector an additional HPLC column to separate sample PFCs and also the

PFCs originating from the system (Wille et al., 2010b). The literature also mentions limited use of mass-labelled analogues as a means to compensate for ionization effects, which can be a contributor to the occurrence of deviations in analytical results. Use of isotopically labelled internal standard for every single PFC requiring quantification is also recommended (Van Leeuwen et al., 2011).

Additionally, there is also the need to consider the occurrence of branched isomers during analysis. Technically PFOS has been reported to contain up to 30% differently branched isomers, which can have (Langlois et al., 2007) varying ionization efficiencies and that in turn can result in systematic quantification bias of PFCs. Examination of PFC isomer profiles can prove to be interesting to analyse the fate of PFCs as well as to distinguish historical contamination from recent contamination. Very recently, Taniyasu et al., 2005 even reported results for HPLC with tandem mass spectrometry (HPLC-MS/MS) wherein isomer-specific quantification of PFCs in water samples was carried out. (Taniyasu et al., 2005).

A diagrammatical representation of the flowchart to be adopted for sample preparation techniques for analysis using LC-HR/MS is detailed in Figure 5.3. As has been elucidated in the flowchart, techniques differ for targeted and non-targeted analysis. Whilst targeted analysis involves quantification for known and specific chemicals, it has the advantage of being supported by reference standards during the course of analysis. On the other hand, non-targeted analysis is predominantly used only for identifying potential (chemical) suspects in a sample. Quantitation cannot be yielded through this branch of analysis as it does not have the support of reference standards.

Different types of CECs, notably pharmaceuticals, PCPs, and EDCs, have been reported in environmental matrices. A number of CECs analysed post collection from different water sources including potable water (surface, well, river, lake, tap) and wastewater have been analysed using different extraction (e.g., SPE, SPME, MDSPME, SPE–DVB, MEPS, μ-SPE, etc.), separation, and analytical separation techniques (including different mobile as well as stationary phases). The use of techniques ranged from GC-MS to UHPLC-QTOF-MS/Orbitrap high-resolution mass spectrometry resulting in precise LOD, LOQ, and recovery percentage. An overview of these published techniques for CEC analysis is given in Table 5.1.

In addition to the published analytical methods for analysis of ECs, one of the authors also optimised the use of GC-MS/MS and LC-HR/MS for analysis of one of the breeds of EC such as pesticides. A detailing of the steps that can be adopted using the two different techniques for said analysis follows.

5.3.5 Optimization of GC-MS/MS

Improved sensitivity and a reduced runtime were the benchmarks based on which the technique was optimised. For this, oven temperature program and flow rate were initially optimized. Post optimization, initial oven temperature was fixed at 70°C. The flow rate of carrier gas helium was assessed at three different rates (i.e., 1.0, 1.2, and 1.5 ml/min) for accomplishing chromatographic separation. Post this, a noticeable optimum was obtained at 1.5 ml/min. Simultaneously the retention time and identification of analyte were established based on mass confirmation of each

Analysis of Emerging Contaminants in Water Samples

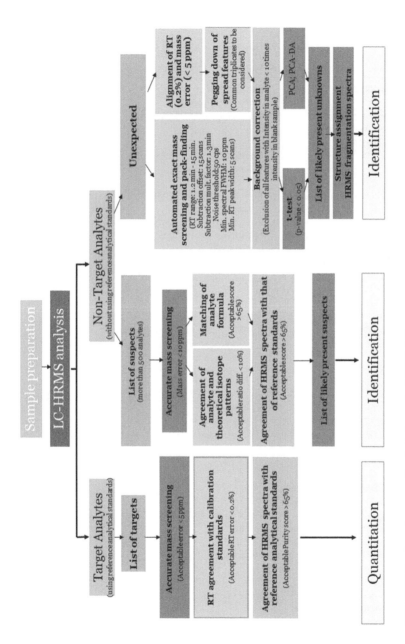

FIGURE 5.3 Sample preparation techniques for LC-HR/MS.

TABLE 5.1
Overview of Analytical Methods for Analysis of Emerging Contaminants

Analyte details	Matrix	Sample Extraction Techniques	Analytical Separation Technique — Mobile Phase/Carrier Gas
Pesticides (n=1607)	Surface water	SPE	-
Pesticides (n=5) viz. Sevin, Fenitrothion, Malathion, Parathion, Diazinon	Real water	MD-SPME	He
Pesticides (n=33)	Ground and well water	SPE–HLB	He
2,4-DMA, Teflubenzuron-Pesticides	River and well water	SPME	He (99.999%)
Atrazine, Simazine, Terbuthylazine, Ametryn, Prometon, Prometryn, Propazine, Hydroxyatrazine and Gesatamin	Lake, River and well water	SPE	$HCOONH_4$, $HCOOH$ and CH_3OH different compositions
Pesticides (n=3) viz. Diazinon, Malathion, Ethion	Tap and well water	MEPS, μ-SPE	He (99.999%)
Diazinon, Profenofos, Chlorpyrifos, Fenitrothion, and Ethion	Water sample	MSPE	He
OC pesticides (n=10) viz. Lindane, Heptachlor, Aldrin, Endosulfan, Chlordane, Dieldrin, Endrin, DDT, Methoxychlor, 1,3-dichloropropene OP pesticides (n=4) viz. Chlorpyrifos, Malathion, Diazinon, Glyphosate Herbicides (n=2) viz. 2-4-D, Alachlor	Raw, treated, drinking and surface water	LLME	-
Pesticides (n=4) Chlorthiamid, Ethyl Parathion, Penconazole and Fludioxonil	Tap, well and lake water	SS-LPE	He (99.999%)
Pesticides (n=4) Dichlorvos, Terbutryn, Cybutryne and Quinoxyfen	Surface water	SPME	He (≥ 99.99%)
39 Multiclass pesticides (n=39) OP, OC, Pyrethroid, Strobilurin, Thiocarbamate, Triazole, Imidazole, and Triazine	Water samples.	DM-SPE	-

Stationary Phase	Analytical Instrument	LOD (µg/L)	LOQ (µg/L)	Recovery (%)	(%) RSD	References
C_{18} and HILIC	LC-QTOF/MS	-	-	-	-	(Becker et al., 2021)
RTX-5MS (30 m × 0.25 mm i.d., 0.25 µm)	GC-MS	0.04–0.07	0.11–0.21	82.9–113.2	< 8.6	(Akbarzade et al., 2018)
5-MS (60 m × 0.25 mm i.d., 0.25 µm)	GC-MS	0.001–0.2	0.0005–0.05	-	-	(Chaza et al., 2018)
VF-5 MS (30 m × 0.25 mm i.d., 0.25 µm)	GC-MS	0.1–0.15	0.5–0.33	91.3–101.4	< 10.3	(Dargahi et al., 2018)
ZB-1 (30 m × 0.25 mm i.d., 0.25 µm)	GC-MS	0.1–5	0.2–17	80–102	1.5–6	(Rimayi et al., 2018)
C_{18}, Biphenyl 100 Å (2.6 µm 100 x 2.1 cm)	LC-MS/MS					
5% phenyl methyl siloxane (30 m × 250 µm i.d., 0.25 µm)	GC-MS	0.1–0.38	-	85.5–103.2	2.8–8.9	(Taghani et al., 2018)
HP-5 (30 m × 0.32 mm i.d., 0.25 µm)	GC-MS	0.02–0.08	0.07–2.27	94.5–107.1	2.6–6.5	(Nasiri et al., 2021)
HP-5 (30 m × 0.25 mm i.d., 0.25 µm)	GC-MS	-	-	-	-	(Kalantary et al., 2021)
HP-5 (30 m × 0.25 mm i.d., 0.25 µm)	GC-MS	0.28–0.73	0.92–2.4	90–105	7.2–9.1	(Bulgurcuoğlu et al., 2021)
DB-5 (30 m × 0.25 mm i.d., 0.25 µm)	GC-MS	0.0007–0.544	0.0023–1.96	90–111	4.3–7.8	(Khademi et al., 2021)
Restek Rtx-5 MS (30 m × 0.25 mm i.d., 0.25 µm)	GC-MS	0.00051–0.0224	0.00179–0.0745	74.2–123	1.5–17.8	(Nascimento et al., 2021)

(continued)

TABLE 5.1 (Continued)
Overview of Analytical Methods for Analysis of Emerging Contaminants

Analyte details	Matrix	Sample Extraction Techniques	Analytical Separation Technique — Mobile Phase/Carrier Gas
LC database: 1600 contaminants including pesticides & pharmaceuticals GC database: 280 compounds including pesticides, PAHs, PCBs, PBDEs	Surface, ground and waste water	SPE, HLB	–
Pharmaceuticals (n=172), OP pesticides	Waste and tap water	SPE	Water and Methanol with different compositions
OP pesticides (n=5) viz. Clofenvinphos Diazinon, Disulfoton, Malathion, and Parathion	River water	CSDF-ME	He (99.9995 %)
OP Pesticides (n=8) viz. Dimethoate, Diazinon, Parathion methyl, Malathion, Fenthion, Parathion ethyl, Methidathion, Fenthion sulfoxide	River, pond, tap and well water	SPE	He (99.999%)
82 pesticides from different classes, including OC and OP	Drinking water	LLME	He
Pyrethroids (n=4) viz. Cyhalothrin, Cyfluthrin, Cypermethrin, Flucythrinate	River and well water	MSPE	He (99.999%)
OC Pesticides Mix (n=508)	River water	SPME, SPE – ENVITM-C_{18}	He (99.999%)
OC Pesticides (n=9)	River, tap and underground water	MSPE	He
Pesticides (n=215)	River water	SPE, HLB	$HCOONH_4$, HCOOH and CH_3OH with different compositions
Pesticides (n=450)	Environmental water	SPE	He
Priority substances (n=24) such as pesticides, hormones and pharmaceuticals	Drinking, surface and effluent waste water	On-line SPE	CH_3OH, HCOOH and NH_4OH with different compositions for negative and positive ion mode

Stationary Phase	Analytical Instrument	Analytical Method Performance				References
		LOD (µg/L)	LOQ (µg/L)	Recovery (%)	(%) RSD	
-	UHPLC-QTOF-MS & GC-TOF-MS	-	-	-	-	(Hernández et al., 2015)
Thermo Hypersil GOLD aQ (50 m × 2.1 mm i.d., 1.9 µm)	Orbitrap HRMS	-	0.5–10	70–120	< 10	Ofrydopoulou et al., 2021)
HP-5MS capillary (30 m × 0.25 mm i.d., 0.25 µm)	GC-MS	0.02–0.30	0.07–1.0	90–110	3.6–5.8	(Moinfar et al., 2021)
HP-5 (30 m × 0.25 mm i.d., 0.25 µm)	GC–NPD/FID	0.05–0.13 and 0.0045–0.0117	-	93.8–104.5	2.9–4.3	(Ballesteros et al., 2004)
Fused-silica (25 m × 0.32 mm i.d., 0.17 µm)	GC-ECD/NPD	-	-	38–114	1.8–19	(Zapf et al., 1995)
RTX-5MS (30 m × 0.25 mm i.d., 0.25 µm)	GC-MS/MS	0.05–0.21	-	88.3–96.9	0.8–8.2	(Liu et al., 2019)
HP 5MS capillary (30 m × 0.25 mm i.d., 0.25)	GC-MS	0.68–16.2	2.04–48.7	42.0	4.93–11.9	(Sifatullah and Tuncel, 2018)
RTX-5MS capillary (30 m × 0.25 mm i.d., 0.25 m)	GC-MS/MS	0.07–1.03	-	75.1–112.7	1.0–8.5	(Huang et al., 2018)
Accucore aQ C_{18} (100 x 2.1 mm, i.d., 2.6 µm), C_{18} guard cartridge (10 x 2.1 mm, i.d., 2.6 µm)	LC-HRMS	-	0.0005–0.1	70–130	5–15	(Casado et al., 2018)
SH-RXI-5SIL MS (30 m × 0.25 mm i.d., 0.25 µm)	GC-MS/MS	-	-	80–130	-	(Zhang et al., 2018)
Kinetex EVO C_{18} (100 x 2.1 mm i.d., 2.6 µm)	LC-MS/MS	0.0001–0.0014	0.0003–0.0048	-	1.1–6.8	(Rubirola and Galceran, 2017)

(continued)

TABLE 5.1 (Continued)
Overview of Analytical Methods for Analysis of Emerging Contaminants

Analyte details	Matrix	Sample Extraction Techniques	Analytical Separation Technique — Mobile Phase/Carrier Gas
Pesticides (n=124)	Drinking Water	SPE	He
Analytes (n=35) including corrosion inhibitors, pesticides and pharmaceuticals	Tap, surface, ground and effluent wastewater	SPE	Water, formic acid and acetonitrile with different compositions Water, HCOONH$_4$, and methanol with different composition
Pesticides (n=160)	Surface water	SPE	He
Targeted pesticides (n=50)	Surface Water	SPE	Methanol, water, and formic acid in different composition
Pesticides (n=50)	Surface and wastewater	-	-
Mycotoxins	Bottled water	SPE	HCOONH$_4$, HCOOH and acetonitrile with different compositions
Pharmaceutical products (n=63)	Natural water	SPE	Water, acetonitrile and formic acid in different composition
Pesticides (n=48) belonging to 12 different chemical groups	River water	SPME	He (99.99%)
Pharmaceutical products (n=44)	Waste water	SPE	Formic acid, ammonium format and methanol with different composition
Pharmaceuticals (n=25)	Hospital wastewater	SPE	HCOOH, NH$_4$OAc, and acetonitrile with different composition

Analysis of Emerging Contaminants in Water Samples

		Analytical Method Performance					
Stationary Phase	Analytical Instrument	LOD (µg/L)	LOQ (µg/L)	Recovery (%)		(%) RSD	References
DB-5MS (30 m × 0.25 mm × 0.25 µm)	GC-MS	0.00102– 0.04821	0.00319– 0.14624	55.54– 121.21		2.31– 27.15	(Schwanz et al., 2019)
Zorbax Extend C_{18} (100mm×2.1mm i.d. × 1.8 µm)	LC-MS/MS						
Luna C_{18} (250 mm × 2.0 mm, i.d., 5 µm)	LC-MS/MS	0.0006 – 0.0078	0.0017 – 0.0255	30 – 128		84 – 144	(Zhang et al., 2018)
RTX-CL Pesticides II (30 m × 0.25 µm i.d., 0.32mm)	GC × GC-ToF/MS	-	-	-		-	(Glinski et al., 2018)
RXI-17Sil MS (2 m × 0.15 µm i.d., 0.15mm)							
Luna $_{18}$ (150 mm × 2.0 mm i.d., 3 µm)	LC-MS/MS	0.00001– 0.0020	-	69 – 95		-	(Aguilar et al., 2017)
-	LC-MS/MS	0.00001– 0.002	-	48–70		-	(Masiá et al., 2013)
Phenomenex Kinetex C_{18} (100 × 2.1 mm i.d., 2.6 µm)	LC-MS/MS	0.10–3.0	0.3–10.0	69–105		-	(Mata et al., 2015)
Zorbax SB-C_{18} (2.1 mm × 100 mm i.d., 3.5 µm)	LC-MS/MS	0.0023– 0.0943	-	68 – 134		1 – 13.8	(Mokh et al., 2017)
VF-5MS (30m × 0.25mm i.d., 0.25 µm)	GC-IT MS/MS	0.01–1.09	0.06–1.43	78.68– 122.16		3–19	(Jabali and El-Hoz, 2019)
Acquity BEH C_{18}	UPLC-MS/MS	0.07–6.7	0.21– 22.37	31–137		0.8–16.4	(Huerta et al., 2016)
C_{18} (50 mm × 2.1 mm i.d., 2.7 µm) endcapped and C_{18} (5 mm × 2.1mm i.d., 2.7 µm)guardcolumn	LC-MS/MS	0.0001– 0.0076	0.0002– 0.0253	≥ 60		-	(Mendoza et al., 2015)

(continued)

TABLE 5.1 (Continued)
Overview of Analytical Methods for Analysis of Emerging Contaminants

Analyte details	Matrix	Sample Extraction Techniques	Analytical Separation Technique — Mobile Phase/Carrier Gas
Pharmaceuticals (n=27)	Waste water	SBSE	Formic acid, water, and acetonitrile in different composition
Paracetamol, Diclofenac, Ibuprofen (3) Pharmaceuticals	Surface and river water	Bar adsorptive micro extraction (BAµE)	Ammonium hydroxide, water, and methanol with different composition
Pharmaceuticals (n=12)	Drinking, running, river, Influent and effluent waste water	DLLME	Formic acid, water and acetonitrile in different composition
OC pesticides (n=22), PCBs (n=28)	Seawater samples	SPE & UAE	He (99.999%)
Pesticides (n=56) including pharmaceuticals	Waste water	SPE & biphenyl cartridge	Formic acid, water, acetonitrile: methanol with different composition
Pesticides (n=24)	Environmental water	SPME	He
Pesticides residues (n=70)	Drinking water	SPE	He
			Ammonium formate, water, methanol with different composition
OC Pesticides (n=6)	River surface water	Magnetic SPE procedure	He
Pesticides (n=48)	Water	DI-SPME	He

Analysis of Emerging Contaminants in Water Samples   103

Stationary Phase	Analytical Instrument	LOD (µg/L)	LOQ (µg/L)	Recovery (%)	RSD (%)	References
Poroshell EC-C$_{18}$ (100 mm × 3.0 mm i.d., 2.7 µm)	LC-MS/MS	-	0.00125–0.0050	-	0.9–20.2	(Klančar et al., 2017)
BEH C$_{18}$ (50 mm × 2.1 mm i.d., 1.7 µm)	UHPLC-MS/MS	0.012–0.6	0.04–2.0	74–118	2–19	(Souza et al., 2018)
BEH Phenyl (50 mm × 2.1 mm i.d., 1.7 µm)	UHPLC-MS/MS	0.006–0.091	0.018–0.281	76.77-99.97	1.6–8.8	(Guan et al., 2016)
DB-5 MS (60 m × 0.25 mm i.d., 0.10 µm)	GC–QqQMS	4×10^{-5} – 343×10^{-5}	-	21–301	1–39	(Zhang and Barry, 2015)
Raptor biphenyl (5.0 mm × 3.0 mm i.d., 2.7 µm)	LC-MS/MS	0.00006–0.533	0.00021–1.777	-	2–55	(Keng Tiong et al., 2020)
DB-5 MS capillary (30 m × 250 µm i.d., 0.25 µm)	GC-MS	0.0002-1.1309	0.0007-3.7320	70-123	7-15	(Valenzuela et al., 2020)
VF-5-MS (30 m × 0.25 mm i.d., 0.25 µm)	GC-MS/MS	0.15	0.5	70–117.3	19.7	(Donato et al., 2015)
UPS Pursuit C$_{18}$ (50 mm × 3.0 mm i.d., 2.4 µm)	LC-MS/MS					
HP-5 MS (30 m × 0.25 mm, i.d., 0.25 µm)	GC–MS/MS	0.00004–0.00035	0.00014–0.00115	79–91	2–7	(Wang et al. 2018)
VF- 5MS (30 m × 0.25 mm i.d., 0.25 µm)	GC-MS/MS	0.001–0.458	0.039–0.732	75.6–137.31	3–19	(Jabali and El-Hoz, 2019)

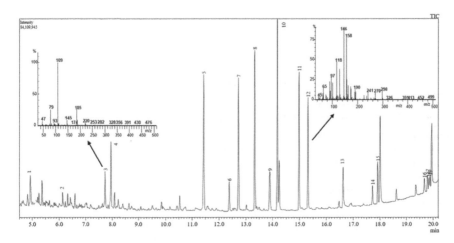

FIGURE 5.4 Total ion chromatogram (TIC) with their corresponding mass spectra (n=2) of the optimized neat standard mixture of fifteen pesticides under scan mode. Peak numbers: (1) dichlorvos; (2) acephate; (3) monocrotophos; (4) phorate; (5) chlorpyrifos; (6) fipronil; (7) quinalphos; (8) butachlor; (9) profenophos; (10) buprofezin; (11) ethion; (12) triazophos; (13) acetamiprid; (14) L-cyhalothrin-1; (15) L-cyhalothrin-2; (16) cypermethrin-1; (17) cypermethrin-2; (18) cypermethrin-3; and (19) cypermethrin-4.

analyte provided through the NIST library search. Figure 5.4 illustrates the sharp as well as resolved peaks obtained for the total ion chromatogram (TIC) for a standard pesticide mixture along with the respective mass spectrum of monocrotophos and triazophos under the SCAN mode, respectively. The precursor ion was identified for each targeted analyte based on abundance, ion cluster, and m/z ratio. The method file was then measured under product ion scan mode to determine collision energy for respective product ion masses. The data acquisition was processed using multiple reaction monitoring (MRM) optimization tool. For obtaining optimum collision energy, a series of fifteen injections of pesticide mixture were made from the same vial, with collision energy in the increasing order of 0 to 45V, tested at 3V interval for each data. After the completion of data acquisition, optimum collision energies of MRM transitions were determined over the defined range.

The method was developed with one target ion and two product ions, each having maximum intensities. According to EU Commission, utilization of 2–3 fragment ions as quantifier ion is deemed to be sufficient for identification of target analytes as each analyte has an individual fragmentation spectrum. These fragmentation spectra were herein employed for routine analyte search and verification using a reference library. Further, the method file was created under MRM mode. Figure 5.5 represents the standard chromatogram of a mixture of pesticides under the MRM mode.

Detailed below is also Figure 5.6, which illustrates a selected ion chromatogram for a water samples containing chlorpyrifos in comparison with blank sample, neat standard and spiked sample at 10 ng/mL and analysed under the MRM mode.

Analysis of Emerging Contaminants in Water Samples

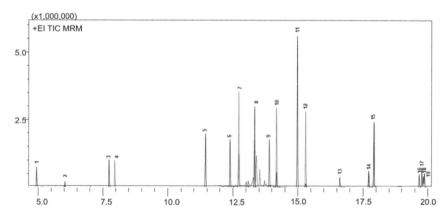

FIGURE 5.5 Total Ion Chromatogram of neat standard mixture under MRM mode. Peak numbers: (1) dichlorvos; (2) acephate; (3) monocrotophos; (4) phorate; (5) chlorpyrifos; (6) fipronil; (7) quinalphos; (8) butachlor; (9) profenophos; (10) buprofezin; (11) ethion; (12) triazophos; (13) acetamiprid; (14) L-cyhalothrin-1; (15) L-cyhalothrin-2; (16) cypermethrin-1, cypermethrin-2, cypermethrin-3, and cypermethrin-4.

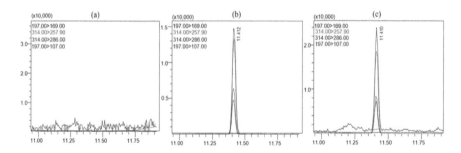

FIGURE 5.6 MRM chromatogram of chlorpyrifos in comparison with (a) blank sample; (b) neat standard; and (c) spiked sample at 10 ng/mL in blood matrix under MRM mode.

5.3.6 Optimization of LC-HR/MS

Optimisation of LC-HR/MS for analysis of EC was done with the prime objective of improved intensity. This was initiated by optimising mobile phase, column flow, cone voltage, and dissolution temperature. Following this the sample was run and TIC was obtained, followed by the extracted ion chromatogram.

During sample analysis mass accuracy was considered a standard measure. In the present study, matrix-matched calibration method along with simultaneous reference standard *(Leucine-enkephalin)* injection was performed for verification and improvement (<1 ppm) in mass accuracy. Therefore, a matrix-matched calibration curve was used for the quantitation of the samples. Malathion, a representative pesticide, was used to evaluate the mass accuracy and elemental composition spiked at 50 µg kg^{-1}. Seven-point matrix-matched calibration curves and simultaneous injection

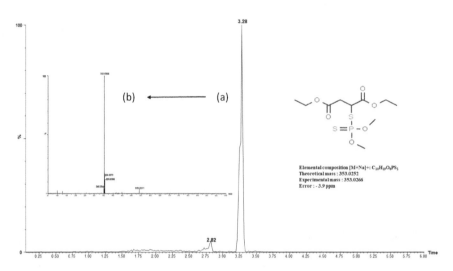

FIGURE 5.7 Extracted Ion Chromatogram, mass spectra and mass accuracy of malathion using LCHR/MS.

of reference standards was used for auto-correction of analyte mass. The experimental mass of malathion was observed to be (m/z) 353.0266 as compared to the calculated/theoretical (m/z) 353.0252. The mass accuracy was therefore observed to be -3.9 ppm, which is within the EU recommended limit of perceived mass accuracy (i.e., <5 ppm). Extracted ion chromatogram and elemental analysis of malathion are depicted in Figures 5.7 and 5.8.

5.4 CONCLUSIONS

Overall, a literature survey in the field of CEC determination in aquatic environments hints at some prominent trends. Primarily it can be concluded that the analysis of EC can be done by both GC and LC coupled with MS. Both these techniques have their own pros and cons. Whilst GC cannot be completely ruled out, due to its ease of operation and it being accorded preference by chemists for separating typical and more hydrophobic CECs such as alkylphenols, steroids, fragrances, and phthalates the fact that LC-MS analysis displays ionization issues cannot be neglected either. And despite this, alternative LC applications for most of these groups continue to be reported. There has also been observed a clear trend towards multi-residue and multi-class methods (Petrovic et al., 2010). Newer technology and advances in instrumentation have enabled simultaneous detection of a multitude of compounds in a single analytical run. Recently instrumentation has also witnessed an emergence of high-resolution LC equipment, which enables the use of sub-2lm particle sizes and high flow rates. Referred to as ultra high performance liquid chromatography (U-HPLC), this technique allows easy resolution of CECs, which in turn save time by analysis in shorter run times. The use of online SPE can also be attempted to shorten analysis times (Petrovic et al., 2010).

Analysis of Emerging Contaminants in Water Samples

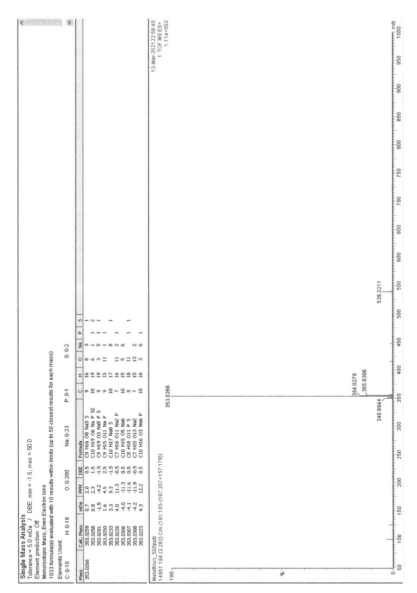

FIGURE 5.8 Elemental analysis of malathion in water samples using LC HR/MS.

The literature also points to an increased popularity of HR full-scan analysis with ToF and Orbitrap instruments proving to be suitable alternatives to QqQ instruments. These in turn have been reported to allow accurate mass-screening of almost an unlimited number of analytes, both targeted as well as untargeted. However, for certain applications, their sensitivity has been proven to be insufficient (Hogenboom et al., 2009). It is in such cases that hybrid mass spectrometers (e.g., triple-quadrupole/linear ion trap mass spectrometer (QTRAP or QLIT), quadrupole/time-of-flight mass spectrometer (QToF), and ion-trap/Orbitrap mass spectrometer) offer significant potential by marrying the sensitivity of triple quadrupole systems with the versatility and identification potential afforded by HRMS.

Further, it is also imperative to note that despite the many evolutions and achievements in analytical approaches, pitfalls also exist within this domain. Primarily, shortfalls in identification and quantification commonly arise due to the want for standardized criteria for identification and confirmation of CECs in environmental matrices. Criterions comparable to European criteria in lines with Commission Decision 2002/657/EC (European Commission, 2002), and concerning determination of analytes in products of animal origin, or (SANCO/10684/2009), with respect to pesticide-residue analysis in food and feed, are missing for environmental matrices. Evaluation of matrix effects is observed to be not always taken into consideration within the validation procedures of newly developed analytical methods, despite the knowledge of the occurrence of matrix interferences being a well-known source of false positives/negatives. And these have in turn been observed to contribute to erroneous quantification when using LC-MS. There exists therefore a greater requirement for standardization of validation procedures for analytical methods developed for environmental applications.

Another issue relating to these modern and versatile instruments, which are today powerful screening and confirmation tools, is the lack of presence of appropriate identification criteria. Criterions are currently observed to be incomplete for the commonly used procedures as prescribed by Commission Decision 2002/657/EC (Hogenboom et al., 2009, European Commission, 2002). It has been observed that criteria for mass resolution have not yet been fully specified for chromatography coupled with MS techniques systems. Several authors over a period of time have even expressed the need for implementation of (Nielen et al., 2007) additional criteria for the use of LC-MS technologies in the standardized validation procedures.

A few critical comments as well as recommendations concerning quantitative data obtained from new analytical methods have been floated in literature by Van Leeuwen et al. (2011), who reported results of an inter-laboratory study on PFC analysis in environmental matrices. They documented some typical sources as being capable of contributing to the variance of analytical data including the occurrence of matrix effects as well. Limited use of mass-labeled internal standards, the necessity for blank as well as recovery correction, and perceived failure in isomer separation (especially for PFCs) were other contributors.

Another important aspect that draws attention is the fact that CECs are usually present at concentration levels extremely close to the Limit of Detection (LODs) of analytical methods being used. This results in a decrease in precise quantification coupled with a higher variance. For methods newly developed, it is recommended to utilize sufficient mass-labelled internal standards and also perform extensive validation at environmentally relevant concentrations. This would facilitate the inclusion of evaluation of matrix effects, natural background levels in control samples and aid in improved accuracy.

Today there is a need for the conduct of international inter-laboratory studies and/or data quality tests, since these may serve as prudent verification tools for ascertaining the reliability of analytical methods. Another shortcoming in the domain of environmental chemistry is the current lack of equivalence of analytical methods and their results; even though literature reviews points at widely differing analytical methodologies for obtaining concentration of CECs in environmental matrices. Thus though there are varied methodologies there is a greater need for standardized state-of-the-art methods for the ever-increasing groups of CECs.

The literature mentions the EPA Method 1694 and ISO 25101/2006 method as being standard methodologies for the analysis of pharmaceuticals and PFCs, respectively, in water samples. There are also currently available EPA methodologies for the analysis of different groups of pesticides as well. However, these are rarely applied in monitoring studies. We believe that the development of advanced and up-to-date standardized methodologies as well as their widespread application would result in the generation of reliable and comparable monitoring data, which is the need of the hour in CEC quantification.

ACKNOWLEDGMENTS

The authors are grateful to the Director-in-charge, ICMR-National Institute of Occupational Health, Ahmedabad-16, for his approval and suggestions and also to Mr Chandrakant G Parmar for his support of laboratory work.

ABBREVIATIONS

BAµE: Bar adsorptive micro extraction
BAµE: Bar Adsorptive Microextraction
BPA: Bisphenol A
CECs: Chemicals of emerging concern
CSDF-ME: Continuous Sample Drop Flow Micro-extraction
DDT: Dichlorodiphenyltrichloroethane
DI-SPME: Direct Immersion solid-phase micro-extraction
DLLME: Dispersive Liquid–Liquid Micro-extraction
DM-SPE: Dispersive Micro solid-phase extraction
DSPE/DSPME: Dispersive solid phase (micro) extraction
ECD: Electron capture detector
EDCs: Endocrine-disrupting compounds
FID: Flame ionization detection
FWHM: Full width half maxima
GC: Gas chromatography
GC-ECD: Gas Chromatography-Electron Capture Detector
GC-IT MS/MS: Gas Chromatography-Ion Trap Tandem Mass Spectrometry

GC-MS: Gas Chromatography Mass Spectrometry
GC-NPD/FID: Gas Chromatography-Nitrogen Phosphorus Detector/Flame Ionisation Detector
GC–QqQMS: Gas Chromatography Triple Quadrupole Mass Spectrometry
GCToF-MS: Gas Chromatography Time-of-Flight Mass Spectrometry
HILIC: Hydrophilic interaction chromatography
HLB: Hydrophilic-lipophilic balance
HPLC: High performance liquid chromatography
HPLC-MS/MS: HPLC-tandem mass spectrometry
HR-GC: High resolution gas chromatography
HRMS: High Resolution Mass Spectrometry
i.d.: inner diameter
IT-MS: Ion-trap mass spectrometry
LC: Liquid chromatography
LC-ESI-Q: Liquid chromatography electrospray ionization quadrupole
LC-QTOF MS: Liquid Chromatography Quadrupole Time-of-Flight Mass Spectrometry
LLE: Liquid-liquid extraction
LLME: Liquid-Liquid Micro-extraction
LOD: Limit of detection
LOQs: Limit of quantitation
MDSPME: Magnetic dispersive solid phase micro-extraction
MEPS: Micro extraction by packed sorbent
μ-SPE: Micro solid-phase extraction
MRM: Multiple reaction monitoring
MS: Mass spectrometry
MSPE: Magnetic solid-phase extraction
Na_2EDTA/Na_4EDTA: di- or tetrasodium ethylenediaminetetraacetate
NIST: National Institute of Standards and Technology
OC: Organochlorine
OP: Organophosphate
PAHs: Polycyclic aromatic hydrocarbons
PBB: Polybrominated biphenyls
PCA: Principal component analysis
PCPs: Personal care products
PEEK: Polyether ether ketone
PFCA: Perfluorinated carboxylic acid
PFCs: Per-fluorinated compounds
PFNA: Perfluorononanoate
PFOA: Perfluorooctanoic acid
PFOS: Perfluorooctanesulfonic acid
PFSA: Perfluorosulfonic acid
PLS-DA: Partial least square discriminant analysis
QqQ-MS: Triple quadrupole-mass spectrometry
QToF-MS: Time-of-flight mass spectrometry
RSD: *Relative standard deviation*

RT: *Retention Time*
SBSE: Stir-bar sorptive extraction
SPE: Solid-phase extraction
SPE HLB: Solid Phase Extraction–Hydrophilic Lipophilic Balance
SPE–DVB: Solid phase extraction–divinylbenzene
SPME: Solid phase microextraction
SS-LPE: Switchable Solvent Liquid Phase Extraction
TIC: Total Ion Chromatogram
UAE: Ultrasound assisted extraction
Ug/l: *microgram* per liter
UHPLC-MS/MS: Ultra high performance liquid chromatography tandem mass spectrometry
UPLC-MS/MS: Ultra performance liquid chromatography tandem mass spectrometry
VOCs: Volatile organic compounds
WWTPs: Wastewater treatment plants

REFERENCES

Akbarzade, Samaneh, Mahmoud Chamsaz, Gholam Hossein Rounaghi, and Mahdi Ghorbani. 2018. "Zero Valent Fe-Reduced Graphene Oxide Quantum Dots as a Novel Magnetic Dispersive Solid Phase Microextraction Sorbent for Extraction of Organophosphorus Pesticides in Real Water and Fruit Juice Samples Prior to Analysis by Gas Chromatography-Mass Spectrometry." *Analytical and Bioanalytical Chemistry* 410 (2): 429–39.

Alavanja, Michael C. R., Jane A. Hoppin, and Freya Kamel. 2004. "Health Effects of Chronic Pesticide Exposure: Cancer and Neurotoxicity." *Annual Review of Public Health* 25 (1): 155–97.

Arthur, Catherine L., and Janusz Pawliszyn. 1990. "Solid Phase Microextraction with Thermal Desorption Using Fused Silica Optical Fibers." *Analytical Chemistry* 62 (19): 2145–48.

Auriol, Muriel, Youssef Filali-Meknassi, Rajeshwar D. Tyagi, Craig D. Adams, and Rao Y. Surampalli. 2006. "Endocrine Disrupting Compounds Removal from Wastewater, a New Challenge." *Process Biochemistry* 41 (3): 525–39.

Azzouz, Abdelmonaim, and Evaristo Ballesteros. 2016. "Determination of 13 Endocrine Disrupting Chemicals in Environmental Solid Samples Using Microwave-Assisted Solvent Extraction and Continuous Solid-Phase Extraction Followed by Gas Chromatography-Mass Spectrometry." *Analytical and Bioanalytical Chemistry* 408 (1): 231–41.

Ballesteros, E., and M. J. Parrado. 2004. "Continuous Solid-Phase Extraction and Gas Chromatographic Determination of Organophosphorus Pesticides in Natural and Drinking Waters." *Journal of Chromatography A* 1029 (1–2): 267–73.

Barceló, D. 1993. "Environmental Protection Agency and Other Methods for the Determination of Priority Pesticides and Their Transformation Products in Water." *Journal of Chromatography A* 643 (1–2): 117–43.

Becker, Raquel W., Débora S. Araújo, Carla Sirtori, Natalie P. Toyama, Diego A. Tavares, Gilcélia A. Cordeiro, Simone F. Benassi, Ana C. Gossen, and Bianca do Amaral. 2021. "Pesticides in Surface Water from Brazil and Paraguay Cross-Border Region: Screening Using LC-QTOF MS and Correlation with Land Use and Occupation through Multivariate

Analysis." *Microchemical Journal., Devoted to the Application of Microtechniques in All Branches of Science* 168: 106502.

Bletsou, Anna A., Junho Jeon, Juliane Hollender, Eleni Archontaki, and Nikolaos S. Thomaidis. 2015. "Targeted and Non-Targeted Liquid Chromatography-Mass Spectrometric Workflows for Identification of Transformation Products of Emerging Pollutants in the Aquatic Environment." *Trends in Analytical Chemistry: TRAC* 66: 32–44.

Bouchard, Maryse F., Jonathan Chevrier, Kim G. Harley, Katherine Kogut, Michelle Vedar, Norma Calderon, Celina Trujillo, et al. 2011. "Prenatal Exposure to Organophosphate Pesticides and IQ in 7-Year-Old Children." *Environmental Health Perspectives* 119 (8): 1189–95.

Brausch, John M., and Gary M. Rand. 2011. "A Review of Personal Care Products in the Aquatic Environment: Environmental Concentrations and Toxicity." *Chemosphere* 82 (11): 1518–32.

Bulgurcuoğlu, Ayşe Evrim, Büşra Yılmaz Durak, Dotse Selali Chormey, and Sezgin Bakırdere. 2021. "Development of a Switchable Solvent Liquid Phase Extraction Method for the Determination of Chlorthiamid, Ethyl Parathion, Penconazole and Fludioxonil Pesticides in Well, Tap and Lake Water Samples by Gas Chromatography Mass Spectrometry." *Microchemical Journal, Devoted to the Application of Microtechniques in All Branches of Science* 168: 106381.

Campbell, Chris G., Sharon E. Borglin, F. Bailey Green, Allen Grayson, Eleanor Wozei, and William T. Stringfellow. 2006. "Biologically Directed Environmental Monitoring, Fate, and Transport of Estrogenic Endocrine Disrupting Compounds in Water: A Review." *Chemosphere* 65 (8): 1265–80.

Casado, J., D. Santillo, and P. Johnston. 2018. "Multi-Residue Analysis of Pesticides in Surface Water by Liquid Chromatography Quadrupole-Orbitrap High Resolution Tandem Mass Spectrometry." *Analytica Chimica Acta* 1024: 1–17.

Cavaliere, Chiara, Anna Laura Capriotti, Francesca Ferraris, Patrizia Foglia, Roberto Samperi, Salvatore Ventura, and Aldo Laganà. 2016. "Multiresidue Analysis of Endocrine-Disrupting Compounds and Perfluorinated Sulfates and Carboxylic Acids in Sediments by Ultra-High-Performance Liquid Chromatography-Tandem Mass Spectrometry." *Journal of Chromatography A* 1438: 133–42.

Chaza, Chbib, Sopheak Net, Mariam Hamzeh, David Dumoulin, Baghdad Ouddane, and Moomen Baroudi. 2018. "Assessment of Pesticide Contamination in Akkar Groundwater, Northern Lebanon." *Environmental Science and Pollution Research International* 25 (15): 14302–12.

Christodoulou, A., and K. Stamatelatou. 2016. "Overview of Legislation on Sewage Sludge Management in Developed Countries Worldwide." *Water Science and Technology: A Journal of the International Association on Water Pollution Research* 73 (3): 453–62.

Dargahi, Rosa, Homeira Ebrahimzadeh, and Reza Alizadeh. 2018. "Polypyrrole Coated ZnO Nanorods on Platinum Wire for Solid-Phase Microextraction of Amitraz and Teflubenzuron Pesticides Prior to Quantitation by GC-MS." *Mikrochimica Acta* 185 (2). https://doi.org/10.1007/s00604-018-2692-y.

Diamanti-Kandarakis, Evanthia, Jean-Pierre Bourguignon, Linda C. Giudice, Russ Hauser, Gail S. Prins, Ana M. Soto, R. Thomas Zoeller, and Andrea C. Gore. 2009. "Endocrine-Disrupting Chemicals: An Endocrine Society Scientific Statement." *Endocrine Reviews* 30 (4): 293–342.

Donato, Filipe F., Manoel L. Martins, Juliana S. Munaretto, Osmar D. Prestes, Martha B. Adaime, and Renato Zanella. 2015. "Development of a Multiresidue Method for

Pesticide Analysis in Drinking Water by Solid Phase Extraction and Determination by Gas and Liquid Chromatography with Triple Quadrupole Tandem Mass Spectrometry." *Journal of the Brazilian Chemical Society.* https://doi.org/10.5935/0103-5053.20150192.
EPA. 2012. *Emerging Contaminants-Perfluorooctane Sulfonate (PFOS) and Perfluorooctanoic Acid (PFOA).* Emerging contaminants fact sheet–PFOS and PFOA.
Erger, Christine, and Torsten C. Schmidt. 2014. "Disk-Based Solid-Phase Extraction Analysis of Organic Substances in Water." *Trends in Analytical Chemistry: TRAC* 61: 74–82.
Espada, M. C. Pablos, A. Garrido Frenich, J. L. Martínez Vidal, and P. Parrilla. 2001. "Comparative Study Using Ecd, Npd, and Ms/Ms Chromatographic Techniques in the Determination of Pesticides in Wetland Waters." *Analytical Letters* 34 (4): 597–614.
European Commission. 2002. "Decision 2002/657/EC." *Off. J. Eur. Commun.* L221: 8.
Farré, Marinella, Damià Barceló, and Damià Barceló. 2013. "Analysis of Emerging Contaminants in Food." *Trends in Analytical Chemistry: TRAC* 43: 240–53.
Farré, Marinella, Sandra Pérez, Carlos Gonçalves, M. F. Alpendurada, and Damià Barceló. 2010. "Green Analytical Chemistry in the Determination of Organic Pollutants in the Aquatic Environment." *Trends in Analytical Chemistry: TRAC* 29 (11): 1347–62.
Ferrer, Imma, Jerry A. Zweigenbaum, and E. Michael Thurman. 2010. "Analysis of 70 Environmental Protection Agency Priority Pharmaceuticals in Water by EPA Method 1694." *Journal of Chromatography A* 1217 (36): 5674–86.
Giulia, D. B. 2009. "Development of Advanced Analytical Methods for the Determination of Emerging Pollutants in Environmental Waters." *Journal of Separation Science* 32: 1249–1261.
Glinski, Donna A., S. Thomas Purucker, Robin J. Van Meter, Marsha C. Black, and W. Matthew Henderson. 2018. "Analysis of Pesticides in Surface Water, Stemflow, and Throughfall in an Agricultural Area in South Georgia, USA." *Chemosphere* 209: 496–507.
González-Barreiro, Carmen, Elena Martínez-Carballo, Andrea Sitka, Sigrid Scharf, and Oliver Gans. 2006. "Method Optimization for Determination of Selected Perfluorinated Alkylated Substances in Water Samples." *Analytical and Bioanalytical Chemistry* 386 (7–8): 2123–32.
Gorchev, H. G., and G. Ozolins. 1984. "WHO Guidelines for Drinking-Water Quality." *WHO Chronicle* 38 (3): 104–8.
Gore, A. C., V. A. Chappell, S. E. Fenton, J. A. Flaws, A. Nadal, G. S. Prins, J. Toppari, and R. T. Zoeller. 2015. "Executive Summary to EDC-2: The Endocrine Society's Second Scientific Statement on Endocrine-Disrupting Chemicals." *Endocrine Reviews* 36 (6): 593–602.
Gorga, Marina, Sara Insa, Mira Petrovic, and Damià Barceló. 2014. "Analysis of Endocrine Disrupters and Related Compounds in Sediments and Sewage Sludge Using On-Line Turbulent Flow Chromatography-Liquid Chromatography-Tandem Mass Spectrometry." *Journal of Chromatography A* 1352: 29–37.
Guan, Jin, Chi Zhang, Yang Wang, Yiguang Guo, Peiting Huang, and Longshan Zhao. 2016. "Simultaneous Determination of 12 Pharmaceuticals in Water Samples by Ultrasound-Assisted Dispersive Liquid–Liquid Microextraction Coupled with Ultra-High Performance Liquid Chromatography with Tandem Mass Spectrometry." *Analytical and Bioanalytical Chemistry* 408 (28): 8099–109.
Handal, Alexis J., Betsy Lozoff, Jaime Breilh, and Siobán D. Harlow. 2006. "Effect of Community of Residence on Neurobehavioral Development in Infants and Young Children in a Flower-Growing Region of Ecuador." *Environmental Health Perspectives* 115 (1): 128–33.

Healy, M. G., O. Fenton, M. Cormican, D. P. Peyton, N. Ordsmith, K. Kimber, and L. Morrison. 2017. "Antimicrobial Compounds (Triclosan and Triclocarban) in Sewage Sludges, and Their Presence in Runoff Following Land Application." *Ecotoxicology and Environmental Safety* 142: 448–53.

Hennion, M-C, V. Pichon, and D. Barceló. 1994. "Surface Water Analysis (Trace-Organic Contaminants) and EC Regulations." *Trends in Analytical Chemistry: TRAC* 13 (9): 361–72.

Hernández, Félix, María Ibáñez, Tania Portolés, María I. Cervera, Juan V. Sancho, and Francisco J. López. 2015. "Advancing towards Universal Screening for Organic Pollutants in Waters." *Journal of Hazardous Materials* 282: 86–95.

Hogenboom, A. C., J. A. van Leerdam, and P. de Voogt. 2009. "Accurate Mass Screening and Identification of Emerging Contaminants in Environmental Samples by Liquid Chromatography-Hybrid Linear Ion Trap Orbitrap Mass Spectrometry." *Journal of Chromatography A* 1216 (3): 510–19.

Huang, Xiaodong, Guangyang Liu, Donghui Xu, Xiaomin Xu, Lingyun Li, Shuning Zheng, Huan Lin, and Haixiang Gao. 2018. "Novel Zeolitic Imidazolate Frameworks Based on Magnetic Multiwalled Carbon Nanotubes for Magnetic Solid-Phase Extraction of Organochlorine Pesticides from Agricultural Irrigation Water Samples." *Applied Sciences (Basel, Switzerland)* 8 (6): 959.

Huerta, B., S. Rodriguez-Mozaz, C. Nannou, L. Nakis, A. Ruhí, V. Acuña, S. Sabater, and D. Barcelo. 2016. "Determination of a Broad Spectrum of Pharmaceuticals and Endocrine Disruptors in Biofilm from a Waste Water Treatment Plant-Impacted River." *The Science of the Total Environment* 540: 241–49.

Jabali, Y., M. Millet, and M. El-Hoz. 2019. "Optimization of a DI-SPME-GC–MS/MS Method for Multi-Residue Analysis of Pesticides in Waters." *Microchemical Journal., Devoted to the Application of Microtechniques in All Branches of Science* 147: 83–92.

Janna, H. 2011. *Occurrence and Removal of Emerging Contaminants in Wastewaters (Ph.D. Thesis)*. London Brunel University.

Jiménez-Díaz, I., F. Vela-Soria, R. Rodríguez-Gómez, A. Zafra-Gómez, O. Ballesteros, and A. Navalón. 2015. "Analytical Methods for the Assessment of Endocrine Disrupting Chemical Exposure during Human Fetal and Lactation Stages: A Review." *Analytica Chimica Acta* 892: 27–48.

Jurewicz, Joanna, and Wojciech Hanke. 2008. "Prenatal and Childhood Exposure to Pesticides and Neurobehavioral Development: Review of Epidemiological Studies." *International Journal of Occupational Medicine and Environmental Health* 21 (2): 121–32.

Kaw, Han Yeong, and Narayanan Kannan. 2017. "A Review on Polychlorinated Biphenyls (PCBs) and Polybrominated Diphenyl Ethers (PBDEs) in South Asia with a Focus on Malaysia." *Reviews of Environmental Contamination and Toxicology* 242: 153–81.

Keng Tiong Ng, Helena Rapp-Wright, Melanie Egli, Alicia Hartmann, Joshua C. Steele, Juan Eduardo Sosa-Hernández, Elda M. Melchor-Martínez, et al. 2020. "High-Throughput Multi-Residue Quantification of Contaminants of Emerging Concern in Wastewaters Enabled Using Direct Injection Liquid Chromatography-Tandem Mass Spectrometry." *Journal of Hazardous Materials* 398: 122933.

Khademi, Seyed Mohammad Seyed, Amir Salemi, Maik Jochmann, Sasho Joksimoski, and Ursula Telgheder. 2021. "Development and Comparison of Direct Immersion Solid Phase Micro Extraction Arrow-GC-MS for the Determination of Selected Pesticides in Water." *Microchemical Journal., Devoted to the Application of Microtechniques in All Branches of Science* 164: 106006.

Klančar, A., M. Zakotnik, R. Roškar, and J. Trontelj. 2017. "Multi-Residue Analysis of Selected Pharmaceuticals in Wastewater Samples by Stir-Bar Sorptive Extraction Followed by Liquid Desorption and Liquid Chromatography-Mass Spectrometry." *Analytical Methods: Advancing Methods and Applications* 9 (36): 5310–21.

Koester, C. J., and R. E. Clement. 1993. "Analysis of Drinking Water for Trace Organics." *Critical Reviews in Analytical Chemistry* 24 (4): 263–316.

Langlois, Ingrid, Urs Berger, Zdenek Zencak, and Michael Oehme. 2007. "Mass Spectral Studies of Perfluorooctane Sulfonate Derivatives Separated by High-Resolution Gas Chromatography." *Rapid Communications in Mass Spectrometry: RCM* 21 (22): 3547–53.

Liu, Huifang, Lihua Jiang, Meng Lu, Guangyang Liu, Tengfei Li, Xiaomin Xu, Lingyun Li, et al. 2019. "Magnetic Solid-Phase Extraction of Pyrethroid Pesticides from Environmental Water Samples Using Deep Eutectic Solvent-Type Surfactant Modified Magnetic Zeolitic Imidazolate Framework-8." *Molecules (Basel, Switzerland)* 24 (22): 4038.

Liu, Jin-Lin, and Ming-Hung Wong. 2013. "Pharmaceuticals and Personal Care Products (PPCPs): A Review on Environmental Contamination in China." *Environment International* 59: 208–24.

Mahara, Bashir M., J. Borossay, and K. Torkos. 1998. "Liquid–Liquid Extraction for Sample Preparation Prior to Gas Chromatography and Gas Chromatography–Mass Spectrometry Determination of Herbicide and Pesticide Compounds." *Microchemical Journal, Devoted to the Application of Microtechniques in All Branches of Science* 58 (1): 31–38.

Manzetti, Sergio, and David van der Spoel. 2015. "Impact of Sludge Deposition on Biodiversity." *Ecotoxicology (London, England)* 24 (9): 1799–814.

Masiá, Ana, Julián Campo, Pablo Vázquez-Roig, Cristina Blasco, and Yolanda Picó. 2013. "Screening of Currently Used Pesticides in Water, Sediments and Biota of the Guadalquivir River Basin (Spain)." *Journal of Hazardous Materials* 263 (Pt 1): 95–104.

Masuo, Yoshinori, and Masami Ishido. 2011. "Neurotoxicity of Endocrine Disruptors: Possible Involvement in Brain Development and Neurodegeneration." *Journal of Toxicology and Environmental Health. Part B, Critical Reviews* 14 (5–7): 346–69.

Mata, A. T., J. P. Ferreira, B. R. Oliveira, M. C. Batoréu, M. T. Barreto Crespo, V. J. Pereira, and M. R. Bronze. 2015. "Bottled Water: Analysis of Mycotoxins by LC-MS/MS." *Food Chemistry* 176: 455–64.

McDonald, Thomas A. 2002. "A Perspective on the Potential Health Risks of PBDEs." *Chemosphere* 46 (5): 745–55.

Mendoza, A., J. Aceña, S. Pérez, M. López de Alda, D. Barceló, A. Gil, and Y. Valcárcel. 2015. "Pharmaceuticals and Iodinated Contrast Media in a Hospital Wastewater: A Case Study to Analyse Their Presence and Characterise Their Environmental Risk and Hazard." *Environmental Research* 140: 225–41.

Metcalfe, C. 2014. "Contamination of Emerging Concern in Effluents from Wastewater Treatment Plants in the Lake Simcoe Watershed." *Environmental and Resource Studies (Trent University, Peterborough)*.

Moinfar, Soleyman, Lazgin Abdi Jamil, Helan Zeyad Sami, and Sorayya Ataei. 2021. "An Innovative Continuous Sample Drop Flow Microextraction for GC–MS Determination of Pesticides in Grape Juice and Water Samples." *Journal of Food Composition and Analysis: An Official Publication of the United Nations University, International Network of Food Data Systems* 95: 103695.

Mokh, Samia, Mohammad El Khatib, Mohamad Koubar, Zeina Daher, and Mohamad Al Iskandarani. 2017. "Innovative SPE-LC-MS/MS Technique for the Assessment of 63

Pharmaceuticals and the Detection of Antibiotic-Resistant-Bacteria: A Case Study Natural Water Sources in Lebanon." *The Science of the Total Environment* 609: 830–41.

Nascimento, Madson Moreira, Gisele Olímpio da Rocha, and Jailson B. de Andrade. 2021. "Customized Dispersive Micro-Solid-Phase Extraction Device Combined with Micro-Desorption for the Simultaneous Determination of 39 Multiclass Pesticides in Environmental Water Samples." *Journal of Chromatography A* 1639: 461781.

Nasiri, Maryam, Hossein Ahmadzadeh, and Amirhassan Amiri. 2021. "Organophosphorus Pesticides Extraction with Polyvinyl Alcohol Coated Magnetic Graphene Oxide Particles and Analysis by Gas Chromatography-Mass Spectrometry: Application to Apple Juice and Environmental Water." *Talanta* 227: 122078.

Nielen, M. W. F., M. C. van Engelen, R. Zuiderent, and R. Ramaker. 2007. "Screening and Confirmation Criteria for Hormone Residue Analysis Using Liquid Chromatography Accurate Mass Time-of-Flight, Fourier Transform Ion Cyclotron Resonance and Orbitrap Mass Spectrometry Techniques." *Analytica Chimica Acta* 586 (1–2): 122–29.

Nohynek, Gerhard J., Christopher J. Borgert, Daniel Dietrich, and Karl K. Rozman. 2013. "Endocrine Disruption: Fact or Urban Legend?" *Toxicology Letters* 223 (3): 295–305.

Ofrydopoulou, Anna, Christina Nannou, Eleni Evgenidou, and Dimitra Lambropoulou. 2021. "Sample Preparation Optimization by Central Composite Design for Multi Class Determination of 172 Emerging Contaminants in Wastewaters and Tap Water Using Liquid Chromatography High-Resolution Mass Spectrometry." *Journal of Chromatography A* 1652: 462369.

Padrón, Ma Esther Torres, Cristina Afonso-Olivares, Zoraida Sosa-Ferrera, and José Juan Santana-Rodríguez. 2014. "Microextraction Techniques Coupled to Liquid Chromatography with Mass Spectrometry for the Determination of Organic Micropollutants in Environmental Water Samples." *Molecules (Basel, Switzerland)* 19 (7): 10320–49.

Pascual Aguilar, Juan Antonio, Vicente Andreu, Julián Campo, Yolanda Picó, and Ana Masiá. 2017. "Pesticide Occurrence in the Waters of Júcar River, Spain from Different Farming Landscapes." *The Science of the Total Environment* 607–608: 752–60.

Patiño, Yolanda, Eva Díaz, and Salvador Ordóñez. 2015. "Performance of Different Carbonaceous Materials for Emerging Pollutants Adsorption." *Chemosphere* 119 Suppl: S124–30.

Petrovic, M. 2003. "Analysis and Removal of Emerging Contaminants in Wastewater and Drinking Water." *Trends in Analytical Chemistry: TRAC* 22 (10): 685–96.

Petrovic, Mira, Marinella Farré, Miren Lopez de Alda, Sandra Perez, Cristina Postigo, Marianne Köck, Jelena Radjenovic, Merixell Gros, and Damia Barcelo. 2010. "Recent Trends in the Liquid Chromatography-Mass Spectrometry Analysis of Organic Contaminants in Environmental Samples." *Journal of Chromatography A* 1217 (25): 4004–17.

Pihlström, Tuija, Anna Hellström, and Victoria Axelsson. 1997. "Gas Chromatographic Analysis of Pesticides in Water with Off-Line Solid Phase Extraction." *Analytica Chimica Acta* 356 (2–3): 155–63.

Rimayi, Cornelius, David Odusanya, Jana M. Weiss, Jacob de Boer, and Luke Chimuka. 2018. "Seasonal Variation of Chloro-s-Triazines in the Hartbeespoort Dam Catchment, South Africa." *The Science of the Total Environment* 613–614: 472–82.

Rivera-Utrilla, José, Manuel Sánchez-Polo, María Ángeles Ferro-García, Gonzalo Prados-Joya, and Raúl Ocampo-Pérez. 2013. "Pharmaceuticals as Emerging Contaminants and Their Removal from Water. A Review." *Chemosphere* 93 (7): 1268–87.

Rubirola, Adrià, Mª Rosa Boleda, and Mª Teresa Galceran. 2017. "Multiresidue Analysis of 24 Water Framework Directive Priority Substances by On-Line Solid Phase Extraction-Liquid Chromatography Tandem Mass Spectrometry in Environmental Waters." *Journal of Chromatography A* 1493: 64–75.

Sanchez-Prado, Lucia, Carmen Garcia-Jares, and Maria Llompart. 2010. "Microwave-Assisted Extraction: Application to the Determination of Emerging Pollutants in Solid Samples." *Journal of Chromatography A* 1217 (16): 2390–414.

Schultz, Melissa M., Douglas F. Barofsky, and Jennifer A. Field. 2006. "Quantitative Determination of Fluorinated Alkyl Substances by Large-Volume-Injection Liquid Chromatography Tandem Mass Spectrometry Characterization of Municipal Waste Waters." *Environmental Science & Technology* 40 (1): 289–95.

Schwanz, Thiago Guilherme, Cristiane Köhler Carpilovsky, Grazielle Castagna Cezimbra Weis, and Ijoni Hilda Costabeber. 2019. "Validation of a Multi-Residue Method and Estimation of Measurement Uncertainty of Pesticides in Drinking Water Using Gas Chromatography-Mass Spectrometry and Liquid Chromatography-Tandem Mass Spectrometry." *Journal of Chromatography A* 1585: 10–18.

Shafrir, Michelle, and Dror Avisar. 2012. "Development Method for Extracting and Analyzing Antibiotic and Hormone Residues from Treated Wastewater Sludge and Composted Biosolids." *Water, Air, and Soil Pollution* 223 (5): 2571–87.

Sifatullah, K. M., and G. S. Tuncel. 2018. "Study of Pesticide Contamination in Soil, Water and Produce Using Gas Chromatography Mass Spectrometry." *J Anal Bioanal Tech* 9 (409): 2.

Söderström, Hanna, Richard H. Lindberg, and Jerker Fick. 2009. "Strategies for Monitoring the Emerging Polar Organic Contaminants in Water with Emphasis on Integrative Passive Sampling." *Journal of Chromatography A* 1216 (3): 623–30.

Souza, M. P. de, T. M. Rizzetti, J. Z. Francesquett, O. D. Prestes, and R. Zanella. 2018. "Bar Adsorptive Microextraction (BAµE) with a Polymeric Sorbent for the Determination of Emerging Contaminants in Water Samples by Ultra-High Performance Liquid Chromatography with Tandem Mass Spectrometry." *Analytical Methods: Advancing Methods and Applications* 10 (7): 697–705.

Souza-Silva, Érica A., Ruifen Jiang, Angel Rodríguez-Lafuente, Emanuela Gionfriddo, and Janusz Pawliszyn. 2015. "A Critical Review of the State of the Art of Solid-Phase Microextraction of Complex Matrices I. Environmental Analysis." *Trends in Analytical Chemistry: TRAC* 71: 224–35.

Stasinakis, Athanasios S. 2012. "Review on the Fate of Emerging Contaminants during Sludge Anaerobic Digestion." *Bioresource Technology* 121: 432–40.

Suffet, I. H., and S. D. Faust. 1972. "The P-Value Approach to Quantitative Liquid-Liquid Extraction of Pesticides from Water. 1. Organophosphates: Choice of PH and Solvent." *Journal of Agricultural and Food Chemistry* 20 (1): 52–56.

Taghani, Abdollah, Nasser Goudarzi, Ghadam Ali Bagherian, Mansour Arab Chamjangali, and Amir Hossein Amin. 2018. "Application of Nanoperlite as a New Natural Sorbent in the Preconcentration of Three Organophosphorus Pesticides by Microextraction in Packed Syringe Coupled with Gas Chromatography and Mass Spectrometry." *Journal of Separation Science* 41 (10): 2245–52.

Taniyasu, Sachi, Kurunthachalam Kannan, Man Ka So, Anna Gulkowska, Ewan Sinclair, Tsuyoshi Okazawa, and Nobuyoshi Yamashita. 2005. "Analysis of Fluorotelomer Alcohols, Fluorotelomer Acids, and Short- and Long-Chain Perfluorinated Acids in Water and Biota." *Journal of Chromatography A* 1093 (1–2): 89–97.

Tremblay, L. A., M. Sterwart, B. M. Peake, and J. B. Gadd. 2011. "Review of the Risks of Emerging Organic Contaminants and Potential Impacts to Hawke's Bay. Prepared for Hawke's Ay Regional Council." *Cawthron Report* 1973: 1–39.

U.S. Environmental Protection Agency (EPA). 2010. *List of Contaminants and Their MCLs*. Washington, DC.

Valenzuela, Eduard F., Fabiano G. F. de Paula, Ana Paula C. Teixeira, Helvécio C. Menezes, and Zenilda L. Cardeal. 2020. "A New Carbon Nanomaterial Solid-Phase Microextraction to Pre-Concentrate and Extract Pesticides in Environmental Water." *Talanta* 217: 121011.

Valsecchi, Sara, Stefano Polesello, Michela Mazzoni, Marianna Rusconi, and Mira Petrovic. 2015. "On-Line Sample Extraction and Purification for the LC–MS Determination of Emerging Contaminants in Environmental Samples." *Trends in Environmental Analytical Chemistry* 8: 27–37.

Van Leeuwen, S. P. J., Strub, M. P., Cofino, W., Lindström, G. and B. van Bavel. 2011. *Third Inter Laboratory Study on Perfluorinated Compounds in Environmental and Human Matrices, Report R-11/04*. IVM Institute for Environmental Studies, VU University, Amsterdam, The Netherlands.

Wang, Hui, Bo Qu, He Liu, Jie Ding, and Nanqi Ren. 2018. "Analysis of Organochlorine Pesticides in Surface Water of the Songhua River Using Magnetoliposomes as Adsorbents Coupled with GC-MS/MS Detection." *The Science of the Total Environment* 618: 70–79.

Wille, K., J. Vanden Bussche, H. Noppe, E. De Wulf, P. Van Caeter, C. R. Janssen, H. F. De Brabander, and L. Vanhaecke. 2010a. "A Validated Analytical Method for the Determination of Perfluorinated Compounds in Surface-, Sea- and Sewagewater Using Liquid Chromatography Coupled to Time-of-Flight Mass Spectrometry." *Journal of Chromatography A* 1217 (43): 6616–22.

Wille, Klaas, Herlinde Noppe, Karolien Verheyden, Julie Vanden Bussche, Eric De Wulf, Peter Van Caeter, Colin R. Janssen, Hubert F. De Brabander, and Lynn Vanhaecke. 2010b. "Validation and Application of an LC-MS/MS Method for the Simultaneous Quantification of 13 Pharmaceuticals in Seawater." *Analytical and Bioanalytical Chemistry* 397 (5): 1797–808.

Yamashita, Nobuyoshi, Sachi Taniyasu, Gert Petrick, Si Wei, Toshitaka Gamo, Paul K. S. Lam, and Kurunthachalam Kannan. 2008. "Perfluorinated Acids as Novel Chemical Tracers of Global Circulation of Ocean Waters." *Chemosphere* 70 (7): 1247–55.

Yoon, M. K. 2010. *Analytical Method Development for Measurement of Unregulated Organic Contaminants in Aqueous Samples Using Liquid Chromatography Tandem Mass Spectrometry (Ph.D. Thesis)*. The State University of New Jersey, New Jersey.

Yu, Rui, Qiang Liu, Jingshuang Liu, Qicun Wang, and Yang Wang. 2016. "Concentrations of Organophosphorus Pesticides in Fresh Vegetables and Related Human Health Risk Assessment in Changchun, Northeast China." *Food Control* 60: 353–60.

Zapf, Andreas, Regine Heyer, and Hans-Jürgen Stan. 1995. "Rapid Micro Liquid-Liquid Extraction Method for Trace Analysis of Organic Contaminants in Drinking Water." *Journal of Chromatography A* 694 (2): 453–61.

Zavala López, Miguel Ángel, and Liliana Reynoso-Cuevas. 2015. "Simultaneous Extraction and Determination of Four Different Groups of Pharmaceuticals in Compost Using Optimized Ultrasonic Extraction and Ultrahigh Pressure Liquid Chromatography-Mass Spectrometry." *Journal of Chromatography A* 1423: 9–18.

Zhang, Hui, Simon Watts, Martin C. Philix, Shane A. Snyder, and Choon Nam Ong. 2018. "Occurrence and Distribution of Pesticides in Precipitation as Revealed by Targeted Screening through GC-MS/MS." *Chemosphere* 211: 210–17.

Zuloaga, O., P. Navarro, E. Bizkarguenaga, A. Iparraguirre, A. Vallejo, M. Olivares, and A. Prieto. 2012. "Overview of Extraction, Clean-up and Detection Techniques for the Determination of Organic Pollutants in Sewage Sludge: A Review." *Analytica Chimica Acta* 736: 7–29.

Zwiener, Christian, and Fritz H. Frimmel. 2004. "LC-MS Analysis in the Aquatic Environment and in Water Treatment Technology – ACritical Review. Part II: Applications for Emerging Contaminants and Related Pollutants, Microorganisms and Humic Acids." *Analytical and Bioanalytical Chemistry* 378 (4): 862–74.

6 Advanced Techniques for Detection of Environmental Pollutants and Recent Progress

Kriti Srivastava, Harsh Prajapati, Gajendra Singh Vishwakarma, Nidhi Verma, and Alok Pandya

CONTENTS

6.1	Introduction	121
6.2	Basics of Traditional Environmental Pollutant Monitoring and Detections	122
6.3	Types of Environmental Pollutions	123
	6.3.1 Air Pollution	123
	6.3.2 Water Pollution	124
6.4	Detection of Environmental Pollutant by Instrumental Technique	125
	6.4.1 Spectroscopy Techniques for Analysis	126
	6.4.2 Separation Techniques for Analysis	127
6.5	Detection of Environmental Pollutants Using Nano-Enabled Sensors	128
	6.5.1 Optical Method	128
	6.5.2 Electrochemical Method	130
	6.5.3 Magnetic Method	130
	6.5.4 Paper-Based Detection Method	131
6.6	Conclusion and Future Perspective	132
Acknowledgments		133
References		133

6.1 INTRODUCTION

The world has seen the negative consequences of the uncontrolled expansion of numerous human activities, such as industries, transportation, agriculture, and urbanization in recent decades. Rising living standards and the demands of consumers have increased the different types of pollutants in air, water, soil, and groundwater pollution. In the case of air pollution, the level of pollutants such as CO_2, CH_4, greenhouse gases, particulate matter, SOx, and NOx is increasing day by day. Similarly, water is also getting polluted due to the deposition of a variety of chemicals, leachate, oil spills, and heavy metals. The fertility and quality of soil is also on the verge of deterioration due to hazardous waste disposal, sewage sludge, mixing of agrochemicals (chemical pesticides and fertilizers) as well as non-biodegradable materials such as glass, computer parts, electronic device, plastics, etc. These pollutants not only create problems in the ecosystem but also in human health. Even trace amounts of these pollutants have the potential to cause acute and chronic disease such as chronic obstructive pulmonary disease, different types of cancer, acute infections respiration, and digestive system disorders. Hence, it is a matter of concern for environmentalists and government bodies globally due to their potential hazard to both human health and the environment. Measures have already been undertaken to decrease such pollutants over time but still with limited success. Although another aspect of such hazardous environmental pollutants is "monitoring," which should be standardized for the accuracy and efficacy.

Currently the monitoring of environmental pollutants is majorly dependent on sophisticated instrumental-based approaches. These methods usually make use of high-end instruments like spectrophotometry, high performance liquid chromatography (HPLC), gas chromatography (GC), mass spectrometry (MS), etc. These methods are still considered as the first choice for the detection of pollutants as they provide results that are highly precise and accurate. However, these methods have tedious sample preparation followed by difficult processing and interpretation. Current research is thus focused on the development of instruments or kits that are able to detect specimens at very low levels at the site of the pollution and would overcome the shortfalls of traditional techniques. In this regard sensor techniques are being explored with the aim of providing results with accuracy in the field.

This chapter discusses recent novel and remarkable sensor strategies reported in recent years for the detection of various environmental pollutants. This chapter also provides details on the types of pollutants followed by a discussion on the traditional techniques used in detection of environmental pollutants. This chapter provides insight on how the field of paper microfluidics can be explored in order to improve the current analytical methods for environmental pollutant detection. The chapter may be of great help to environmentalists and governmental bodies to help develop new methods for detection.

6.2 BASICS OF TRADITIONAL ENVIRONMENTAL POLLUTANT MONITORING AND DETECTIONS

Environmental monitoring includes a set of instruments and procedures that are used to examine an environmental pollutant, assess its quality, and analyse various environmental factors in order to precisely measure the influence of a certain activity on the environment. Air, water, and soil are the three main components of the human environment that must be monitored on a regular basis in order to ensure safety of human health and environment (Halmann 1996). The sampling of these components begins with absorption, condensation, grab sampling, and composite sampling, followed by storage of the samples in impingers, sample bottles, or sampling medium. These samples are treated by digestion, filtration, sedimentation, and electrostatic techniques prior to analysis. Finally, these samples are analysed using a variety of sophisticated instruments such as GCMS, ICP-MS, and HPLC. The data acquired from various environmental monitoring systems are then be entered into a database management system (DBMS), where it can be categorized, analysed, displayed, and suitable information generated. In the case of water monitoring physical parameters like pH, TDS, and conductivity are monitored through probe-based hand haled meter, while toxic chemicals and heavy metals are analysed through instrumental techniques such as chromatography and spectrometry. In the case of air pollution, the environmental data collected from multiple environmental networks and institutes using specialized observation tools such as sensor networks and geographic information system (GIS) models is integrated into air dispersion models, which combine emissions, meteorological, and topographic data to detect and predict concentrations of air pollutants. Similarly in the case of soil pollution monitoring the process includes grab or random sampling followed by the processing and analysis of sampling through high-end equipment. All in all, in traditional monitoring systems the information provided regarding the actual source or position of a pollutant is not very clear, therefore *in situ* monitoring of pollutants in the environment is a serious concern. Hence, quick sensing technologies are necessary to solve this issue.

6.3 TYPES OF ENVIRONMENTAL POLLUTIONS

In general pollution can be defined as any undesirable change in the physical, chemical, and biological characteristic of air, water, or land that may or will harmfully affect human life or that of various species, industrial processes, living conditions, and cultural assets. Environmental pollution is divided into different types based on the different components of the environment. Among the various types of environmental pollution the main ones are air pollution, soil pollution, water pollution, noise pollution, thermal pollution, radiation pollution, etc. Some significant types of environmental pollution are discussed in brief in the section below and described in Figure 6.1.

6.3.1 Air Pollution

It is any undesirable change in the quality of air that harmfully affects people's well-being. Atmosphere, the gaseous layer called 'air,' is the life blanket of Earth, the essential ingredient for all living things. It is composed of a mixture of gases (Edward 1998). Therefore, preserving or maintaining the air in its pure form becomes necessary, since polluted air may profoundly affect health and bring in other consequences. The atmosphere can cleanse itself of the impurities firstly by its vertical and horizontal mixing (dispersion—favoured by wind) and secondly washing out the pollutants to a certain extent by rain. But the presence of high concentrations of pollutants in air such as sulphur oxides (SOx), nitrogen oxides (NOx), hydrocarbons (HCs), carbon monoxide (CO), and particulates affects the natural cleansing mechanism of the atmosphere, altering the percentage of concentration of gases naturally present in the

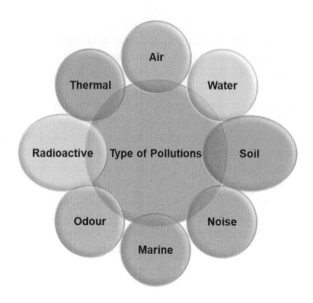

FIGURE 6.1 Types of environmental pollutants.

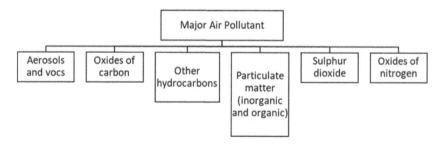

FIGURE 6.2 Major air pollutants.

atmosphere and causing air pollution. The major air pollutants and their sources are shown in Figure 6.2 and in Table 6.1, respectively (Fa et al., 2018).

6.3.2 WATER POLLUTION

Water pollution refers to one or more substances building up in aquatic bodies to a degree that negatively affects life forms (Abel, 2002; Aldwell, Burdon, and Sherwood, 1983). Water pollution usually arises due to discharging of s e w a g e and wastewater in natural water bodies. Annually, more than 400 billion tons of industrial waste are produced globally, the bulk of which is discharged untreated into the streams, rivers, seas, oceans, and other water bodies. Major water pollutants and their sources of are shown in Figure 6.3 and Figure 6.4, respectively (Barcelo and Petrovic, 2007).

TABLE 6.1
Summary of the Various Classes of Air Pollutants and Their Source

Nos.	Pollutant	Source/Cause
1.	Carbon monoxide	Primarily produced by incomplete combustion; automobile exhaust, atmospheric photo- chemical reactions, biological oxidation by marine organism, etc
2.	Carbon dioxide (CO_2), though not a pollutant at normal level and is harmful in excess levels.	Fossil fuels combustion, forests diminution (forests take away excess carbon dioxide and help in maintaining the O_2 and CO_2 ratio).
3.	Sulphur dioxide (SO_2)	Industrial processes, fossil fuel combustion, wild fire, thermal power plants, smelters, industrial boilers, oil refineries and volcanic eruptions.
4.	Polynuclear Aromatic Compounds (PACs) and Polynuclear Aromatic Hydrocarbons (PAHs)	Vehicular exhaust and industries, leaky fuel tanks, leaching of noxious waste from dumping sites, coal tar lining of some water supply pipelines.
5.	Chlorofluorocarbons (CFCs)	Refrigerators, air conditioning machines, foam shaving cream, spray cans and cleaning solvents.
6.	Nitrogen Oxides	Vehicular exhaust, fossil fuel combustion, wild fires, thermal power plants, smelters, industrial boilers, oil refineries and volcanoes.
7.	Nitrous oxide (N_2O)	Nitrogenous fertilizers, deforestation and biomass burning.
8.	Peroxy Acetyl Nitrate (PAN)	Chemical reaction of hydrocarbons and nitrogen oxides in presence of sunlight.
9.	Particulate matter lead halides (Lead pollution)	Combustion of leaded gasoline products
10.	Asbestos particles	Mining activities
11.	Silicon dioxide	Cutting of stones, cutting, crushing and grinding, cement industries, making of pottery and glass.
12.	Biological matters like pollen grains, fungal spores, bacteria, and virus.	Flowers and microbes
13.	Dioxins and furans	From combustion of chlorinated substance.

6.4 DETECTION OF ENVIRONMENTAL POLLUTANT BY INSTRUMENTAL TECHNIQUE

Analytical techniques that provide precise estimates of the quantitative and qualitative composition of pollutants aid in the study of environmental pollution. They also aid in determining the composition variation of pollutants across time and space, as well as in assessing the source and intensity of emissions. Pollution management techniques rely on accurate data and numbers that show the precise degree of pollution. As a result, detection of environmental pollution is the most effective approach to choose which pollution control technique to use, based on the level of pollution in a given region, and assure the overall economic feasibility of the control measure.

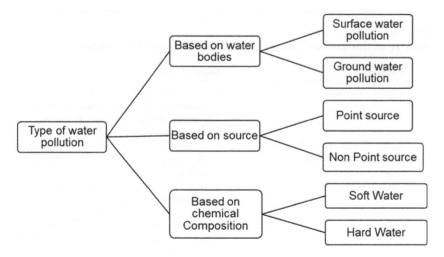

FIGURE 6.3 Major types of water pollutants.

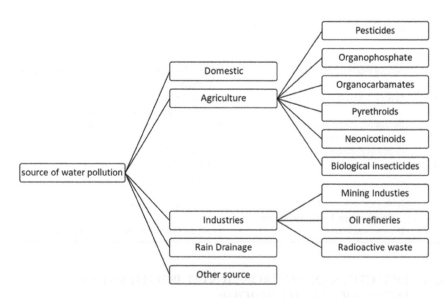

FIGURE 6.4 Source of water pollution.

6.4.1 Spectroscopy Techniques for Analysis

Spectroscopy plays an important role in the analysis and quantification of various pollutants in the environmental samples (Singh et al., 2019; Alula, Mengesha, and Mwenesongole, 2018; Armenta and de la Guardia, 2014) (Figure 6.5). The choice of appropriate spectroscopic technique and detection system has broadened the

Advanced Techniques for Detecting Environmental Pollutants

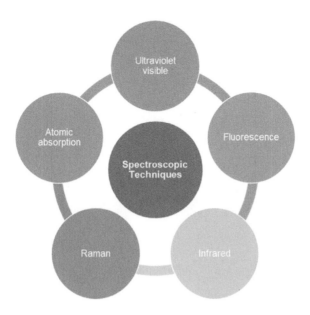

FIGURE 6.5 Spectroscopic techniques for analysis.

application capabilities of these techniques for the efficient and sensitive analysis of target pollutants such as heavy metals, pesticides, herbicides, and polycyclic aromatic hydrocarbons (PAHs) in environmental samples (i.e., wastewater, soil, and air). The careful choice and application of sampling techniques and preconcentration techniques allow the development of the sensitivity of these spectroscopic analyses. Impressive progress in the limit of detection for the target environmental pollutants, mostly because of the design and development of chromatographic systems with sensitive detectors, has pushed the target concentrations to very low levels (in the range between µg and ng). With the rapid progress in the instrumentation of spectroscopy, extraction methods have also become facile, rapid, environmentally friendly, and low cost ensuring the preconcentration of target pollutants in environmental samples (Carstea et al., 2016; Dominguez et al., 2020; Esch et al., 2019; Halvorson and Vikesland, 2010; Mesquita et al., 2017; Mirnaghi et al., 2019).

6.4.2 SEPARATION TECHNIQUES FOR ANALYSIS

Chromatography plays a crucial role in the detection and quantification of various pollutants in environmental samples (Figure 6.6). The choices of suitable detection systems and the types of columns have broadened the capabilities of this technique for the sensitive analysis of target environmental pollutants. The careful applications of sampling techniques and sample treatment approaches have allowed the development of the sensitivity of these chromatographic analyses.

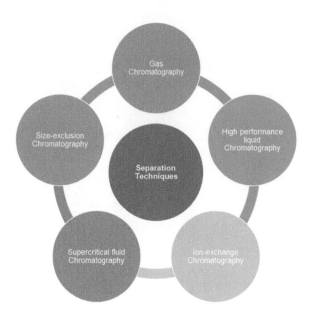

FIGURE 6.6 Separation techniques for analysis.

6.5 DETECTION OF ENVIRONMENTAL POLLUTANTS USING NANO-ENABLED SENSORS

Owing to the multiple advantages of nano sensors like low cost, higher sensitivity and selectivity, real-time detection, high efficiency, portability, and several other features, nano-enabled sensors are also being explored for the detection of environmental contaminants. Nano-enabled sensors still have a lot more potential to achieve improved sensing abilities by using various metals, biological functional groups like aptamers, antibodies and enzymes, and chemical treatments. They are widely researched for the detection of pathogens and heavy metals from several environmental samples.

Nanomaterial-enabled sensors generally are made up of three crucial components: a nanomaterial, a recognition element to detect the analyte, and a signal transduction method that confirms the detection of the target analyte. Sensors can be specifically designed to detect a single analyte or multiple analytes. The presence of target analyte is confirmed using several signal transduction techniques after the interaction of the sensor and target. The commonly used methods are optical transduction, electrochemical transduction, magnetic transduction, and paper-based detection. Illustration of the components of a nano sensor and signal transduction methods are shown in Figure 6.7.

6.5.1 Optical Method

Biosensors use optical signal transduction mechanisms, as they provide higher sensitivity to detect lower concentrations, are not interfered with by electromagnetic noise,

Advanced Techniques for Detecting Environmental Pollutants

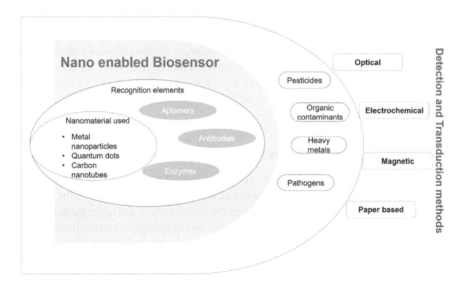

FIGURE 6.7 Illustration of the components of a nano sensor and signal transduction methods.

and they provide widespread potential for further advancements. Optical methods monitor the change in an optical signal that occurs after a recognition event between a functionalized nanomaterial and contaminants like pathogens, organic contaminants, and heavy metals. The optical signals are monitored using a variety of spectroscopic methods. Optical transduction is based on the interaction of a sensing element with electromagnetic radiation. The analytical techniques measure the emission or absorption by the sample under irradiation by ultraviolet, visible, or infrared light (Willner and Vikesland, 2018). The commonly used methods include surface plasmon resonance-enabled spectroscopies, fluorescence, and colorimetric analysis.

Detection of pathogens like *Salmonella typhimurium* and *Pseudomonas aeruginosa* has been achieved by immobilizing different types of thiolated aptamers on a multispot gold-capped NPs array (MG-NPA) chip (Mustafa, Hassan, and Andreescu, 2017; Yoo, Kim, and Lee, 2015). The chip was fabricated from a dielectric layer of a thin gold layer on silica NPs over a glass slide. Detection was achieved by measuring change in the localized surface plasmon resonance (LSPR) upon the binding of bacteria. An aptamer/graphene interdigitated gold piezoelectric sensor was fabricated using mercaptobenzenediazonium tetrafluoroborate (MBDT) attached to graphene on a gold surface through thiol chemistry for the detection of *S. aureus* (Lian et al., 2015).

Quantum dots (QDs) are photostable, fluorescent semiconductor (e.g., CdSe, CdTe, CdS, ZnSe, PbS, PbTe) nanocrystals that have emission wavelengths that scale with particle size and chemical composition (Vikesland and Wigginton, 2010). QDs have been functionalized with antibodies and aptamers for the fluorescent detection of pathogens as diverse as respiratory syncytial virus (RSV) (Agrawal et al., 2005; Bentzen et al., 2005), *E. coli* O157:H7 (Hahn, Keng, and Krauss, 2008), *Bacillus thurinigensis* (Ikanovic et al., 2007), and protozoan parasites such as *Cryptosporidium* and *Giardia* (Zhu, Ang, and Liu, 2004).

Functionalised electrospun nanofibers for colorimetric detection of various metal ions such as copper, mercury, lead, cadmium, chromium, iron, zinc, nickel, cesium, aluminum, and silver are achieved following a direct blending and the surface immobilization of functional agents for metal ion interaction (Balusamy, Senthamizhan, and Uyar, 2020).

6.5.2 Electrochemical Method

Electrochemical detection methods work by detecting the change in current or potential that result from the interaction between an analyte and an electrode. In the electrochemical sensors, electron transfer between the sensing layer and target contaminants can produce an electrochemical response that results in voltage or current modulations (Khanmohammadi et al., 2020). For this, a sensing layer is placed at the surface of a conductive substrate for the detection of targeted analyte. Electrochemical sensors are dependent on the electron transfer between the surface of the sensor and target analyte/intermediate A variety of techniques have been used to observe these changes and include of cyclic voltammetry, chronoamperometry, chronopotentiometry, impedance spectroscopy, and various field-effect transistor based methods (Willner and Vikesland ,2018). Nano-enabled sensor designs can involve modification of the solid electrode (e.g., platinum, gold, silver, graphite) with nano-carbons (e.g., carbon nanotubes, graphene) or functionalization with recognition elements (e.g., antibodies, aptamers) (Sadik, Aluoch, and Zhou, 2009). Direct spatial contact between the nanoscale architecture of the electrode and the recognition element gives rise to large signal amplification and improved signal-to-noise ratios compared to traditional electrochemical techniques (Sadik, Aluoch, and Zhou, 2009; Grieshaber et al., 2008; Sanvicens et al., 2009).

Electrochemical sensors include graphene for bacterial sensing (Mannoor et al., 2012); carbon nanotubes for ammonium, Co^{2+}, and pesticides (Bandodkar et al., 2016; Gou et al., 2013; Yu et al., 2015); copper nanowire electrodes for nitrate (Stortini et al., 2015); polymeric nanocomposite membranes for Ag^+, Hg^{2+}, and Cu^{2+} (Li, Zhang, and Huang, 2014); and reduced graphene oxide/gold nanoparticle nanocomposite for organophosphates (Konwarh, Gollavelli, and Babu, 2020). Methyl parathion pesticide has been determined in groundwater using $CuO–TiO_2$ as the sensing layer by a nonenzymatic electrochemical sensor (Tian et al., 2018).

6.5.3 Magnetic Method

Magnetic transduction method is often preferred for detection in biological samples due to the low background magnetic signal from the sample. Often, the use of magnetic nanoparticles to concentrate, separate, and purify the analyte of interest in the detection zone is termed magnetic transduction (Koets et al., 2009).

The principle for detection mechanism is the clustering/agglomeration of individual nanomagnetic probes into larger assemblies following interaction with a target (Willner and Vikesland, 2018). Analyte binding with the sensor probes results in the formation and enhanced dephasing of the spins of the surrounding water protons. The changes in the spin–spin relaxation is often analysed by the technique magnetic resonance relaxometry (Reddy et al., n.d.; Kaittanis, Naser, and Perez, 2007). Magnetic relaxation switches have been used to detect proteins (Zhang et al., 2013) and viruses (Perez et al., 2003).

Biomolecules such as antibodies (Abs) or aptamers (APTs), which exhibit good specificity, can be modified on metal nanoparticles to improve the efficiency of bacterial capture (Hu et al., 2019). Aptamer-modified MNPs have been successfully applied for the selective capture and sensitive detection of *Pseudomonas aeruginosa, Salmonella typhimurium*, and many other foodborne pathogens (Duan et al., 2012; Zhong and Ma, 2018; Tian et al., 2021). Magnetic nanoparticles (MNPs) have been used with biological binding probes to concentrate and separate bacterial pathogens inexpensively and rapidly with good control using a magnet (Lee, Chen, and Lee, 2021). Epoxy-organosilane modified lysine and its functionalization on metal nanoparticles on SiO_2 is found effective for the simultaneous removal of heavy metals from real environmental systems (Plohl et al., 2021).

Multi-walled carbon nanotubes (MWCNTs) are more often used for the purpose of detection and removal of heavy metals and dyes. Magnetic nanoparticles embedded with MWCNTs are usually produced by using the chemical composition of λ-Fe_2O_3, or Fe_2O_3 onto covalently modified MWCNTs (Saleem et al., 2021). Fe_3O_4 functionalised with ZrO_2 nanoparticles magnetic solid phase extraction coupled with flame atomic absorption spectrometry has been tested for separation of Cr(III) in environmental and biological samples (López-García, Marín-Hernández, and Hernández-Córdoba 2020; Bagherzadeh et al., 2021).

6.5.4 Paper-Based Detection Method

Paper-based platforms have become widely known for the development of analytical devices. Paper-based sensors are the most convenient new technology for fabricating simple, low-cost, portable, and disposable analytical devices for many application areas including clinical diagnosis, food quality control, and environmental monitoring. The properties of paper like passive liquid transport and compatibility with chemicals/biochemicals is one of the features that has been widely exploited for the development of environmental sensors. The most common paper-based sensors are lateral flow immunoassays (LFIA) and lateral flow dipsticks (LFD).

Paper-based detection has been described for specific *E. coli* O157:H7 using LFIA; an assay employing AuNPs labelled with specific anti-*E. coli* O157:H7 antibodies provide the detector for the presence of the specific antigen (i.e., *E. coli* O157:H7) (Sun et al., 2020). Another method used for detection has been developed as a multiplex loop-mediated isothermal amplification (mLAMP) assay for the simultaneous detection of *Salmonella spp. and Cronobacter spp.* in powder infant formulas as a foodborne pathogen. The invasion protein (invA) gene (*Salmonella spp.*) and the 16Se23S rRNA internal transcribed spacer (ITS) sequence (*Cronobacter spp.*) were amplified by labelled specific primers in the mLAMP reaction. Then, the modified amplicons were visually monitored through the accumulation of gold nanoparticles (AuNPs) on the LFD (Yang et al., 2021).

For heavy metal detection, a high-sensitivity lateral flow immunoassay device (LFID) for Cd^{2+} detection in aqueous samples using the Cd–EDTA–BSA conjugate labelled with gold nanoparticles as signal producer tool has been described. The developed LFID were based on the recognition of Cd–EDTA complex (formed after Cd^{2+} analyte complexation) by using the specific 2A81G5 monoclonal antibody (mAb) (López, Blake, and Merkoc 2013). In another work, ultra-sensitive surface enhanced Raman scattering (SERS) LFIA with silver nanoparticles (AgNPs) was shown to trace chromium in hard-to-detect surface water and polluted environmental water samples (Liang et al., 2014). Another research group developed a lateral flow immunoassay device (LFID) with the combination of the gold nanoparticles (AuNP)-based visual test using AuNP probes of carboxylic modified protein (COOH-BSA) for the field detection of Hg^{2+} (Chao et al., 2012).

Microfluidic paper-based device (μPADs) and a screen-printed electrochemical paper-based device (ePADs) have been seen as effective alternative tools for paper-based detection in environmental applications (Sgobbi, Henry, and Coltro 2021). The dual colorimetric and electrochemical approaches on paper-based microfluidic platforms have been researched by a few groups for the detection of heavy metals in water samples. The data reported suggests the use of μPADs for colorimetric detection or ePADs for electrochemical detection of select heavy metals including iron, copper, nickel, mercury, lead, chromium, zinc, and cadmium (Wu et al., 2019; Rattanarat et al., 2014; Mettakoonpitak, Volckens, and Henry 2020).

6.6 CONCLUSION AND FUTURE PERSPECTIVE

Despite the fact that environmental pollutants are found in miniscule amounts in soil, air, and water bodies, their negative impacts on living creatures cannot be overstated owing to their continual discharge into the environment. Due to the complexity of contaminants in samples, removing and detecting environmental pollutants and their transformed products in natural environments is a difficult task. However, due to the immense efforts and time invested by numerous scientists working in many scientific disciplines, considerable progress has been achieved on the evaluation of both biological and chemical contaminants. With regard to the existing technologies, the detection has been achieved but more focus needs to be directed towards development of paper-based electrochemical and microfluidics platforms as they have advantages of being biodegradable and low cost. Electrochemical sensors can help in detection of trace pollutants in the nick of time. However, not much has been investigated in this area. Future developments in the detection of emerging pollutants will rely on more cost-effective and point-of-testing approaches since they can overcome the disadvantages of existing techniques. However, the robustness of field-deployable sensors needs to be evaluated for on-ground applications. The detection of degraded compounds along with parent compounds is of the utmost importance. Exhaustive research is already in progress on the detection of biological pollutants, but emergence of new biological threats is always expected. In short, feasible techniques need to be in developed to detect contaminants within regulated levels and more research is necessary to discover new recognition agents for continuously monitoring pollutants in real applications.

ACKNOWLEDGMENTS

We gratefully acknowledge Department of Science and Technology, Govt of India (DST/TMD(EWO)/OWUIS-2018/RS-20(G)) for financial support. Dr. Gajendra Singh Vishwakarma gratefully acknowledge GSBTM, Govt of Gujarat, India (GSBTM/JDR&D/604-2019/307) for financial support.

REFERENCES

Abel, P.D., 2002. *Water Pollution Biology*, 0 ed. London: CRC Press. https://doi.org/10.1201/9781482295368

Aldwell, C.R., D.J. Burdon, and M. Sherwood. 1983. "Impact of Agriculture on Groundwater in Ireland." *Environmental Geology* 5 (1): 39–48. doi:10.1007/BF02381300

Alula, Melisew Tadele, Zebasil Tassew Mengesha, and Ellen Mwenesongole. 2018. "Advances in Surface-Enhanced Raman Spectroscopy for Analysis of Pharmaceuticals: A Review." *Vibrational Spectroscopy* 98 (September): 50–63. doi:10.1016/j.vibspec.2018.06.013

Agrawal, Amit, Ralph A. Tripp, Larry J. Anderson, and Shuming Nie. 2005. "Real-Time Detection of Virus Particles and Viral Protein Expression with Two-Color Nanoparticle Probes." *Journal of Virology* 79 (13): 8625–28. doi:10.1128/jvi.79.13.8625-8628.2005.

Armenta, Sergio, and Miguel de la Guardia. 2014. "Vibrational Spectroscopy in Soil and Sediment Analysis." *Trends in Environmental Analytical Chemistry* 2 (May): 43–52. doi:10.1016/j.teac.2014.05.001

Bagherzadeh, Mojtaba, Zahra Hadizadeh, Zakyeh Akrami, and Zahra Shams. 2021. "Materials Science in Semiconductor Processing Electrochemical Detection of Cr (III) and Cr (VI) in Solution by Using ZrO_2 Modified Magnetic Nanoparticles by Redox Probes." *Materials Science in Semiconductor Processing* 131 (April): 105840. doi:10.1016/j.mssp.2021.105840

Balusamy, Brabu, Anitha Senthamizhan, and Tamer Uyar. 2020. "Functionalized Electrospun Nanofibers as a Versatile Platform for Colorimetric Detection of Heavy Metal Ions in Water: A Review." *Materials* 13: 1–44.

Bandodkar, Amay J., Itthipon Jeerapan, Jung Min You, Rogelio Nuñez-Flores, and Joseph Wang. 2016. "Highly Stretchable Fully-Printed CNT-Based Electrochemical Sensors and Biofuel Cells: Combining Intrinsic and Design-Induced Stretchability." *Nano Letters* 16 (1): 721–27. doi:10.1021/acs.nanolett.5b04549

Barceló, D., and M. Petrovic. 2007. "Challenges and Achievements of LC-MS in Environmental Analysis: 25 Years On." *TrAC Trends in Analytical Chemistry* 26: 2–11. https://doi.org/10.1016/j.trac.2006.11.006

Bentzen, Elizabeth L., Frances House, Thomas J. Utley, James E. Crowe, and David W. Wright. 2005. "Progression of Respiratory Syncytial Virus Infection Monitored by Fluorescent Quantum Dot Probes." *Nano Letters* 5 (4): 591–95. doi:10.1021/nl048073u

Carstea, Elfrida M., John Bridgeman, Andy Baker, and Darren M. Reynolds. 2016. "Fluorescence Spectroscopy for Wastewater Monitoring: A Review." *Water Research* 95 (May): 205–19. doi:10.1016/j.watres.2016.03.021

Chao, Cheng Han, Chung Shu Wu, Chung Chih Huang, Jie Chian Liang, Hsiao Ting Wang, Pin Ting Tang, Lih Yuan Lin, and Fu Hsiang Ko. 2012. "A Rapid and Portable Sensor Based on Protein-Modified Gold Nanoparticle Probes and Lateral Flow Assay for Naked Eye Detection of Mercury Ion." *Microelectronic Engineering* 97. Elsevier B.V.: 294–96. doi:10.1016/j.mee.2012.03.015

Domínguez, I., A. Garrido Frenich, and R. Romero-González. 2020. "Mass Spectrometry Approaches to Ensure Food Safety." *Analytical Methods* 12: 1148–62. https://doi.org/10.1039/C9AY02681A

Duan, Nuo, Shijia Wu, Changqing Zhu, Xiaoyuan Ma, Zhouping Wang, Ye Yu, and Yuan Jiang. 2012. "Dual-Color Upconversion Fluorescence and Aptamer-Functionalized Magnetic Nanoparticles-Based Bioassay for the Simultaneous Detection of Salmonella Typhimurium and Staphylococcus Aureus." *Analytica Chimica Acta* 723: 1–6. doi:10.1016/j.aca.2012.02.011

Edward, King. 1998. "Air Pollution." in : Jeffrey Peirce, J., Ruth, F. Weiner, Aarne Vesilind, P. (Eds.), *Environmental Pollution and Control*, 245–69. Elsevier. doi:10.1016/B978-075069899-3/50019-5

Esch, Elisabeth Von Der, Philipp Anger, Thomas Baumann, Martin Elsner, Reinhard Niessner, and Natalia P. Ivleva. 2019. "Raman Microspectroscopy as a Tool for Microplastic Particle Analysis." Unpublished. doi:10.13140/RG.2.2.15927.47524

Fa, Y., Y. Yu, F. Li, F. Du, X. Liang, and H. Liu. 2018. "Simultaneous Detection of Anions and Cations in Mineral Water by Two Dimensional Ion Chromatography." *Journal of Chromatography A* 1554: 123–27. https://doi.org/10.1016/j.chroma.2018.04.017

Gou, Pingping, Nadine D. Kraut, Ian Matthew Feigel, and Alexander Star. 2013. "Rigid versus Flexible Ligands on Carbon Nanotubes for the Enhanced Sensitivity of Cobalt Ions." *Macromolecules* 46 (4): 1376–83. doi:10.1021/ma400113m

Grieshaber, Dorothee, Robert MacKenzie, Janos Vörös, and Erik Reimhult. 2008. "Electrochemical Biosensors—Sensor Principles and Architectures." *Sensors* 8 (3): 1400–1458. doi:10.3390/s8031400

Hahn, Megan A., Peter C. Keng, and Todd D. Krauss. 2008. "Flow Cytometric Analysis to Detect Pathogens in Bacterial Cell Mixtures Using Semiconductor Quantum Dots." *Analytical Chemistry* 80 (3): 864–72. doi:10.1021/ac7018365

Halmann, M.M., 1996. *Photodegradation of Water Pollutants*. CRC Press, Boca Raton.

Halvorson, Rebecca A., and Peter J. Vikesland. 2010. "Surface-Enhanced Raman Spectroscopy (SERS) for Environmental Analyses." *Environmental Science & Technology* 44 (20): 7749–55. doi:10.1021/es101228z

Ikanovic, Milada, Walter E. Rudzinski, John G. Bruno, Amity Allman, Maria P. Carrillo, Sulatha Dwarakanath, Suneetha Bhahdigadi, Poornima Rao, Johnathan L. Kiel, and Carrie J. Andrews. 2007. "Fluorescence Assay Based on Aptamer-Quantum Dot Binding to Bacillus Thuringiensis Spores." *Journal of Fluorescence* 17 (2): 193–99. doi:10.1007/s10895-007-0158-4

Kaittanis, Charalambos, Saleh A. Naser, and J. Manuel Perez. 2007. "One-step, nanoparticle-mediated bacterial detection with magnetic relaxation". *Nano Letters*, 7(2), 380–383.

Khanmohammadi, Akbar, Arash Jalili, Ghazizadeh Pegah, Hashemi Abbas, and Afkhami Fabiana. 2020. "An Overview to Electrochemical Biosensors and Sensors for the Detection of Environmental Contaminants." *Journal of the Iranian Chemical Society*. Springer Berlin Heidelberg. doi:10.1007/s13738-020-01940-z

Koets, M., T. van der Wijk, J.T.W.M. van Eemeren, A. van Amerongen, and M.W.J. Prins. 2009. "Rapid DNA Multi-Analyte Immunoassay on a Magneto-Resistance Biosensor." *Biosensors and Bioelectronics* 24 (7): 1893–98. doi:10.1016/j.bios.2008.09.023

Konwarh, Rocktotpal, Ganesh Gollavelli, and Suresh Babu. 2020. *Designing of Novel Nanosensors for Environmental Aspects. Nanofabrication for Smart Nanosensor Applications*. Amsterdam, Netherlands: Elsevier Inc. doi:10.1016/B978-0-12-820702-4.00003-9

Lee, Seon-yeong, Feixiong Chen, and Tae Yoon Lee. 2021. "Tryptamine-Functionalized Magnetic Nanoparticles for Highly Sensitive Detection of Salmonella Typhimurium." *Analyst*, 146(8), 2559–2566. doi:10.1039/d0an02458a

Leo M. L. Nollet (Ed.). 2006. *Chromatographic Analysis of the Environment*, 3rd Edition (Hogeschool Gent, Ghent, Belgium), CRC Press/Taylor & Francis Group: Boca Raton, FL. 2006. ISBN 0-8247-2629-4. 8095–8095. https://doi.org/10.1021/ja069720h

Li, Xin Gui, Jia Li Zhang, and Mei Rong Huang. 2014. "Chemical Response of Nanocomposite Membranes of Electroactive Polydiaminonaphthalene Nanoparticles to Heavy Metal Ions." *Journal of Physical Chemistry C* 118 (22): 11990–99. doi:10.1021/jp5016005

Lian, Yan, Fengjiao He, Huan Wang, and Feifei Tong. 2015. "A New Aptamer/Graphene Interdigitated Gold Electrode Piezoelectric Sensor for Rapid and Specific Detection of Staphylococcus Aureus." *Biosensors and Bioelectronics* 65. Elsevier: 314–19. doi:10.1016/j.bios.2014.10.017

Liang, Jiajie, Hongwu Liu, Caifeng Lan, Qiangqiang Fu, Caihong Huang, Zhi Luo, Tianjiu Jiang, and Yong Tang. 2014. "Silver Nanoparticle Enhanced Raman Scattering-Based Lateral Flow Immunoassays for Ultra-Sensitive Detection of the Heavy Metal Chromium." *Nanotechnology* 25 (49). IOP Publishing. doi:10.1088/0957-4484/25/49/495501

López, Adaris M., Diane A. Blake, and Arben Merkoc. 2013. "Biosensors and Bioelectronics High Sensitive Gold-Nanoparticle Based Lateral Flow Immunodevice for Cd^{2+} Detection in Drinking Waters" 47: 190–98. doi:10.1016/j.bios.2013.02.031

López-García, Ignacio, Juan José Marín-Hernández, and Manuel Hernández-Córdoba. 2020. "Speciation of Chromium in Waters Using Dispersive Micro-Solid Phase Extraction with Magnetic Ferrite and Graphite Furnace Atomic Absorption Spectrometry." *Scientific Reports* 10 (1): 1–8. doi:10.1038/s41598-020-62212-7

Mannoor, Manu S., Hu Tao, Jefferson D. Clayton, Amartya Sengupta, David L. Kaplan, Rajesh R. Naik, Naveen Verma, Fiorenzo G. Omenetto, and Michael C. McAlpine. 2012. "Graphene-Based Wireless Bacteria Detection on Tooth Enamel." *Nature Communications* 3. Nature Publishing Group: 763–68. doi:10.1038/ncomms1767

Mesquita, Daniela P., Cristina Quintelas, A. Luís Amaral, and Eugénio C. Ferreira. 2017. "Monitoring Biological Wastewater Treatment Processes: Recent Advances in Spectroscopy Applications." *Reviews in Environmental Science and Bio/Technology* 16 (3): 395–424. doi:10.1007/s11157-017-9439-9

Mettakoonpitak, Jaruwan, John Volckens, and Charles S. Henry. 2020. "Janus Electrochemical Paper-Based Analytical Devices for Metals Detection in Aerosol Samples." *Analytical Chemistry* 92 (1): 1439–46. doi:10.1021/acs.analchem.9b04632

Mirnaghi, Fatemeh S., Natalie P. Pinchin, Zeyu Yang, Bruce P. Hollebone, Patrick Lambert, and Carl E. Brown. 2019. "Monitoring of Polycyclic Aromatic Hydrocarbon Contamination at Four Oil Spill Sites Using Fluorescence Spectroscopy Coupled with Parallel Factor-Principal Component Analysis." *Environmental Science: Processes & Impacts* 21 (3): 413–26. doi:10.1039/C8EM00493E

Mustafa, Fatima, Rabeay Y.A. Hassan, and Silvana Andreescu. 2017. "Multifunctional Nanotechnology-Enabled Sensors for Rapid Capture and Detection of Pathogens." *Sensors (Switzerland)* 17 (9). doi:10.3390/s17092121

Perez, J. Manuel, F. Joseph Simeone, Yoshinaga Saeki, Lee Josephson, and Ralph Weissleder. 2003. "Viral-Induced Self-Assembly of Magnetic Nanoparticles Allows the Detection of Viral Particles in Biological Media." *Journal of the American Chemical Society* 125 (34): 10192–93. doi:10.1021/ja036409g

Plohl, Olivija, Marjana Simonič, Ken Kolar, Sašo Gyergyek, and Lidija Fras Zemljič. 2021. Magnetic Nanostructures Functionalized with a Derived Lysine Coating Applied to Simultaneously Remove Heavy Metal Pollutants from Environmental Systems. *Science and Technology of Advance Materials* 22 (1): 55–71. doi: 10.1080/14686996.2020.1865114.

Rattanarat, Poomrat, Wijitar Dungchai, David Cate, John Volckens, Orawon Chailapakul, and Charles S. Henry. 2014. "Multilayer Paper-Based Device for Colorimetric and Electrochemical Quantification of Metals." *Analytical Chemistry* 86 (7): 3555–62. doi:10.1021/ac5000224

Reddy, L. Harivardhan, Jose L. Arias, Julien Nicolas, and Patrick Couvreur. (2012). Magnetic nanoparticles: design and characterization, toxicity and biocompatibility, pharmaceutical and biomedical applications. Chemical reviews, 112(11), 5818–5878.

Sadik, Omowunmi A., Austin O. Aluoch, and Ailing Zhou. 2009. "Status of Biomolecular Recognition Using Electrochemical Techniques." *Biosensors and Bioelectronics* 24 (9): 2749–65. doi:10.1016/j.bios.2008.10.003

Saleem, Fahad, Ahmed Khan, Nabisab Mujawar, and Yie Hua. 2021. "A Comprehensive Review on Magnetic Carbon Nanotubes and Carbon Nanotube-Based Buckypaper for Removal of Heavy Metals and Dyes." *Journal of Hazardous Materials* 413 (November 2020). Elsevier B.V.: 125375. doi:10.1016/j.jhazmat.2021.125375

Sanvicens, Nuria, Carme Pastells, Nuria Pascual, and M. Pilar Marco. 2009. "Nanoparticle-Based Biosensors for Detection of Pathogenic Bacteria." *TrAC—Trends in Analytical Chemistry* 28 (11). Elsevier Ltd: 1243–52. doi:10.1016/j.trac.2009.08.002

Sgobbi, Lívia F., S. Henry, and Wendell K.T. Coltro. 2021. "For Multiplexed Detection of Metals†." *Royal Society of Chemistry*. doi:10.1039/d1an00176k

Silva-Neto, H. A., Cardoso, T. M., McMahon, C. J., Sgobbi, L. F., Henry, C. S., & Coltro, W. K. (2021). Plug-and-play assembly of paper-based colorimetric and electrochemical devices for multiplexed detection of metals. *Analyst*, 146(11), 3463–3473.

Singh, Priyanka, Manish Kumar Singh, Younus Raza Beg, and Gokul Ram Nishad. 2019. "A Review on Spectroscopic Methods for Determination of Nitrite and Nitrate in Environmental Samples." *Talanta* 191 (January): 364–81. doi:10.1016/j.talanta.2018.08.028

Stortini, A.M., L.M. Moretto, A. Mardegan, M. Ongaro, and P. Ugo. 2015. "Arrays of Copper Nanowire Electrodes: Preparation, Characterization and Application as Nitrate Sensor." *Sensors and Actuators, B: Chemical* 207 (Part A). Elsevier B.V.: 186–92. doi:10.1016/j.snb.2014.09.109

Sun, Yu-ling, Chiu-mei Kuo, Chung-lun Lu, and Chih-sheng Lin. (2021). Review of recent advances in improved lateral flow immunoassay for the detection of pathogenic Escherichia coli O157: H7 in foods. *Journal of Food Safety*, 41(1), e12867.

Tian, Xike, Lin Liu, Yong Li, Chao Yang, Zhaoxin Zhou, Yulun Nie, and Yanxin Wang. 2018. "Nonenzymatic Electrochemical Sensor Based on CuO-TiO$_2$ for Sensitive and Selective Detection of Methyl Parathion Pesticide in Ground Water." *Sensors and Actuators, B: Chemical* 256: 135–42. doi:10.1016/j.snb.2017.10.066

Tian, Yu, Xueke Li, Rui Cai, Kang Yang, Zhenpeng Gao, and Yahong Yuan. 2021. "Aptamer Modified Magnetic Nanoparticles Coupled with Fluorescent Quantum Dots for Efficient Separation and Detection of Alicyclobacillus Acidoterrestris in Fruit Juices." *Food Control* 126 (November 2020): 108060. doi:10.1016/j.foodcont.2021.108060

Vikesland, Peter J., and Krista R. Wigginton. 2010. "Nanomaterial Enabled Biosensors for Pathogen Monitoring—A Review" *Environmental Science & Technology* 44 (10): 3656–69.

Willner, Marjorie R., and Peter J. Vikesland. 2018. "Nanomaterial Enabled Sensors for Environmental Contaminants Prof Ueli Aebi, Prof Peter Gehr." *Journal of Nanobiotechnology* 16 (1). BioMed Central: 1–16. doi:10.1186/s12951-018-0419-1

Wu, Jing, Miaosi Li, Hua Tang, Jielong Su, Minghui He, Guangxue Chen, Liyun Guan, and Junfei Tian. 2019. "Portable Paper Sensors for the Detection of Heavy Metals Based on Light Transmission-Improved Quantification of Colorimetric Assays." *Analyst* 144 (21): 6382–90. doi:10.1039/c9an01131e

Yang, Xinyan, Yujun Jiang, Yang Song, Xue Qin, Yuwei Ren, Yueming Zhao, Chaoxin Man, and Wei Zhang. 2021. "Point-of-Care and Visual Detection of *Salmonella* Spp. and *Cronobacter* Spp. by Multiplex Loop-Mediated Isothermal Amplification Label-Based Lateral Flow Dipstick in Powdered Infant Formula." *International Dairy Journal* 118. Elsevier Ltd: 105022. doi:10.1016/j.idairyj.2021.105022

Yoo, Seung Min, Do Kyun Kim, and Sang Yup Lee. 2015. "Aptamer-Functionalized Localized Surface Plasmon Resonance Sensor for the Multiplexed Detection of Different Bacterial Species." *Talanta* 132. Elsevier: 112–17. doi:10.1016/j.talanta.2014.09.003

Yu, Guangxia, Weixiang Wu, Qiang Zhao, Xiaoyun Wei, and Qing Lu. 2015. "Efficient Immobilization of Acetylcholinesterase onto Amino Functionalized Carbon Nanotubes for the Fabrication of High Sensitive Organophosphorus Pesticides Biosensors." *Biosensors and Bioelectronics* 68. Elsevier: 288–94. doi:10.1016/j.bios.2015.01.005

Zhang, Yuting, Yongkun Yang, Wanfu Ma, Jia Guo, Yao Lin, and Changchun Wang. 2013. "Uniform Magnetic Core/Shell Microspheres Functionalized with Ni^{2+}-Iminodiacetic Acid for One Step Purification and Immobilization of His-Tagged Enzymes." *ACS Applied Materials and Interfaces* 5 (7): 2626–33. doi:10.1021/am4006786

Zhong, Jessie S., and Lynn Ma. 2018. "Interview with Prof. Evan Martin Bloch—Public and Clinical Awareness of Babesiosis Is Critical." *Annals of Blood* 3 (February): 8–11. doi:10.21037/aob.2018.01.05

Zhu, Liang, Simon Ang, and Wen Tso Liu. 2004. "Quantum Dots as a Novel Immunofluorescent Detection System for Cryptosporidium Parvum and Giardia Lamblia." *Applied and Environmental Microbiology* 70 (1): 597–98. doi:10.1128/AEM.70.1.597-598.2004

7 Impact of Heavy Metals on Different Ecosystems

Onila Lugun, Richa Singh, Shambhavi Jha, and Alok Kumar Pandey

CONTENTS

7.1 Introduction 139
7.2 What Are HMs? 140
7.3 Major HMs in the Environment 140
    7.3.1 Cadmium (Cd) 140
    7.3.2 Chromium (Cr) 141
    7.3.3 Lead (Pb) 141
    7.3.4 Mercury (Hg) 142
    7.3.5 Arsenic (As) 142
7.4 HMs and Their Effects on Different Types of Ecosystem 143
    7.4.1 Forest Ecosystem 143
    7.4.2 Grassland Ecosystem 144
    7.4.3 Fresh Water Ecosystem 145
    7.4.4 Marine Ecosystem 146
    7.4.5 Mangrove Ecosystem 147
    7.4.6 Coral Reefs 148
7.5 Sources of HMs 150
7.6 Responsible Factors 152
    7.6.1 Bio-availability 152
    7.6.2 Uptake 152
    7.6.3 Bioaccumulation 153
    7.6.4 Biomagnification 154
    7.6.5 Tropic Transfer 154
7.7 Conclusion 155
Acknowledgments 155
References 155

7.1 INTRODUCTION

Metals and metalloids having density > 5 g/cm^3 are referred to as heavy metals (HMs) and have characteristic properties in nature and environmental behaviour (Oves et al., 2012). HM pollution is a serious problem in the environment due to their ubiquitous occurrence and toxicity. HM contamination poses a serious threat to the environment because these metals are persistent and present in the environment for a long period of time and can enter into the food chain and cause different health problems (Ali et al., 2019). These HMs are released into the environment through various processes such as natural or anthropogenic sources like weathering of rocks, volcanic eruption, industrial emission, pesticides, and fertilizers that contaminate the natural water bodies, sediments, soils, air, and living organisms (Yadav et al., 2017). HMs are natural constituents of the environment and are present in very low concentration but due to recent human activities HM levels in the environment have increased dramatically. Some HMs are beneficial at one concentration and essential for life forms while at other concentrations are highly toxic. Once excess of these HMs enters into the pedosphere, it can spread to different ecosystems and can pose health risks to human beings. This chapter describes different HMs existing in diverse ecosystems along with their potential sources of origin. It also discusses HM pollution level and associated risks to ecosystems. How unintended release of these HMs is impacting different types of ecosystem components is also outlined here (Figure 7.1).

7.2 WHAT ARE HMs?

HMs are high-density metals that are toxic even at relatively low concentrations. Some HMs act as essential elements for various life forms, while others are nonessential trace elements and are highly toxic in nature. HMs are found naturally in the earth's surface and some remain in the environment for a long period of time. Some due to their bioaccumulative nature neither break down in the environment nor are easily metabolized by organisms and accumulate in the ecological food chain. Once taken up in the food chain they are transferred to the upper tropic level and accumulate through consumption at consumer level. HMs are highly toxic and dangerous at low concentrations, and they cause serious health problems to humans and ecosystems (Oyetibo et al., 2017; Wuana et al., 2011). HMs and metalloids accumulate slowly in plants, animals, and humans through air, water, and the food chain (Briffa et al., 2020). Examples of HMs include lead (Pb), arsenic (As), mercury (Hg), cadmium (Cd), zinc (Zn), silver (Ag), copper (Cu), iron (Fe), chromium (Cr), nickel (Ni), palladium (Pd), and platinum (Pt). These HMs are released into the environment from both natural and anthropogenic sources, but due to industrialization and urbanization anthropogenic contributions of HMs are increasing in the biosphere (Table 7.1).

7.3 MAJOR HMs IN THE ENVIRONMENT

7.3.1 Cadmium (Cd)

Cd, a natural ore, is hazardous and discharged into the environment due to recent agricultural and industrial processes. Cd is present in trace amounts in nature in the air, soil, and water; however, because of its bioaccumulative nature it accumulates in the ecological food chain (Dziubanek et al., 2015; Järup and Åkesson, 2009; Queiroz and Waissmann, 2006). There are several ways through which Cd can enter the environment, among which hydrospheric Cd is a major contributor. In nature aqueous Cd comes from the dissolution of solid Cd (minerals), the deposition of atmospheric Cd, and bedrock weathering in watershed, soils, and sediments (Shine et al., 2002). Cd is mostly used as a stabilizer in different products like colour pigments, alloys, and in Polyvinyl chloride (PVC)-related products (Järup et al., n.d.). Other than these Cd can enter into the system through food and tobacco as plants like vegetables, grains, and tobacco take Cd directly from soil (Alloway et al., 1990). Due to large-scale commercial product production and anthropogenic activities, Cd has increased in the environment and workplaces (Järup and Åkesson, 2009). It was observed that Cd enters into aquatic macro invertebrates and fishes through biofilms from water and sediments. Once Cd enters an organism it bioaccumulates in various tissues.

7.3.2 Chromium (Cr)

Cr is a naturally occurring steel-gray element found in two valence states in the earth's crust: trivalent Cr(III) and hexavalent Cr(VI). The trivalent Cr(III) and hexavalent Cr(VI) species are the most stable forms of Cr; Cr(III) has a lower toxicity than Cr(VI). Cr is introduced by both natural and anthropogenic mechanisms into the environment. According to valence states Cr III levels rise as a result of leather, textile, and steel manufacturing, whereas Cr VI levels rise as a result of industrial applications such as electro painting and chemical manufacturing (Bakshi and Panigrahi, 2018). From soil Cr can be introduced into the surface water system (Ranieri et al., 2016; 2013). Huge amounts of Cr are released into streams as outflows from the metal mining and electroplating industries, making Cr in marine ecosystems a major concern. Through sedimentation of Cr-containing rocks, direct release from industrial applications and soil leaching Cr reaches natural streams. Cr levels in soil can rise largely due to various anthropogenic depositions, such as air deposition or the dumping of Cr-bearing liquids and solid wastes such as Cr wastes, ferrochromium slag, or Cr painting (Oliveira, 2012). Reports suggest people develop health issues when consuming drinking water contaminated with Cr HM. They showed adverse effects on physiology of kidney, digestive system, circulatory system, neurological system, and other organs and systems throughout the body (Sankhla et al., 2019).

7.3.3 Lead (Pb)

Among all the HMs reported Pb is one of the most highly toxic HMs in the environment. It occurs naturally in the earth's crust in the form of bluish grey metal with unique properties like softness, high malleability, ductility, and low melting point. Due to its persistent nature it poses a major threat to human health in several aspects. Pb is mostly found in air, water, and soil, and is predominantly derived from a variety of products such as leaded gasoline, paints, ceramics, solders, water pipes, hair colour, cosmetics, aeroplanes, farm equipment, x-ray machine shielding, etc. (Boskabady et al., 2018). These anthropogenic sources impart effects on the terrestrial, aquatic organisms and human health. In aquatic ecosystems fish are used as a bioindicator for the aquatic ecosystem as they show detrimental effects in the presence of Pb. Aquatic bodies are mostly contaminated by Pb metals, which do not degrade biologically and disrupt the ecological balance by interfering with various physiological, biochemical, and cellular processes (Paul et al., 2019). Due to its persistent nature Pb accumulates in different organisms both directly and indirectly from water and enters the food chain.

7.3.4 MERCURY (HG)

Hg is one of the deadliest HMs on the globe having atomic weight 200.59 with an average plethora of only 0.08 PPM (parts per million). It is a non-transition metal that is extremely rare in the earth's crust. After reaching surface waters or soil microorganisms, Hg is converted to Methyl mercury (MeHg), a chemical that is quickly absorbed. In aquatic food systems, it accumulates and multiplies to a fatal level. Hg^0 is oxidised in the air to its inorganic forms (Hg^+ and Hg^{2+}), which are then deposited in soil or in the water of rivers, lakes, and seas during precipitation (Alex, 2018). Its most common type of organic Hg found in ecosystems is MeHg. MeHg can easily be transferred into aquatic habitats via water. It is lipid soluble because of its limited water-soluble nature. Smaller species can easily pierce MeHg, which goes up in the food chain and has a proclivity for bioaccumulation in fish. Fish poisoning appears to be the most common cause of MeHg poisoning in humans. Hg-contaminated soil or water are the main sources of contamination through which Hg enter the food chain through plants and livestock. Once Hg enters the food chain, it bioaccumulates, posing a health concern to humans. The specific mechanism(s) by which Hg enters different food chains is unknown, and it is likely to differ amongst ecosystems (Hsu-Kim et al., 2013). Soil is another source of Hg, and it plays a significant part in biogeochemical Hg circulation by accumulating the element and serving as a source for other environmental components. In this process vegetation on the ground surface affects environmental factors by reducing solar radiation, temperature, and wind velocity and acts as a surface for Hg uptake. The main sources of Hg in leaves are from polluted air with Hg^0 rather than from soil contamination. Hg emission in the atmosphere is still one of the most serious environmental issues in the modern world (Gworek et al., 2020).

7.3.5 ARSENIC (AS)

As is a naturally occurring element of the earth's crust and is widely spread in the environment. It is found in air, soil, water, plants, animals, and rocks and is predominantly released into the environment through weathering, rock erosion, volcanic eruption, and forest fires. As is a naturally occurring metalloid and one of the few metals that can be metabolized inside the human body. Around 200 million people are exposed to potentially dangerous levels of As worldwide, and significant resources have been spent on research and potential mitigation strategies (Chen and Costa, 2021). Most environmental As contamination originates from anthropogenic activities, and such contamination leads to life-threatening complications to millions of people. Another source of As in soil is due to continuous use of the arsenical pesticides (Islam et al., 2014). As in ground water, As affects the agricultural products irrigated by contaminated water. As exists in arsenite As(III) or arsenate As(V) but

TABLE 7.1
Sources of Different HMs in Ecosystem

Ni	Forest fire, volcano activities, weathering of rocks, automobile batteries, surgical instruments, industrial effluents, landfills, etc.
Cd	Burning of fossil fuels such as coal and oil, incineration of municipal waste, metal smelting and refining, industrial activities
Cu	Electrical manufacture, domestic and agricultural use of pesticide and fungicide, leather processing, volcanic eruption, forest fire, copper mining activities
Pb	Lead based paint, disposal of lead acid batteries, smelting,mining, urban soil waste, processing of ore
Cr	Leather tanning, wood polishing, volcanic dust, waste treatment, cement manufacturing
Hg	Deposition from atmosphere, coal plant, cement plant, urban sewage system,waste incineration and cremation, small-scale gold mining
As	Weathering of rock and mineral, paint, tannery, irrigation water, cattle waste, industrial emissions, fertilizers, pesticides
Zn	Electrical waste, paints, pigments, mining, refining, smelting.

As(III) is the most abundant form in ground water due to underground reducing conditions (Kesici, 2016). The consumption of inorganic As-contaminated foods could increase the worldwide cancer burden (Oberoi et al., 2014). As contamination has been declared as the twentieth-to-twenty-first-century tragedy by researchers and officials (Alka et al., 2021).

7.4 HMS AND THEIR EFFECTS ON DIFFERENT TYPES OF ECOSYSTEM

7.4.1 Forest Ecosystem

The forest ecosystem is a diverse ecosystem comprising a large diversity of flora and fauna. Due to rapid industrialization in the last decade, components of this ecosystem are primarily affected. When forest tree rings were analysed for HMs, Cu, Zn, and Pb were found in an elevated amount, indicating recent increased anthropogenic activities in the past decade (Xu et al., 2017). Similarly, the moso bamboo forest near the Pb-Zn mining area exhibited high HM concentration in the forest soil and plant shoots (Yan et al., 2015). Pb concentration exceeded the standard limits in the bamboo shoots and exhibited a spatial correlation with the soil HM concentration. With increasing age, HM depositions in various plant parts are found to be increased. When pine plant needles were explored for their HM content in the pine forest near smelter power plants of southern Poland, HM content was increased with the age of needles (Kandziora-Ciupa et al., 2016). Due to high Cd deposition the Beech forests of Bulgaria are at risk of damages by Cd pollution (Damyanova, 2010). The efficiency of pollutant transfer from soil to different plant parts is a characteristic of that plant and varies among different plant species. Plants developing on waste heaps left by Zn-Pb ore mining in the Beech Forest region showed active translocation of these HM pollutants from soil to plant organs (roots and shoots) and their species dependence (Stefanowicz et al., 2016).

Microbial populations in forest ecosystems have shown a negative impact of HM contamination on the microbial dynamics. With increased Zn content above 220 mg/kg and Pb above 100 mg/kg in the soil, microbial dehydrogenase and urease activities are noted to be reduced (Pająk et al., 2016). The diversity and structure of the bacterial community were negatively affected by high concentrations of HM pollution in the forest soils (Chodak et al., 2013). Boreal forest moss-associated cyanobacteria displayed a reduced propensity of biological nitrogen fixation potential when moved closer towards the roadside, indicative of the detrimental effect of HM contamination (Scott et al., 2018). Edible mushrooms collected from Yunnan province in China had elevated concentrations of toxic HMs like As, Pb, and Cd, values higher than China's national standard values in most cases (Liu et al., 2015).

In forest ecosystems, HMs interact with different plant systems and alter the physiochemical and biochemical response. Oxidative stress, reactive oxygen species generation, genotoxicity induction, and interference with various signalling pathways has also been reported for HM plant interaction (Srivastava et al., 2017). In the tropical forests of West Bengal, dust fall on leaves and foliar transfer of HMs resulted in increased peroxidase and catalase activities and the generation of a high level of reactive oxygen species (Karmakar and Padhy, 2019). Similarly, glutathione levels in *Pinussylvestris L.* plants tend to decrease when the Mn level rises (Kandziora-Ciupa et al., 2016). With the increase in Cd, Pb, and Zn concentration *Vvitis-idaea* leaves showed a growing trend in ascorbic acid content and superoxide dismutase activity, whereas increased Mn accumulation decreased antioxidant response (Kandziora-Ciupa et al., 2017). Soil quality degradation due to HM contamination profoundly affects woody plant growth patterns in the forest ecosystem. Reports suggest it affects forest community structure by lowering the species richness and by reducing tree generation to change future ecosystem productivity (Trammell et al., 2011).

7.4.2 Grassland Ecosystem

Grassland ecosystems comprise vegetation dominated by grasses and other herbaceous (non-woody) plants with few or no trees in the area. In one of the grassland ecosystems of Northeast China, Cd and Hg are amongst the potential factors responsible for posing a risk due to soil contamination (Chai et al., 2015). Similarly, Cd, Cu, and Fe are reported to be the most significant pollutants in the grassland ecosystem when soils and plants of former polymetallic ore mining and smelting area was examined (Demková et al., 2017). Soil characteristics are considered as a major factor responsible for the accumulation, distribution, and mobility of HMs in the grassland ecosystem (Liu et al., 2016). Pb and Hg derived from the anthropogenic origin are accumulated in the soils of Tibetan grassland. Hg showed a positive correlation with the average annual rainfall amount and may be contributed by its precipitation into the soil. In a similar study transfer factor value was used to measure HM uptake from soil to plants (Ratko et al., 2013). As compared to Zn, Cu, and Cd, Pb showed a comparatively lower transfer factor and thus was present in nonmobile fractions, which facilitates easy availability for plants and easy uptake.

Similarly, in three species-rich calcareous grassland ecosystems, HM contamination was increased as we moved towards high traffic-density regions (Lee et al.,

2012). Roads and exhaust emissions increase Cu, Pb, and Zn concentration, which ultimately declines plant species richness and abundance of forb and moss species. In a study, when two grassland ecosystems of north Belgium were evaluated for the effect of HM contamination on microbial activity, the diversity index decreased dramatically (Boshoff et al., 2014). The feeding activity of macrofauna, mesofauna, and microarthropods was reduced from 90 to 10% when moving plots near to source pollution sites. Feeding activities and functional indices are negatively impacted by As, Cu, and Pb and are correlated significantly. In Romanian grasslands, numeric densities of soil mite communities are decreased dramatically when HM pollutants are increased, indicative of damaged grasslands (Manu et al., 2017b). Distinguishable soil mite communities were observed concerning HM contamination in the grassland ecosystems of Transylvania, indicating species sensitivity towards different HM contaminations (Manu et al., 2017a). HMs like Zn, Pb, and Cd affect the soil microbial community by reducing microbial biomass, functional diversity, bioactivity diversity, and increasing metabolic quotient (Stefanowicz et al., 2010; Xie et al., 2016). Soil actinobacterial community composition were found to be affected by HMs like Cu and Pb in the grassland soils along a gradient of mixture of HMs (Větrovský and Baldrian, 2015).

7.4.3 Fresh Water Ecosystem

Analysing the water and sediments is the most direct approach for detection of environmental HM pollution. Thus various studies are concerned with the direct measurement of HMs in water and sediments of freshwater ecosystems to assess and monitor environmental pollution. China's primary freshwater lake sediments presented various HM contaminations and posed nationwide potential ecological risk (Cheng et al., 2015). Sediment dating records from two lakes in China showed an increase in total As level to 21.4-fold relative to pre-1950 (Chen et al., 2015). In the sediments and benthic fauna, when explored for HMs, Pb was found at a level exceeding the threshold effective level (Enuneku et al., 2018). Cu, Cd, and Pb were observed in benthic fauna and showed a significant health risk with non-carcinogenic effects. When muscle, gill, and intestines were examined for HMs in two fish species of the Mbaa river in southeastern Nigera, appreciable bioaccumulation was observed in the fish tissues (Ajima et al., 2015). The bioconcentration factor in fish has exceeded the limit recommended by the WHO and FEPA, hinting that drinking water and fish are not safe for human consumption. Similarly, water and African catfish samples collected near vehicle mechanical industrial area were heavily contaminated with Cr, Fe, Zn, and Pb than other sampling areas (Golam Mortuza and Al-Misned, 2015). In fish livers, HMs like Cd and Cu are mostly accumulated due to metallothionin protein, whereas in gills, the main HMs (Mn, Pb) uptake pathway was directly through water (Jia et al., 2017). Cd and other HMs are bioaccumulated from sediments to macrobenthos in the Yellow river delta wetland, creating ecological risk (Li et al., 2016). In some cases, natural phenomena like mineralogy and lithogenic sources are also responsible for substantial HM pollution apart from anthropogenic sources. In some lentic compartments, HMs like Zn and Fe are elevated and are closely associated with natural incidence (Meena et al., 2017).

Though direct detection of HMs it is possible to determine the status of environment pollution, although it does not provide information about these metal species' interaction and toxicity status towards different biota present in the ecosystem. Studies concerning the different levels of biological organization present in freshwater ecosystems and their interaction with different HM pollution are on the rise as in current times there is more industrial actviity. A relatively high As level in a freshwater ecosystem documented a >10-fold loss of crustacean zooplankton and the development of highly metal tolerant alga up to > 5-fold, cascading through events causing potential catastrophic consequences (Chen et al., 2015).

When edible freshwater cyprinids from Punjab, India, were examined for HM contamination and their responses on various fish organs, various toxic effects were observed in liver, kidney, and muscles (Kaur et al., 2018). The histopathological observations indicated various hepatic lesions like cytoplasmic degradation, necrosis, melano-macrophage centres, and infiltration of leukocytes, pyknosis, and nuclear degradation. In muscle tissue, shortening and elongation of muscle bundles and variation in kidney architecture were observed in response to high concentrations of HMs like As, Cr, Cd, Mn, and Pb, which were detected above the WHO/FAO recommended permissible limit. Similarly in response to metal toxicity, fish gills showed an association between metal concentration and gill injury (Fonseca et al., 2017).

All HMs examined are taken up readily by the filamentous green algae *D. swartzii*, a biomass producer in acid peat bogs, and sequestered in the cell walls and intracellular compartments (Andosch et al., 2015). This resulted in effects on cell ultrastructure, disturbed photosynthesis activity, and biomass production resulting in threat to the ecosystem. Significant microbial populations performing crucial ecological functions are harmed by HM pollutants generated from mine waste (Jackson et al., 2015). Reduction in overall productivity, biomass, biodiversity of microflora, and development of tolerant species are observed in response to HM contamination. As a toxicological impact, HMs are reported to alter hematological and glycogen status in fishes (Javed and Usmani, 2014).

7.4.4 Marine Ecosystem

Sediments from the South East Coast of Tamil Nadu displayed HM contamination with major pollutants containing Ti, V, Cr, Mn, Ca, and Pb (Devanesan et al., 2017). Coastal water, sediments, and soft tissues of *S cucullata* exhibited high Zn, Cu, and Mn accumulation in the Gulf of Chabahar, Oman Sea (Bazzi, 2014). It was observed that different marine lifeforms show different metal accumulation behaviors. Results from a recent work indicate that mollusks comprising seasnail, benthic bivalves, and oysters accumulated a relatively higher amount of HMs than other groups. They were followed by mollusks, crustaceans (crab and shrimp) accumulated relatively fewer HMs in their total biomass. Fishes comparatively accumulate low concentrations of all HMs in the marine ecosystem (Q. Liu et al., 2019). When five microalgae from the Roumanian Black Sea were compared for their bioaccumulation potential, a considerable variation was observed within the species (Trifan et al., 2015). Red algae show the highest capacity to accumulate and concentrate HMs from surrounding water and sediments. HM content in bivalves, starfish, gammaridiean amphipods, and heart urchin from waste disposal tip sites at Brown Bay were comparatively higher than those of non-impacted sites (Chu et al., 2019). HMs in different species are mainly due to contaminated nearshore water containing contaminants like Pb, Cu, and Zn. Among invertebrates, bivalves are severely affected due to HM contamination, whereas metal transfer through food chain is evidenced in starfish. Similarly, penguins from the Antarctic ecosystem showed high Cd levels, indicating high exposure to this HM (Jerez et al., 2013). Apart from this, several other biotic communities like soft-shelled clam, fishes, mosses, lichens, and cryptograms also suffered from high HM exposure (Chu et al., 2019). There is also evidence of tropic transfer and bioaccumulation through the food web in the Antarctic ecosystem. Seagrasses and wetland macrophytes showed bioaccumulation and translocation from sediments in a species-specific manner (Bonanno et al., 2017). However, high levels of HMs are found to be more accumulated in the below-ground organs than the other organs. HMs, after translocation, compartmentalize in different organs of seagrasses.

Benthic diatom community composition is strongly and negatively associated with HM concentrations (Chu et al., 2019). Reduction in cellular lipid content was a more prominent indicator of Cd, Ni, and Pb toxicity index than growth inhibition in the case of Antarctic marine microalgae (Koppel et al., 2017). In a similar set of experiments, electron flow in the reaction centers of chlorophyll a was found to be affected by the HM exposure in microalgae (Gissi et al., 2015). Not only the electron transfer increased but granularity showed distinct cellular changes in various organs like lysosomes, starch granules, vacuoles, polyphosphate bodies, and nucleus of marine Antarctic microalgae. The accumulation of Cd was reported mainly in the cell walls and induced decreased intracellular levels of other metals like Ca, and inhibition in the growth, photosynthesis, and ultra-structural damage to the algal species (Chu et al., 2019). As a whole organism response, HMs affect growth, morphology, reproduction, recruitment, and behavior in the marine ecosystem (Langston, 2018).

7.4.5 Mangrove Ecosystem

Mangrove ecosystems are mostly confined to the tropical and subtropical regions of earth, where they transit between marine and terrestrial environment. When explored in different urban functional zoning, urban mangroves showed highest the HM contamination at a proximity near point source discharges of rivers (Chai et al., 2019). Shajing mangroves are predicted to be at higher ecological risk than in industrial districts with metal product industries. Zn, Ni, and Cr were detected at elevated amounts in the fractions collected with anthropogenic sources of origin. In the temperate mangrove ecosystem of New Zealand, when sedimentary column was compared between inside and outside mangrove strands, HM build-up was observed inside mangroves (Bastakoti et al., 2018). Road run-off into the sedimentary column of mangrove shows high Pb content, indicating the anthropogenic origin of HMs. Not only in the sediments but plant tissues are also reported to accumulate Pb due to their high translocation and bioaccumulation potentials in rhizophores of mangrove plants (Ganeshkumar et al., 2019). This reported concentration was several folds higher than the WHO recommended level for plant protection, indicating potential risk. Plant species in the mangrove ecosystem plays an essential role in altering soil quality as it was observed that different plant species could sequester HMs to different extent (Chai et al., 2018). Apart from sediments and rhizophores other plant parts like branch, bark, root, and leaves are also found to accumulate high HM content. *Avicenia marina* collected from the Muthupet mangrove forest were found to accumulate Cd and Pb in various plant parts and showed high bioaccumulation and translocation potential (Arumugam et al., 2018). In a similar set of research, very high Cd was bioaccumulated above the threshold effect level and posed a risk towards biota present in the sediments of the mangrove ecosystem along the Red Sea coast of Saudi Arabia (Alzahrani et al., 2018). At the Sundarbans mangrove forest, some macrobenthic fauna reported having high Zn and Fe in their dry bodyweight (Ahmed et al., 2011). In the edible portion of macrobenthos, Pb was accumulated over the international permissible limits certified by the WHO. Fe and Cd were found to be highly bioaccumulated and magnified in the benthic fauna. As mangrove plants are well known for their metal bioaccumulation potential, metal recycling into the sediments was observed when leaves fall during senescence (Almahasheer et al., 2018).

The mangrove soils of the Mahanadi delta HMs showed a negative impact on the microbial load (Behera et al., 2013). In response to high trace metal contamination, unusual breeding mechanisms have been observed in the organisms present in mangrove ecosystem (Bayen, 2012). When examined under controlled environmental conditions in the presence of metals mangrove plants exhibited adverse outcomes like reduced photosynthesis, reduced growth and biomass, and sometimes mortalities (Bayen, 2012). In-field studies are complex as it is impossible to control dose exposure to the mangrove-associated biota. Due to these limitations of dose-response relationship and fixed ecological endpoints, studies are mainly concerned with the potential risk assessment in the presence of different HMs. In the case of microorganisms, diversity and structure are known to be affected by metal contaminations. Under a controlled environment, mangrove fauna exposed to HM pollutants imparted enzyme activity inhibition, behavioural changes, and sometimes mortality (Bayen, 2012). Low water quality due to pollutants has also been associated with declining population density and reduced diversity indices (Bayen, 2012).

7.4.6 CORAL REEFS

Due to their pristine environment, tropical seas and oceans are suitable for growing the most productive and diverse coral ecosystems. Coral reefs provide food and shelter for various aquatic lifeforms and play a significant role in tourism and coastline protection from floods and storms (Beck et al., 2018). Coral being sedentary best represents the study site environment. Due to recent human developmental activities near coastal areas like costal mining, industrial and domestic effluent discharge, these species are proceeding towards an unintended decline or extinction. HMs, due to their prevalence, toxicity, non-biodegradable properties, long half-lives, and accumulative nature, pose a severe threat to seawater quality and the health of aquatic organisms. Therefore, studies concerning the detection and effect of HM levels in the coral cytoskeletons are essential for their conservation.

Surface sediments of the coral ecosystem also constitute a crucial role as a significant part of this ecosystem. These surface sediments directly indicate anthropogenic activities and can monitor environmental changes (Li et al., 2020). These sediment depositions can impact the coral ecosystem by smothering of corals. It has been reported that polluted sediments could serve as a primary pollution source in the water ecosystem (R. Wang et al., 2020; X. Wang et al., 2020). Sediment has been identified as the main reservoir of HM pollution, and their monitoring and regulation can directly determine risk towards the coral ecosystem. A renowned tourist destination located at the Gulf of Aqaba, Red Sea in the Western Indo-Pacific region, overflowing with the magnificent coral reef ecosystem, was explored for its susceptibility to metal pollution (Al-Rousan et al., 2012). Five living coral species and their fossil counterparts collected from the Jordanian Coast showed a heavy load of HM contamination. The area hosting intense development and human costal activities showed a majority of HM contamination in coral skeletons collected. Cd and Cu concentrations are elevated in the sites and collected species. Lead was found at the highest concentration in all the coral species reported. Among all coral species, *Porites* species showed accumulation of highest metal concentration. Source apportionment study suggests that the primary sources of Cu in corals collected from Dazhou Island, Hainan, China is from non-crustal material generated from anthropogenic sources (Yang et al., 2019).

Satellite imaging shows that the Magdalena River, which is mainly responsible for the continental flux into the Caribbean Sea and the Rosario Island National Park, is putting strains on the coral reef ecosystem by dispensing HM pollutants into the coral zone (Restrepo et al., 2016). Further geochemical data also suggests that suspended sediment load over the coral reefs has increased during the last decade due to increased industrial activities and reduced coral cover and declined water quality. In the Persian Gulf region, the distribution of HMs across different coral islands' coastal reef sediments suggests that HMs, especially Cd and Hg, are on the rise due to industrial development and oil refineries (Ranjbar Jafarabadi et al., 2017).

In the Caribbean region, when coral tissues were examined for the HM content, Cu and Zn were found abundantly in the tissues (Berry et al., 2013). Some coral species like *P. furcate* were found to bioaccumulate high Cu and Hg content in their tissues. This suggests that the nearshore coral reefs are more at risk from Cu, Zn, and Hg than the off-shore reefs. These four out of five current sites exceeded the "effects range low" (ERL) guideline value for Cu in sediments, suggesting possible health risks towards existing biota inhabiting these reef ecosystems. Further lower hard coral abundance was also found to be affected by the change in environmental conditions. Coral reef abundance and diversity were declining towards proximity to ports, indicating the degraded environmental condition due to human activities.

For instance, due to recent anthropogenic activities, metal contamination was observed gradually when moved from the bottom of the sediment core upwards. When the ambient sediments from 10 coral reef ecosystems in the Persian Gulf of Iran were analyzed for metal pollution, highly exchangeable and mobile fractions of metal pollutants were observed indicative of high bioavailability (Ranjbar Jafarabadi et al., 2020). Locations with high anthropogenic activities showed considerable HM contaminations (As, Cd, and Hg) and exhibited the highest potential ecological risk.

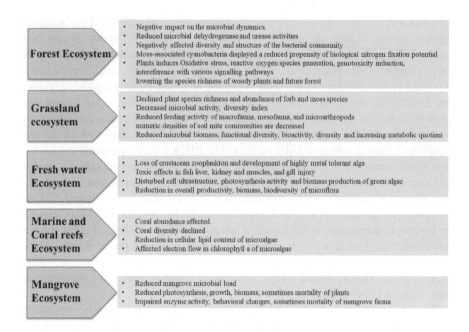

FIGURE 7.1 HMs and their adverse effects on different biotic components of ecosystem.

Areas with low anthropogenic activities are mainly contaminated with river runoff and urban emissions. Not only HMs but these coral skeletons are also explored for trace metal contaminations. The impact of trace metals and HM contamination on seven Pleistocene and recent coral species of Egyptian Red Sea coast coral skeletons were analyzed for Co, Cr, Cu, Fe, Mn, Ni, Pb, and Zn content (El-Sorogy et al., 2012). Except for Mn and Ni, all metals showed increased concentrations in recent coral skeleton species. Variations were observed in the bioaccumulation of metal concentrations among different coral species depending on the microstructure and microarchitecture.

7.5 SOURCES OF HMs

There are two possible ways known through which HMs may be introduced into the agro ecosystem or bioecosystem by natural processes or anthropogenic activities (Figure 7.2). Various studies have indicated that anthropogenic activities lead to higher HMs in the ecosystem than natural processes (Dixit et al., 2015). The main way of origination of HMs are from igneous rock, as 95% of the earth's crust is made up of igneous rock, and 5% is sedimentary rock (Sarwar et al., 2017). Generally, it is known that igneous rock is comprised of HMs such as Ni, Cu, Cd, and Co, whereas laminated sedimentary rock, which consists of clay-sized particles known as shale, consists of HMs such as Pb, Cu, Zn, Mn, and Cd in higher amounts. Naturally, rocks are the core source of HMs, so it is clear that several natural processes such as weathering, sedimentation, volcanic processes, erosion, leaching, and biogenic processes lead to the addition of HMs in ecosystems (Muradoglu et al., 2015).

Impact of Heavy Metals on Different Ecosystems

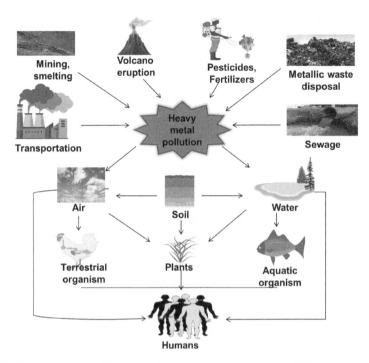

FIGURE 7.2 Sources of HMs and their routes of exposure to different environmental ecosystems.

Source: Created by biorender.com.

TABLE 7.2
Different Sources of HMs

Natural Sources of HMs	Anthropogenic Sources of HMs
Parent Material	Mining
Volcanic eruption	Smelting
Weathering of metal enriched rock	Fossil Fuel Combustion
Forest fire	Waste disposal
Aerosol Formation over sea	Warfare
Soil Erosion	Electronics
Leaching	Organic and inorganic Fertilizer
Surface Wind	Sewage sludge Pollution
Biogenic	Energy Intensive Industries
	Modern intensive Agriculture

Today, this natural cycle of addition of HMs to ecosystems is gradually getting disturbed due to several anthropogenic processes such as industrial activities, advancement in agricultural sectors, urbanization, and mining, which have led to the addition of one or more HMs in soil ecosystems (Li et al., 2019). Based on the sources, HMs can be divided into natural and anthropogenic, as summarized in Table 7.2.

7.6 RESPONSIBLE FACTORS

7.6.1 Bio-availability

With the alarming rate of increasing urbanization and industrialization, HM deposition into the soil layer and water bodies is of great concern to the environment, which eventually becomes an essential aspect for human and animal health. For example, industrial wastewater is used for irrigation purposes in China, which eventually contaminates millions of tons of grain in vast land areas (Oves et al., 2012). The bioavailability of HMs to the soil and water bodies is the leading cause of HMs entering into the ecosystem's natural food chain of lower-level and higher-level organisms (Balabanova et al., 2015). Many authors have reported that different plant parts act differently for absorption of HMs (Balabanova et al., 2015). According to Danielly et al. bioavailability of metals to freshwater ecosystem depends on the speciation of metals (i.e., the existence of any element in different forms). Several factors influence speciation in the aquatic ecosystem such as hardness of water, pH of water, ionic strength of medium, amount of organic matter, and many other factors. When these factors combine, it provides a favorable condition for forming toxic metal species, which becomes biologically available to the ecosystem (de Paiva Magalhães et al., 2015). For example, arsenite, the As(III) form of As, is less mobile but has been recognized as more toxic than the arsenate, As(V) form (Yadav et al., 2019). For a typical pH range in any soil type, stability of metals can be outlined as Ni, Zn, and Cd being less stable and Pb, Cu, and Cr metals being highly stable (Kim et al., 2015). There is one term known as environmental bioavailability of HMs from pore water to the plant roots, which depends on the rate of transfer of HM ions and is controlled by physiological factors, uptake, and transfer specificity of plant and HMs (Kim et al., 2015). In soil with slightly alkaline pH, HM ions are absorbed explicitly in hydroxyl sites of oxides and clay minerals (Shahid et al., 2017). In one of the studies on aquatic organisms a higher level of Zn and Cu was found from the Luoyang area of China in 2010 (Li et al., 2018). According to Moore et. al., methylmercury is more readily and highly available to plants than inorganic Hg, and humans readily consume it as it accumulates into the aquatic food chain very quickly (Moore et al., 2014).

7.6.2 Uptake

A process in which HM contamination is migrated or translocated from the outer environment like air, water, and supplemented nutrients by the plant's rhizosphere to the plant parts like stem, leaves, and fruits is called uptake of HMs. HMs may be present in the groundwater, and migration through plant's roots is a common way to uptake HMs from soil. Pesticides and other nutrient supplements provided to the plants also contain HMs and are a good source of HMs. Many more anthropogenic and natural processes make such HMs available to the plant or any other living organisms like bacteria, fungi, and crustaceans. According to Egbenda et al., different soil properties such as pH, organic content, and redox potential also influence the uptake of HMs from environment. HMs are cumulative, so long-term exposure to it in the soil makes it readily available for uptake by plants and other microbes living in the soil environment. Thus, it becomes a matter of concern for human health (Hao et al., 2019). Some studies have noted that plant roots take up metal ions by active transport (metabolic path) or passive pathway (i.e., diffusion or mass flow of metal ions). For example, ions of Cd, Ni, Cr(III), and Pb metals are primarily absorbed by active transport, whereas Cu and Zn ions can be absorbed via both active and passive uptake (Kim et al., 2015). Mobility of metal ions and organic complex varies as alkalinity of soil reduces at acidic pH and increases at alkaline pH (Shahid et al., 2017).

7.6.3 BIOACCUMULATION

Bioaccumulation is a major factor in the concentration of absorbed elements into any living system through normal biological processes such as absorption and food intake. Bioaccumulated metals may get localized in a particular organ or may get dispersed into the system, and it may remain in the system or may get passed on. The term known as BAF (bioaccumulation factor) is the ability of any organism to digest any pollutant and eliminate it by its physiological function. Low BAF suggests high bioaccumulation ability and vice versa.

BAF = C organism/C sediment or C seawater

De Paiva Magalhães et al. (2015) reported that HMs present in the environment primarily come into contact with the cell wall of microorganisms, which is composed of different biopolymers like lipids, proteins, and polysaccharides. These polymers contain functional groups such as imidazole, hydroxyl, phosphate, phenol, and others. The presence of these functional groups enables metal absorption by making a complex with metal ions present in the extracellular environment of microbial cells (de Paiva Magalhães et al., 2015). HMs such as Hg, Pb, Ag, and Ni are highly bioaccumulative as these metals neither get broken down nor metabolized by the organism. Several studies have been conducted to precisely understand how HMs accumulate in the aquatic system. Generally, marine organisms absorb HMs from surrounding sources, sediments, and food as primary sources (Bosch et al., 2016). One of the studies in China by Hao et al. showed accumulation of Cd, Cr, Cu, Pb, Hg, and Zn metals in different marine organisms (Hao et al., 2019). During exposure of HMs to organisms, organs act differently for accumulation of different HMs. In one study it was observed that liver is the primary site of accumulation of Cu and Hg, kidney acts as the primary site of accumulationfor Cd and Zn, and spleen and gall bladder are the primary organs of accumulation for Cr (Wei et al., 2014). Bioaccumulation of different HMs was studied on vegetables and soil in old mining areas of copper and iron by selecting different sampling sites, and different factors like translocation factor, BAF, and BCF were analyzed. As, Cd, Cu, and Pb were found in different samples (Balabanova et al., 2015). Daphnia Magna was used as a model organism to analyze the bioaccumulative nature of some HMs. Levels of some HMs were found to be higher than the permissible limits in samples collected from aquatic source of the Xiaolangdi Reservoir in the Yellow River like water and suspended sediments (SPS). These samples determined speciation and concentration for Cu, Ni, Cd, and Zn (Dong et al., 2018).

There is another factor known as water hardness that reduces the rate of bioaccumulation of HMs in aquatic organisms as hard water contains Ca^{2+} and Mg^{2+}

ions in the form of carbonate, which specifically does not allow binding of metal ions to organism binding sites as these ions acts as blocking agents and compete with divalent metal ions. Calcium ion also bind to fish gills to confer positive charge (Saglam et al., 2013). Intake of HMs is beneficial to human life in small concentrations. For example, Zn and Cu helps maintain normal function of physiology (Gu et al., 2017).

7.6.4 Biomagnification

HMs are accumulated by plants or other organisms like bacteria, fungi, and other higher organisms through the food chain. When their concentrations reach a threshold level in the tissues of organisms, they get accumulated there. This is called the biomagnification of HM in the food chain. In a study by Shilla et al., at the Rufiji estuary in Tanzania, biomagnification of trace metals was studied in aquatic food web. Different samples from three different sites of Rufiji estuary such as sediments, fish, and invertebrates were collected and concentration of different trace metals were analyzed, quantified, and evaluated by BAF and BMF. Trophic magnification factor was also analyzed, and it was found that concentration of Zn and As increased with trophic level (Shilla et al., 2019). A terrestrial food chain was created as a system to study biomagnification of Cerium (two types, nano CeO_2 and bulk CeO_2, were selected). For this study, plants were grown on soil with these compounds with particular concentration for 36 days, and it was observed that this compound was present in tissue of beetles and bugs that fed upon this plant. Thus it was concluded that despite Ce's particle size, it accumulated in roots and leaves of the plant, then in beetles and then to bugs and hence was biomagnified (Majumdar et al., 2016). Liu et al. (2019) performed a study to observe trophic level transfer and biomagnification of four known HMs (i.e., Cu, Cr, Cd and Hg) in food web comprised of 43 representative species of the main trophic level. It was found that Hg and Cu were efficiently biomagnified between the trophic levels along the food chain in the food web. In comparision, Cu was diluted when passing on to higher trophic level, and Cd did not show any trend in biomagnification or trophic level transfer (J. Liu et al., 2019). Another such study was performed to analyze the biomagnification and biodilution of HMs in the marine food web globally by meta-analysis. For this study various studies between the years 2000–2019 were re-compiled involving 33 countries, 154 sampling sites, 61 studies, and 9929 biological samples. From this analysis it was concluded that Cu and Cd were biomagnified from secondary producer to secondary consumer at trophic level and other metals showed more or less biodilution with increasing trophic level. It was also found in this study that concentration of Hg was more noticeable among the trophic level food web of developed countries than of developing countries (Sun et al., 2020).

7.6.5 Tropic Transfer

When any element or pollutant gets transferred from lower to higher level of the ecological food chain, it is known as tropic transfer. In the case of HMs this theory is also applicable, since there are several HMs that are present in many forms in air, water, and soil ecosystems. Due to several natural and anthropogenic activities, which interact with producers these HMs easily reach to these organisms living in different environments. Gradually with time these metals are accumulated into the plants and organisms and get transferred to the consumers (i.e., to higher level organisms of food chain, which is called tropic level transfer of HMs). With successive transfer of HM at each tropic level these HMs act differently in the organism's biological makeup. But there are some metals like Hg, Pb, Ni, and Ag that are neither broken down by environmental factors (e.g., humidity, pH, temperature, location) nor get metabolized by the organism. Thus such metals are accumulated in the ecological food chain as these metals are taken up by primary producers such as microalgae, phytoplankton, and soil microbe, and then transferred to consumers like animals, fishes, etc. (Nagajyoti et al., 2010).

7.7 CONCLUSION

It is evident from several studies that HMs have profound effects on plants, animals, and other living organisms in different ecosystems. A number of analytical techniques are used to evaluate the concentrations of HMs in different tropic levels of ecosystem. However, accurate and reliable measurements of their toxic effects are important in understanding their biological roles. General studies suggest that HM concentrations are within the allowable range in various organisms of ecosystems and pose no threat to public health. But in some places studies have reported HM concentrations exceeding the WHO permissible limits. Judicious use of HMs in the current era of industry and urbanization is important to maintain ecological harmony of the globe. Further research programmes and better understanding are needed with regard to various aspects of ecosystem components and their interactions with different HMs. Therefore, regular monitoring and health risk assessment studies are important to ensure safety conditions.

ACKNOWLEDGMENTS

This work was supported by grants from the Council of Scientific and Industrial Research (CSIR) funded network projects. We are grateful to the Institutional Manuscript Review Committee of CSIR-IITR for reviewing and allotting communication number IITR/SCE/2021-2022/43 to the manuscript.

REFERENCES

Ahmed, K., Mehedi, Y., Haque, R., and Mondol, P., 2011. Heavy metal concentrations in some macrobenthic fauna of the Sundarbans mangrove forest, south west coast of Bangladesh. *Environ. Monit. Assess.* 177, 505–514. https://doi.org/10.1007/S10 661-010-1651-9

Ajima, M.N.O., Nnodi, P.C., Ogo, O.A., Adaka, G.S., Osuigwe, D.I., and Njoku, D.C., 2015. Bioaccumulation of heavy metals in Mbaa River and the impact on aquatic ecosystem. *Environ. Monit. Assess.* 187, 1–9. https://doi.org/10.1007/S10661-015-4937-0

Al-Rousan, S., Al-Shloul, R., Al-Horani, F., and Abu-Hilal, A., 2012. Heavy metals signature of human activities recorded in coral skeletons along the Jordanian coast of the Gulf of Aqaba, Red Sea. *Environ. Earth Sci.* 67, 2003–2013. https://doi.org/10.1007/S12665-012-1640-0

Alex, S., 2018. Mercury and its associated impacts on environment and human health: a review. *J Env. Heal. Sci* 4, 37–43. https://doi.org/10.15436/2378-6841.18.1906

Ali, H., Khan, E., and Ilahi, I., 2019. Environmental chemistry and ecotoxicology of hazardous heavy metals: environmental persistence, toxicity, and bioaccumulation. *J. Chem.* 2019. https://doi.org/10.1155/2019/6730305

Alka, S., Shahir, S., Ibrahim, N., Ndejiko, M.J., Vo, D.V.N., and Manan, F.A., 2021. Arsenic removal technologies and future trends: a mini review. *J. Clean. Prod.* 278, 123805. https://doi.org/10.1016/J.JCLEPRO.2020.123805

Alloway, B.J., Jackson, A.P., and Morgan, H., 1990. The accumulation of cadmium by vegetables grown on soils contaminated from a variety of sources. *Sci. Total Environ.* 91, 223–236. https://doi.org/10.1016/0048-9697(90)90300-J

Almahasheer, H., Serrano, O., Duarte, C.M., and Irigoien, X., 2018. Remobilization of heavy metals by mangrove leaves. *Front. Mar. Sci.* 0, 484. https://doi.org/10.3389/FMARS.2018.00484

Alzahrani, D.A., Selim, E.M.M., and El-Sherbiny, M.M., 2018. Ecological assessment of heavy metals in the grey mangrove (Avicennia marina) and associated sediments along the Red Sea coast of Saudi Arabia. *Oceanologia* 60, 513–526. https://doi.org/10.1016/J.OCEANO.2018.04.002

Andosch, A., Höftberger, M., Lütz, C., and Lütz-Meindl, U., 2015. Subcellular sequestration and impact of heavy metals on the ultrastructure and physiology of the multicellular freshwater alga *Desmidium swartzii*. *Int. J. Mol. Sci* 16, 10389–10410. https://doi.org/10.3390/IJMS160510389

Arumugam, G., Rajendran, R., Ganesan, A., and Sethu, R., 2018. Bioaccumulation and translocation of heavy metals in mangrove rhizosphere sediments to tissues of Avicenia marina—a field study from tropical mangrove forest. *Environ. Nanotechnology, Monit. Manag.* 10, 272–279. https://doi.org/10.1016/J.ENMM.2018.07.005

Bakshi, A., and Panigrahi, A.K., 2018. A comprehensive review on chromium induced alterations in fresh water fishes. *Toxicol. Reports* 5, 440–447. https://doi.org/10.1016/J.TOXREP.2018.03.007

Balabanova, B., Stafilov, T., and Bačeva, K., 2015. Bioavailability and bioaccumulation characterization of essential and heavy metals contents in R. acetosa, S. oleracea and U. dioica from copper polluted and referent areas. *J. Environ. Heal. Sci. Eng.* 13, 1–13. https://doi.org/10.1186/S40201-015-0159-1

Bastakoti, U., Robertson, J., and Alfaro, A.C., 2018. Spatial variation of heavy metals in sediments within a temperate mangrove ecosystem in northern New Zealand. *Mar. Pollut. Bull.* 135, 790–800. https://doi.org/10.1016/J.MARPOLBUL.2018.08.012

Bayen, S., 2012. Occurrence, bioavailability and toxic effects of trace metals and organic contaminants in mangrove ecosystems: a review. *Environ. Int.* 48, 84–101. https://doi.org/10.1016/J.ENVINT.2012.07.008

Bazzi, A.O., 2014. Heavy metals in seawater, sediments and marine organisms in the Gulf of Chabahar, Oman Sea. *J. Oceanogr. Mar. Sci.* 5, 20–29. https://doi.org/10.5897/JOMS2014.0110

Beck, M.W., Losada, I.J., Menéndez, P., Reguero, B.G., Díaz-Simal, P., and Fernández, F., 2018. The global flood protection savings provided by coral reefs. *Nat. Commun.* 9, 1–9. https://doi.org/10.1038/s41467-018-04568-z

Behera, B., Mishra, R., Patra, J., Sarangi, K., Dutta, S.K., and Thatoi, H.N., 2013, 2013. Impact of heavy metals on bacterial communities from mangrove soils of the Mahanadi Delta (India). *Chemistry and Ecology Taylor Fr.* 29, 1–8. https://doi.org/10.1080/02757540.2013.810719

Berry, K.L.E., Seemann, J., Dellwig, O., Struck, U., Wild, C., and Leinfelder, R.R., 2013. Sources and spatial distribution of heavy metals in scleractinian coral tissues and sediments from the Bocas del Toro Archipelago, Panama. *Environ. Monit. Assess.* 185, 9089–9099. https://doi.org/10.1007/S10661-013-3238-8

Bonanno, G., Borg, J.A., and Di Martino, V., 2017. Levels of heavy metals in wetland and marine vascular plants and their biomonitoring potential: a comparative assessment. *Sci. Total Environ.* 576, 796–806. https://doi.org/10.1016/J.SCITOTENV.2016.10.171

Bosch, A.C., O'Neill, B., Sigge, G.O., Kerwath, S.E., and Hoffman, L.C., 2016. Heavy metals in marine fish meat and consumer health: a review. *J. Sci. Food Agric.* 96, 32–48. https://doi.org/10.1002/JSFA.7360

Boshoff, M., De Jonge, M., Dardenne, F., Blust, R., and Bervoets, L., 2014. The impact of metal pollution on soil faunal and microbial activity in two grassland ecosystems. *Environ. Res.* 134, 169–180. https://doi.org/10.1016/J.ENVRES.2014.06.024

Boskabady, M., Marefati, N., Farkhondeh, T., Shakeri, F., Farshbaf, A., and Boskabady, M.H., 2018. The effect of environmental lead exposure on human health and the contribution of inflammatory mechanisms, a review. *Environ. Int.* 120, 404–420. https://doi.org/10.1016/J.ENVINT.2018.08.013

Briffa, J., Sinagra, E., and Blundell, R., 2020. Heavy metal pollution in the environment and their toxicological effects on humans. *Heliyon* 6, e04691. https://doi.org/10.1016/J.HELIYON.2020.E04691

Chai, M., Li, R., Ding, H., and Zan, Q., 2019. Occurrence and contamination of heavy metals in urban mangroves: a case study in Shenzhen, *China. Chemosphere* 219, 165–173. https://doi.org/10.1016/J.CHEMOSPHERE.2018.11.160

Chai, M., Li, R., Tam, N.F.Y., and Zan, Q., 2018. Effects of mangrove plant species on accumulation of heavy metals in sediment in a heavily polluted mangrove swamp in Pearl River Estuary, *China. Environ. Geochemistry Heal.* 41, 175–189. https://doi.org/10.1007/S10653-018-0107-Y

Chai, Y., Guo, J., Chai, S., Cai, J., Xue, L., and Zhang, Q., 2015. Source identification of eight heavy metals in grassland soils by multivariate analysis from the Baicheng–Songyuan area, Jilin Province, *Northeast China. Chemosphere* 134, 67–75. https://doi.org/10.1016/J.CHEMOSPHERE.2015.04.008

Chen, G., Shi, H., Tao, J., Chen, L., Liu, Y., Lei, G., Liu, X., and Smol, J.P., 2015. Industrial arsenic contamination causes catastrophic changes in freshwater ecosystems. *Sci. Reports* 5, 1–7. https://doi.org/10.1038/srep17419

Chen, Q.Y., and Costa, M., 2021. Arsenic: a global environmental challenge. *Annual Review of Pharmacology and Toxicology* 61, 47–63. https://doi.org/10.1146/annurev-pharmtox-030220-013418

Cheng, H., Li, M., Zhao, C., Yang, K., Li, K., Peng, M., Yang, Z., Liu, F., Liu, Y., Bai, R., Cui, Y., Huang, Z., Li, L., Liao, Q., Luo, J., Jia, S., Pang, X., Yang, J., and Yin, G., 2015. Concentrations of toxic metals and ecological risk assessment for sediments of major freshwater lakes in China. *J. Geochemical Explor.* 157, 15–26. https://doi.org/10.1016/J.GEXPLO.2015.05.010

Chodak, M., Gołebiewski, M., Morawska-Płoskonka, J., Kuduk, K., and Niklińska, M., 2013. Diversity of microorganisms from forest soils differently polluted with heavy metals. *Appl. Soil Ecol.* 64, 7–14. https://doi.org/10.1016/J.APSOIL.2012.11.004

Chu, W.L., Dang, N.L., Kok, Y.Y., Ivan Yap, K.S., Phang, S.M., and Convey, P., 2019. Heavy metal pollution in Antarctica and its potential impacts on algae. *Polar Sci.* 20, 75–83. https://doi.org/10.1016/J.POLAR.2018.10.004

Damyanova, S., 2010. Comparative risk assessment studies of heavy metal pollutions in beech forests. *Forestry Ideas* 16(2), 40.

de Paiva Magalhães, D., da Costa Marques, M.R., Baptista, D.F., and Buss, D.F., 2015. Metal bioavailability and toxicity in freshwaters. *Environ. Chem. Lett.* 13, 69–87. https://doi.org/10.1007/S10311-015-0491-9

Demková, L., Árvay, J., Bobuľská, L., Tomáš, J., Stanovič, R., Lošák, T., Harangozo, L., Vollmannová, A., Bystrická, J., Musilová, J., and Jobbágy, J., 2017. Accumulation and environmental risk assessment of heavy metals in soil and plants of four different ecosystems in a former polymetallic ores mining and smelting area (Slovakia). *Journal of Environmental Science and Health, Part A* 52, 479–490. https://doi.org/10.1080/10934529.2016.1274169

Devanesan, E., Suresh Gandhi, M., Selvapandiyan, M., Senthilkumar, G., and Ravisankar, R., 2017. Heavy metal and potential ecological risk assessment in sedimentscollected from Poombuhar to Karaikal Coast of Tamilnadu using Energy dispersive X-ray fluorescence (EDXRF) technique. *Beni-Suef Univ. J. Basic Appl. Sci.* 6, 285–292. https://doi.org/10.1016/J.BJBAS.2017.04.011

Dixit, R., Wasiullah, Malaviya, D., Pandiyan, K., Singh, U.B., Sahu, A., Shukla, R., Singh, B.P., Rai, J.P., Sharma, P.K., Lade, H., and Paul, D., 2015. Bioremediation of heavy metals from soil and aquatic environment: an overview of principles and criteria of fundamental processes. *Sustain.* 7, 2189–2212. https://doi.org/10.3390/SU7022189

Dong, J., Xia, X., Zhang, Z., Liu, Z., Zhang, X., and Li, H., 2018. Variations in concentrations and bioavailability of heavy metals in rivers caused by water conservancy projects: insights from water regulation of the Xiaolangdi Reservoir in the Yellow River. *J. Environ. Sci.* 74, 79–87. https://doi.org/10.1016/J.JES.2018.02.009

Dziubanek, G., Piekut, A., Rusin, M., Baranowska, R., and Hajok, I., 2015. Contamination of food crops grown on soils with elevated heavy metals content. *Ecotoxicol. Environ. Saf.* 118, 183–189. https://doi.org/10.1016/J.ECOENV.2015.04.032

Egbenda, P.O., Thullah, F., Kamara, I., 2015. A physico-chemical analysis of soil and selected fruits in one rehabilitated mined out site in the Sierra Rutile environs for the presence of heavy metals: Lead, Copper, Zinc, Chromium and Arsenic. *African J. Pure Appl. Chem.* 9, 27–32. https://doi.org/10.5897/AJPAC2015.0606

El-Sorogy, A.S., Mohamed, M.A., and Nour, H.E., 2012. Heavy metals contamination of the Quaternary coral reefs, Red Sea coast, Egypt. *Environ. Earth Sci.* 67, 777–785. https://doi.org/10.1007/S12665-012-1535-0

Enuneku, A., Omoruyi, O., Tongo, I., Ogbomida, E., Ogbeide, O., and Ezemonye, L., 2018. Evaluating the potential health risks of heavy metal pollution in sediment and selected benthic fauna of Benin River, Southern Nigeria. *Appl. Water Sci.* 8, 1–13. https://doi.org/10.1007/S13201-018-0873-9

Fonseca, A.R., Sanches Fernandes, L.F., Fontainhas-Fernandes, A., Monteiro, S.M., and Pacheco, F.A.L., 2017. The impact of freshwater metal concentrations on the severity of histopathological changes in fish gills: a statistical perspective. *Sci. Total Environ.* 599–600, 217–226. https://doi.org/10.1016/J.SCITOTENV.2017.04.196

Ganeshkumar, A., Arun, G., Vinothkumar, S., and Rajaram, R., 2019. Bioaccumulation and translocation efficacy of heavy metals by Rhizophora mucronata from tropical mangrove ecosystem, Southeast coast of India. *Ecohydrol. Hydrobiol.* 19, 66–74. https://doi.org/10.1016/J.ECOHYD.2018.10.006

Gissi, F., Adams, M.S., King, C.K., and Jolley, D.F., 2015. A robust bioassay to assess the toxicity of metals to the Antarctic marine microalga Phaeocystis antarctica. *Environ. Toxicol. Chem.* 34, 1578–1587. https://doi.org/10.1002/ETC.2949

Golam Mortuza, M., and Al-Misned, F.A., 2015. Heavy metal concentration in two freshwater fishes from Wadi Hanifah (Riyadh, Saudi Arabia) and evaluation of possible health hazard to consumers. *Pakistan J. Zool* 47, 839–845.

Gu, Y.G., Lin, Q., Huang, H.H., Wang, L. gen, Ning, J.J., and Du, F.Y., 2017. Heavy metals in fish tissues/stomach contents in four marine wild commercially valuable fish species from the western continental shelf of South China Sea. *Mar. Pollut. Bull.* 114, 1125–1129. https://doi.org/10.1016/J.MARPOLBUL.2016.10.040

Gworek, B., Dmuchowski, W., and Baczewska-Dąbrowska, A.H., 2020. Mercury in the terrestrial environment: a review. *Environ. Sci. Eur.* 32, 1–19. https://doi.org/10.1186/S12302-020-00401-X

Hao, Z., Chen, L., Wang, C., Zou, X., Zheng, F., Feng, W., Zhang, D., and Peng, L., 2019. Heavy metal distribution and bioaccumulation ability in marine organisms from coastal regions of Hainan and Zhoushan, China. *Chemosphere* 226, 340–350. https://doi.org/10.1016/J.CHEMOSPHERE.2019.03.132

Hsu-Kim, H., Kucharzyk, K.H., Zhang, T., and Deshusses, M.A., 2013. Mechanisms regulating mercury bioavailability for methylating microorganisms in the aquatic environment: a critical review. *Environ. Sci. Technol.* 47, 2441–2456. https://doi.org/10.1021/ES304370G

Islam, M.S., Ahmed, M.K., Habibullah-Al-Mamun, M., and Masunaga, S., 2014. Trace metals in soil and vegetables and associated health risk assessment. *Environ. Monit. Assess.* 186, 8727–8739. https://doi.org/10.1007/S10661-014-4040-Y

Jackson, T.A., Vlaar, S., Nguyen, N., Leppard, G.G., and Finan, T.M., 2015. Effects of bioavailable heavy metal species, arsenic, and acid drainage from mine tailings on a microbial community sampled along a pollution gradient in a freshwater ecosystem. *Geomicrobiology Journal* 32, 724–750. https://doi.org/10.1080/01490451.2014.969412

Järup, L., and Åkesson, A., 2009. Current status of cadmium as an environmental health problem. *Toxicol. Appl. Pharmacol.* 238, 201–208. https://doi.org/10.1016/J.TAAP.2009.04.020

Järup, L., Berglund, M., Elinder, C., Nordberg, G., and Vahter, M., 1998. Health effects of cadmium exposure—a review of the literature and a risk estimate. JSTOR *Scandinavian Journal of Work, Environment & Health* 1:1–51.

Javed, M., and Usmani, N., 2014. Impact of heavy metal toxicity on hematology and glycogen status of fish: a review. *Proc. Natl. Acad. Sci. India Sect. B Biol. Sci.* 85, 889–900. https://doi.org/10.1007/S40011-014-0404-X

Jerez, S., Motas, M., Benzal, J., Diaz, J., and Barbosa, A., 2013. Monitoring trace elements in Antarctic penguin chicks from South Shetland Islands, Antarctica. *Mar. Pollut. Bull.* 69, 67–75. https://doi.org/10.1016/J.MARPOLBUL.2013.01.004

Jia, Y., Wang, L., Qu, Z., Wang, C., and Yang, Z., 2017. Effects on heavy metal accumulation in freshwater fishes: species, tissues, and sizes. *Environ. Sci. Pollut. Res.* 24, 9379–9386. https://doi.org/10.1007/S11356-017-8606-4

Kandziora-Ciupa, M., Ciepał, R., Nadgórska-Socha, A., and Barczyk, G., 2016. Accumulation of heavy metals and antioxidant responses in Pinus sylvestris L. needles in polluted and non-polluted sites. *Ecotoxicol.* 25, 970–981. https://doi.org/10.1007/S10646-016-1654-6

Kandziora-Ciupa, M., Nadgórska-Socha, A., Barczyk, G., and Ciepał, R., 2017. Bioaccumulation of heavy metals and ecophysiological responses to heavy metal stress in selected populations of *Vaccinium myrtillus* L. and *Vaccinium vitis-idaea* L. *Ecotoxicol.* 26, 966–980. https://doi.org/10.1007/S10646-017-1825-0

Karmakar, D., and Padhy, P.K., 2019. Metals uptake from particulate matter through foliar transfer and their impact on antioxidant enzymes activity of S. robusta in a Tropical Forest, West Bengal, India. *Arch. Environ. Contam. Toxicol.* 76, 605–616. https://doi.org/10.1007/S00244-019-00599-9

Kaur, S., Singh Khera, K., Kaur Kondal, J., and Saravpreet Kaur, C., 2018. Heavy metal induced histopathological alterations in liver, muscle and kidney of freshwater cyprinid, *Labeo rohita* (Hamilton). *J. Entomol. Zool. Stud.* 6, 2137.

Kesici, G.G., 2016. Arsenic ototoxicity. *J. Otol.* 11, 13–17. https://doi.org/10.1016/J.JOTO.2016.03.001

Kim, R.-Y., Yoon, J.-K., Kim, T.-S., Yang, J.E., Owens, G., and Kim, K.-R., 2015. Bioavailability of heavy metals in soils: definitions and practical implementation—a critical review. *Environ. Geochemistry Heal.* 37, 1041–1061. https://doi.org/10.1007/S10653-015-9695-Y

Koppel, D.J., Gissi, F., Adams, M.S., King, C.K., and Jolley, D.F., 2017. Chronic toxicity of five metals to the polar marine microalga Cryothecomonas armigera—application of a new bioassay. *Environ. Pollut.* 228, 211–221. https://doi.org/10.1016/J.ENVPOL.2017.05.034

Langston, W.J., 2018. Toxic effects of metals and the incidence of metal pollution in marine ecosystems. *Heavy Met. Mar. Environ.* 101–120. https://doi.org/10.1201/9781351073158-7

Lee, M.A., Davies, L., and Power, S.A., 2012. Effects of roads on adjacent plant community composition and ecosystem function: an example from three calcareous ecosystems. *Environ. Pollut.* 163, 273–280. https://doi.org/10.1016/J.ENVPOL.2011.12.038

Li, C., Sanchez, G.M., Wu, Z., Cheng, J., Zhang, S., Wang, Q., Li, F., Sun, G., and Meentemeyer, R.K., 2020. Spatiotemporal patterns and drivers of soil contamination with heavy metals during an intensive urbanization period (1989–2018) in southern China. *Environ. Pollut.* 260, 114075. https://doi.org/10.1016/J.ENVPOL.2020.114075

Li, C., Zhou, K., Qin, W., Tian, C., Qi, M., Yan, X., and Han, W., 2019. A review on heavy metals contamination in soil: effects, sources, and remediation techniques. *Soil and Sediment Contamination: An International Journal* 28, 380–394. https://doi.org/10.1080/15320383.2019.1592108

Li, J., Zhang, C., Lin, J., Yin, J., Xu, J., and Chen, Y., 2018. Evaluating the bioavailability of heavy metals in natural-zeolite-amended aquatic sediments using thin-film diffusive gradients. *Aquac. Fish.* 3, 122–128. https://doi.org/10.1016/J.AAF.2018.05.003

Li, M., Yang, W., Sun, T., and Jin, Y., 2016. Potential ecological risk of heavy metal contamination in sediments and macrobenthos in coastal wetlands induced by freshwater releases: a case study in the Yellow River Delta. *China. Mar. Pollut. Bull.* 103, 227–239. https://doi.org/10.1016/J.MARPOLBUL.2015.12.014

Liu, B., Huang, Q., Cai, H., Guo, X., Wang, T., and Gui, M., 2015. Study of heavy metal concentrations in wild edible mushrooms in Yunnan Province. *China. Food Chem.* 188, 294–300. https://doi.org/10.1016/J.FOODCHEM.2015.05.010

Liu, J., Cao, L., and Dou, S., 2019. Trophic transfer, biomagnification and risk assessments of four common heavy metals in the food web of Laizhou Bay, the Bohai Sea. *Sci. Total Environ.* 670, 508–522. https://doi.org/10.1016/J.SCITOTENV.2019.03.140

Liu, Q., Xu, X., Zeng, J., Shi, X., Liao, Y., Du, P., Tang, Y., Huang, W., Chen, Q., and Shou, L., 2019. Heavy metal concentrations in commercial marine organisms from Xiangshan Bay, China, and the potential health risks. *Mar. Pollut. Bull.* 141, 215–226. https://doi.org/10.1016/J.MARPOLBUL.2019.02.058

Liu, Y.-R., He, Z.-Y., Yang, Z.-M., Sun, G.-X., and He, J.-Z., 2016. Variability of heavy metal content in soils of typical Tibetan grasslands. *RSC Adv.* 6, 105398–105405. https://doi.org/10.1039/C6RA23868H

Majumdar, S., Trujillo-Reyes, J., Hernandez-Viezcas, J.A., White, J.C., Peralta-Videa, J.R., and Gardea-Torresdey, J.L., 2016. Cerium biomagnification in a terrestrial food chain: influence of particle size and growth stage. *Environ. Sci. Technol.* 50, 6782–6792. https://doi.org/10.1021/ACS.EST.5B04784

Manu, M., Băncilă, R.I., Iordache, V., Bodescu, F., and Onete, M., 2017a. Impact assessment of heavy metal pollution on soil mite communities (Acari: Mesostigmata) from Zlatna Depression—Transylvania. *Process Saf. Environ. Prot.* 108, 121–134. https://doi.org/10.1016/J.PSEP.2016.06.011

Manu, M., Onete, M., Florescu, L., Bodescu, F., and Iordache, V., 2017b. Influence of heavy metal pollution on soil mite communities (Acari) in Romanian grasslands. *J. Zool.* 13, 200–210.

Meena, R.A.A., Sathishkumar, P., Ameen, F., Yusoff, A.R.M., and Gu, F.L., 2017. Heavy metal pollution in immobile and mobile components of lentic ecosystems—a review. *Environ. Sci. Pollut. Res.* 25, 4134–4148. https://doi.org/10.1007/S11356-017-0966-2

Moore, K.L., Chen, Y., Meene, A.M.L. van de, Hughes, L., Liu, W., Geraki, T., Mosselmans, F., McGrath, S.P., Grovenor, C., and Zhao, F.-J., 2014. Combined NanoSIMS and synchrotron X-ray fluorescence reveal distinct cellular and subcellular distribution patterns of trace elements in rice tissues. *New Phytol.* 201, 104–115. https://doi.org/10.1111/NPH.12497

Muradoglu, F., Gundogdu, M., Ercisli, S., Encu, T., Balta, F., Jaafar, H.Z., and Zia-Ul-Haq, M., 2015. Cadmium toxicity affects chlorophyll a and b content, antioxidant enzyme activities and mineral nutrient accumulation in strawberry. *Biol. Res.* 48, 1–7. https://doi.org/10.1186/S40659-015-0001-3

Nagajyoti, P.C., Lee, K.D., and Sreekanth, T.V.M., 2010. Heavy metals, occurrence and toxicity for plants: a review. *Environ. Chem. Lett.* 8, 199–216. https://doi.org/10.1007/S10311-010-0297-8

Oberoi, S., Barchowsky, A., and Wu, F., 2014. The global burden of disease for skin, lung, and bladder cancer caused by arsenic in food. *Cancer Epidemiol. Prev. Biomarkers* 23, 1187–1194. https://doi.org/10.1158/1055-9965.EPI-13-1317

Oliveira, H., 2012. Chromium as an environmental pollutant: insights on induced plant toxicity. *J. Bot.* 2012. https://doi.org/10.1155/2012/375843

Oves, M., Khan, M.S., Zaidi, A., and Ahmad, E., 2012. Soil contamination, nutritive value, and human health risk assessment of heavy metals: an overview. *Toxic. Heavy Met. to Legum. Bioremediation* 9783709107300, 1–27. https://doi.org/10.1007/978-3-7091-0730-0_1

Oyetibo, G.O., Miyauchi, K., Huang, Y., Chien, M.F., Ilori, M.O., Amund, O.O., and Endo, G., 2017. Biotechnological remedies for the estuarine environment polluted with heavy metals and persistent organic pollutants. *Int. Biodeterior. Biodegradation* 119, 614–625. https://doi.org/10.1016/J.IBIOD.2016.10.005

Pająk, M., Błońska, E., Frąc, M., and Oszust, K., 2016. Functional diversity and microbial activity of forest soils that are heavily contaminated by lead and zinc. *Water, Air, Soil Pollut.* 227, 1–14. https://doi.org/10.1007/S11270-016-3051-4

Paul, S., Mandal, A., Bhattacharjee, P., Chakraborty, S., Paul, R., and Kumar Mukhopadhyay, B., 2019. Evaluation of water quality and toxicity after exposure of lead nitrate in fresh water fish, major source of water pollution. *Egypt. J. Aquat. Res.* 45, 345–351. https://doi.org/10.1016/J.EJAR.2019.09.001

Queiroz, E.K.R. de, and Waissmann, W., 2006. Occupational exposure and effects on the male reproductive system. *Cad. Saude Publica* 22, 485–493. https://doi.org/10.1590/S0102-311X2006000300003

Ranieri, E., Fratino, U., Petrella, A., Torretta, V., and Rada, E.C., 2016. Ailanthus Altissima and Phragmites Australis for chromium removal from a contaminated soil. *Environ. Sci. Pollut. Res.* 23, 15983–15989. https://doi.org/10.1007/S11356-016-6804-0

Ranieri, E., Fratino, U., Petruzzelli, D., and Borges, A.C., 2013. A comparison between *Phragmites australis* and *Helianthus annuus* in chromium phytoextraction. *Water, Air, Soil Pollut.* 224, 1–9. https://doi.org/10.1007/S11270-013-1465-9

Ranjbar Jafarabadi, A., Mitra, S., Raudonytė-Svirbutavičienė, E., and Riyahi Bakhtiari, A., 2020. Large-scale evaluation of deposition, bioavailability and ecological risks of the potentially toxic metals in the sediment cores of the hotspot coral reef ecosystems (Persian Gulf, Iran). *J. Hazard. Mater.* 400, 122988. https://doi.org/10.1016/J.JHAZMAT.2020.122988

Ranjbar Jafarabadi, A., Riyahi Bakhtiyari, A., Shadmehri Toosi, A., and Jadot, C., 2017. Spatial distribution, ecological and health risk assessment of heavy metals in marine surface sediments and coastal seawaters of fringing coral reefs of the Persian Gulf, Iran. *Chemosphere* 185, 1090–1111. https://doi.org/10.1016/J.CHEMOSPHERE.2017.07.110

Ratko, K., Snežana, B., Dragica, O., Ivana, B., and Nada, D., 2013. Assessment of heavy metal content in soil and grasslands in national park of the lake plateau of the N. P. "Durmitor" Montenegro. *African J. Biotechnol.* 10, 5157–5165. https://DOI: 10.5897/AJB10.2129.

Restrepo, J.D., Park, E., Aquino, S., and Latrubesse, E.M., 2016. Coral reefs chronically exposed to river sediment plumes in the southwestern Caribbean: Rosario Islands, Colombia. *Sci. Total Environ.* 553, 316–329. https://doi.org/10.1016/J.SCITOTENV.2016.02.140

Saglam, D., Atli, G., and Canli, M., 2013. Investigations on the osmoregulation of freshwater fish (Oreochromis niloticus) following exposures to metals (Cd, Cu) in differing hardness. *Ecotoxicol. Environ. Saf.* 92, 79–86. https://doi.org/10.1016/J.ECOENV.2013.02.020

Sankhla, M.S., Kumar, R., and Prasad, L., 2019. Distribution and contamination assessment of potentially harmful element chromium in water. *SSRN Electron. J.* 2(3), November 2019. https://doi.org/10.2139/SSRN.3492307

Sarwar, N., Imran, M., Shaheen, M.R., Ishaque, W., Kamran, M.A., Matloob, A., Rehim, A., and Hussain, S., 2017. Phytoremediation strategies for soils contaminated with heavy metals: modifications and future perspectives. *Chemosphere* 171, 710–721. https://doi.org/10.1016/J.CHEMOSPHERE.2016.12.116

Scott, D.L., Bradley, R.L., Bellenger, J.P., Houle, D., Gundale, M.J., Rousk, K., and DeLuca, T.H., 2018. Anthropogenic deposition of heavy metals and phosphorus may reduce biological N_2 fixation in boreal forest mosses. *Sci. Total Environ.* 630, 203–210. https://doi.org/10.1016/J.SCITOTENV.2018.02.192

Shahid, M., Dumat, C., Khalid, S., Schreck, E., Xiong, T., and Niazi, N.K., 2017. Foliar heavy metal uptake, toxicity and detoxification in plants: a comparison of foliar and root metal uptake. *J. Hazard. Mater.* 325, 36–58. https://doi.org/10.1016/J.JHAZMAT.2016.11.063

Shilla, D., Pajala, G., Routh, J., Dario, M., and Kristoffersson, P., 2019. Trophodynamics and biomagnification of trace metals in aquatic food webs: the case of Rufiji estuary in Tanzania. *Appl. Geochemistry* 100, 160–168. https://doi.org/10.1016/J.APGEOCHEM.2018.11.016

Shine, J.P., Ika, R.V., and Ford, T.E., 2002. Multivariate statistical examination of spatial and temporal patterns of heavy metal contamination in new bedford harbor marine sediments. *Environ. Sci. Technol.* 29, 1781–1788. https://doi.org/10.1021/ES00007A014/ASSET/ES00007A014.FP.PNG_V03

Srivastava, V., Sarkar, A., Singh, S., Singh, P., de Araujo, A.S.F., and Singh, R.P., 2017. Agroecological responses of heavy metal pollution with special emphasis on soil health and plant performances. *Front. Environ. Sci.* 0, 64. https://doi.org/10.3389/FENVS.2017.00064

Stefanowicz, A.M., Niklinska, M., Kapusta, P., and Szarek-lukaszewska, G., 2010. Pine forest and grassland differently influence the response of soil microbial communities to metal contamination. *Sci. Total Environ.* 408, 6134–6141. https://doi.org/10.1016/J.SCITOTENV.2010.08.056

Stefanowicz, A.M., Stanek, M., and Woch, M.W., 2016. High concentrations of heavy metals in beech forest understory plants growing on waste heaps left by Zn-Pb ore mining. *J. Geochemical Explor.* 169, 157–162. https://doi.org/10.1016/J.GEXPLO.2016.07.026

Sun, T., Wu, H., Wang, X., Ji, C., Shan, X., and Li, F., 2020. Evaluation on the biomagnification or biodilution of trace metals in global marine food webs by meta-analysis. *Environ. Pollut.* 264, 113856. https://doi.org/10.1016/J.ENVPOL.2019.113856

Trammell, T.L.E., Schneid, B.P., and Carreiro, M.M., 2011. Forest soils adjacent to urban interstates: Soil physical and chemical properties, heavy metals, disturbance legacies, and relationships with woody vegetation. *Urban Ecosyst.* 14, 525–552. https://doi.org/10.1007/S11252-011-0194-3

Trifan, A., Breabăn, I.G., and Bucur, L., 2015. Heavy metal content in macroalgae from Roumanian Black Sea *Rev. Roum. Chim.* 60(9), 915–920.

Větrovský, T., and Baldrian, P., 2015. An in-depth analysis of actinobacterial communities shows their high diversity in grassland soils along a gradient of mixed heavy metal contamination. *Biol. Fertil. Soils* 51, 827–837. https://doi.org/10.1007/S00374-015-1029-9

Wang, R., Zhang, C., Huang, X., Zhao, L., Yang, S., Struck, U., and Yin, D., 2020. Distribution and source of heavy metals in the sediments of the coastal East China sea: geochemical controls and typhoon impact. *Environ. Pollut.* 260, 113936. https://doi.org/10.1016/J.ENVPOL.2020.113936

Wang, X., Fu, R., Li, H., Zhang, Y., Lu, M., Xiao, K., Zhang, X., Zheng, C., and Xiong, Y., 2020. Heavy metal contamination in surface sediments: a comprehensive, large-scale evaluation for the Bohai Sea, *China. Environ. Pollut.* 260, 113986. https://doi.org/10.1016/J.ENVPOL.2020.113986

Wei, Y.H., Zhang, J.Y., Zhang, D.W., Tu, T.H., and Luo, L.G., 2014. Metal concentrations in various fish organs of different fish species from Poyang Lake, *China. Ecotoxicol. Environ. Saf.* 104, 182–188. https://doi.org/10.1016/J.ECOENV.2014.03.001

Wuana, R.A., Okieimen, F.E., Montuelle, B., and Steinman, A.D., 2011. Heavy metals in contaminated soils: a review of sources, chemistry, risks and best available strategies for remediation. *Int. Sch. Res. Netw. ISRN Ecol.* 2011, 20. https://doi.org/10.5402/2011/402647

Xie, Y., Fan, J., Zhu, W., Amombo, E., Lou, Y., Chen, L., and Fu, J., 2016. Effect of heavy metals pollution on soil microbial diversity and bermudagrass genetic variation. *Front. Plant Sci.* 0, 755. https://doi.org/10.3389/FPLS.2016.00755

Xu, J., Jing, B., Zhang, K., Cui, Y., Malkinson, D., Kopel, D., Song, K., and Da, L., 2017. Heavy metal contamination of soil and tree-ring in urban forest around highway in Shanghai, *China. Human and Ecological Risk Assessment: An International Journal* 23, Oct 2017, 1745–1762. https://doi.org/10.1080/10807039.2017.1340826

Yadav, A., Chowdhary, P., Kaithwas, G., and Bharagava, R.N., 2017. Toxic metals in the environment: threats on ecosystem and bioremediation approaches. *Handb. Met. Interact. Bioremediation* CRC Press, 2017 Apr 7 128–141. https://doi.org/10.1201/9781315153353-11

Yadav, K.K., Kumar, V., Gupta, N., Kumar, S., Rezania, S., and Singh, N., 2019. Human health risk assessment: study of a population exposed to fluoride through groundwater of Agra city, India. *Regul. Toxicol. Pharmacol.* 106, 68–80. https://doi.org/10.1016/J.YRTPH.2019.04.013

Yan, W., Mahmood, Q., Peng, D., Fu, W., Chen, T., Wang, Y., Li, S., Chen, J., and Liu, D., 2015. The spatial distribution pattern of heavy metals and risk assessment of moso bamboo forest soil around lead–zinc mine in Southeastern China. *Soil Tillage Res.* 153, 120–130. https://doi.org/10.1016/J.STILL.2015.05.013

Yang, T., Cheng, H., Wang, H., Drews, M., Li, S., Huang, W., Zhou, H., Chen, C.M., and Diao, X., 2019. Comparative study of polycyclic aromatic hydrocarbons (PAHs) and heavy metals (HMs) in corals, surrounding sediments and surface water at the Dazhou Island, *China. Chemosphere* 218, 157–168. https://doi.org/10.1016/J.CHEMOSPHERE.2018.11.063

8 Metabolism of Environmental Pollutants with Specific Reference to Pesticides, Endocrine Disrupters, and Mutagenic Pollutants

Simran Nasra and Ashutosh Kumar

CONTENTS

8.1 Introduction 165
8.2 Cytochrome P450 167
8.3 Pesticides as an Environmental Pollutant 168
    8.3.1 Adverse Effects of Pesticides 168
    8.3.2 Routes of Entry for Pesticides 169
    8.3.3 Metabolism of Pesticides 169
8.4 Metabolism of Endocrine Disruptors 172
8.5 Mutagenic Pollutants 172
8.6 Conclusion 173
References 174

8.1 INTRODUCTION

A significant decline in environmental health has been observed due to high-scale industrialization, urbanization, and growing population burden, and is one of the foremost concerns for developing countries like India. The pollution of the environment by various toxic pollutants discharged from various natural and anthropogenic human activities and its undesirable consequences on living creatures demands attention [1]. Environmental pollution is caused by the introduction of toxic materials like gaseous pollutants, harmful metals, and particulate matter (PM) into the atmosphere; sewage, industrial effluents, agricultural runoffs, and electronic wastes into water bodies; and activities such as mining, deforestation, landfills, and prohibited dumping of litter that cause land pollution [2]. The challenges of environmental pollutants like plastics and agricultural pollutants like pesticides have become a question of debate in recent times. This type of pollution harms the ecosystems, retards biodiversity, and eventually has the capability to disturb the lives of birds, fishes, crabs, turtles, and marine creatures [3]. Additionally, pollution negatively affects genetic variability and the biodiversity of the natural population. Findings have suggested that genomes in fish residing in polluted environments retain substantially complex ribosomal sequences. A systematic surge in the number of copies of the ribosomal DNA is seen, which appears as a reaction of alterations in environmental settings. This occurs as these sequences occur predominantly in the conservation of genome integrity [4], [5].

The Global Burden of Disease (GBD), an extensive local and global disease burden research programme that evaluates mortality and disability from serious illnesses, accidents, and associated risk factors, reported that one element of ambient pollution, PM, with an aerodynamic size of 2.5 mm (PM 2.5), is the fifth highest cause of death in the world, accounting for 4.2 million deaths and more than 103 million disability accommodated life years lost in 2015 [6]. It was also reported that shorter newborn telomere length is correlated with maternal exposures to PM 2.5, PM 10, CO, and SO2 during the third trimester of pregnancy [7]. This infers that not only mankind is in danger of these pollutants, but also that they cause severe susceptible threats to unborn children. Early embryonic development may be an especially vulnerable time for exposure to PAHs, leading to an increase in PAH-DNA adducts and alterations in DNA methylation. It has been revealed that several pollutants such as persistent organic pollutants (POPs) and polycyclic aromatic hydrocarbons (PAHs) are associated with PM, mainly PM 2.5, and cause cardiovascular diseases, respiratory diseases, cancer, and noncancer effects in humans [8]. Airborne pollutants tend to move a longer distance and create more hazardous effects as they reach the population either through breathing, drinking water, or exposed foods, thereby polluting them. There are several studies describing rising applications of engineered nanomaterials

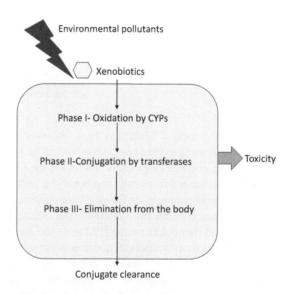

FIGURE 8.1 Metabolic process in hepatocytes.

[9] in daily use and food products that report their trophic transmit and resulting impact on ecology [10], [11], [12]. The growing use of artificial nanoparticles raises concerns about their harmful impacts on the environment and human health [13], [14], [15], [16].

Therapeutic drugs, environmental pollutants, cosmetics, and substances present in our food like additives are a part of an extensive category of xenobiotics, which can be harmless, but tend to be toxic in many cases. Xenobiotics are metabolized and subsequently removed via the urine, bile, and faeces, along with some minor routes through expiration and sweat. However, without efficient detoxification and successive excretion, various substances may become toxic and interfere with cellular homeostasis, causing cellular and tissue impairment, with damaging effects on health [17], [18], [19]. Four stages can be classified in the course of absorption, metabolism, and cellular excretion of xenobiotics: (1) influx by transporter enzymes, biotransformation in (2) Phases I and (3) Phase II, mediated by drug-metabolizing enzymes (DMEs), followed by (4) Phase III, the excretion facilitated generally through transporter enzyme as explained in Figure 8.1 [20], [21].

8.2 CYTOCHROME P450

Amongst the enzymes of phase I biotransformation, cytochrome P450 (CYPs) exhibit noteworthy catalytic diversity. This enzyme superfamily, prevailing across 50 forms, is the majorly crucial enzyme system involved in the biotransformation of many components including pesticides. CYPs are unevenly dispersed in different tissues and can be found in more or less all tissues and organs. Physiological xenobiotic substrates contain drugs, herbal toxins, and pesticides. CYPs largely catalyse oxidative reactions, which consisits of insertion of one oxygen atom into the substrate ($RH + O_2 + 2e- + 2H+ \rightarrow ROH + H_2O$), although the activated oxygen atom may not necessarily be incorporated but used in different types of reactions.

CYPs are primarily divided into three groups [22]. The first contains CYP families 5 to 51 with an increased affinity for endogenous substrates, which are well conserved throughout evolution. The second group contains CYP families 1 to 3, which have a lower affinity for their substrates and have been less conserved in evolution. The third subset includes CYP family 4, which is known to metabolize fatty acids, related substrates, and some xenobiotics. CYP families 1 to 3 are responsible for 70% to 80% of all phase I- dependent metabolism number of xenobiotic substances [23]. Largely, the enzymes in CYP families 1 to 3 show inter-individual irregularity in catalytic activity because of genetic polymorphisms or unpredictability in expression levels. Every CYP isoform contains its individual set of metabolized substrates. The single xenobiotic is proficient at being metabolized by various isoforms into the same or distinct metabolic products. Enzymes categorised in CYP families 1 to 4 are shown to have intersecting substrate specificity.

The essential enzymes for pesticide metabolism are CYP2C9*, CYP2C19, CYP2D6, and CYP3A4 and are the most significant isoforms responsible for the biotransformation of pesticides. In CYP families 2 and 3, new genes like CYP2R1, CYP2S1, CYP2U1, and CYP3A43 are involved. There is no noticeable association between the number of hepatic CYPs and their fundamental importance in the metabolism of pesticides. This could imply that substantially expressed CYPs show a vital function in food metabolism and also play a predominant role in pesticide metabolism. Most CYP enzymes are favorably expressed in the centrilobular area of the liver. Most CYPs participating in the biotransformation of pesticide can be preferentially induced [24]. CYP2D6 is an exemption where numerous gene copies are responsible for expanded detoxifying potential of the enzyme [25]. Induction is a major adaptive reaction against environmental pollutants. CYP expression can be constrained at various levels like transcriptional, mRNA, translational, and posttranslational stages. Transcriptional control is extremely imperative and three crucial cytosolic receptors sense the concentration of environmental xenobiotics: pregnane X receptor (PXR), constitutive androgen receptor (CAR) and aryl hydrocarbon receptor (AhR). AhR controls CYP1A1, CYP1A2, and CYP2S1; PXR controls CYP2C9 and CYP3A4; and CAR controls CYP2B6, CYP2C9, and CYP3A4. Activation of CYPs as well as phase II and phase III proteins is stimulated by increased cellular concentration of pesticides that may result in enhanced protein expression and consequently in lower concentration of pesticides. It is distinct that these transcriptional factors are involved in the control of most human pesticide-metabolizing CYPs.

8.3 PESTICIDES AS AN ENVIRONMENTAL POLLUTANT

Exposure to pesticide is a global threat. Today human susceptibility to the carcinogenic effects of pesticides has become a major research goal. Persistent exposure to low levels of pesticide can lead to mutations. The long-established toxicity demands understanding of the associations connecting chemicals in various doses and tissue pathology. Various studies are fabricated to deliver the potential impact that pesticide exposure may have on farm workers and agricultural helpers. Moreover, agricultural pesticides are employed in different formulations depending on the time of the growing season [26]. This makes the exposures complicated, and the bio-monitoring of the particular compound for exposure evaluation might become challenging. The possible combined lethal effects of such complex exposures are not normally understood [27]. The more important feature in pesticide effect is the proficiency of an organism to metabolize it and active ingredients or formulates alone is not enough to estimate the risk of detrimental health outcomes from pesticide exposure [28]. A basic dogma of toxicology is that toxic effects are a function of the amount of the bioactive form of a chemical in a recipient organ. Consequently, the degree and duration of a toxic response are dependent on how much of the bioactive moiety moves to its desired site and the time it remains there. This is a function of the magnitude of the chemical's system absorption, distribution, metabolism, interaction with cellular components, and elimination [26].

8.3.1 ADVERSE EFFECTS OF PESTICIDES

Organophosphate pesticides (OP) triesters of phosphoric acid are a widely used group of pesticides in the world. OPs have progressed to a more economical substitute for organochlorines. This group also includes malathion, parathion, and dimethoate, which are recognised for their undesirable properties in the role of cholinesterase enzyme [29], ability to reduce insulin secretion, disruption of normal cellular metabolism of proteins, carbohydrates and fats [30], and also their genotoxic effects [31], [32] and influences on mitochondrial function, affecting cellular oxidative stress and creating challenges for the nervous and endocrine systems. OPs also trigger serious health effects such as cardiovascular diseases [33], undesirable effects on the male reproductive system [34] and on the nervous system, and also a possible augmented risk for non-Hodgkin's lymphoma. In addition, prenatal exposure to OPs has been associated with diminished gestational duration and neurological problems appearing in children [35].

8.3.2 Routes of Entry for Pesticides

The chemical state of the pollutant, namely, solid, liquid, or gas, alters the probabilities of pesticide introduction into the body [36]. Liquid or gas substances can enter the body via all routes of admission, whereas solids tend to have a lower chance of entry through the lungs. Also, if solid particles of the pesticide are minute enough or if they linger on the skin long enough, diffusion into the body can occur in a similar means as those of liquids or gases. The most frequent pathway for pesticide poisoning among regular users is absorption via skin [37]. Absorption via skin may take place as a result of splashes and spills while handling pesticides. To a minimal degree, dermal absorption may occur from exposure to a large amount of residue. The degree of hazard by dermal absorption varies based on the toxicity of the pesticide to the skin, the extent of the exposure, the pesticide formulation, and the body part contaminated. Powders, dusts, and granular pesticides are not absorbed as easily through the skin and other body tissues as are the liquid formulations. On the other hand, liquid pesticides containing solvents like organic solvents and oil-based pesticides typically are absorbed more rapidly than dry pesticides. For instance, the emulsifiable concentrates, comprising a large percentage of the toxic substances in a comparatively small amount of solvent, are quickly absorbed by the skin. Specific body areas are more predisposed to absorption of pesticides than other areas.

8.3.3 Metabolism of Pesticides

Metabolic enzymes are categorised into two groups: Phase I and Phase II enzymes [38]. As described in Figure 8.2 phase I reactions are facilitated primarily by cytochrome P450 family of enzymes, but other enzymes like flavin monooxygenases, peroxidases, amine oxidases, dehydrogenases, and xanthine oxidases also catalyse the oxidation of certain functional groups. In addition to the oxidative reactions, there are different types of hydrolytic reactions catalysed by enzymes like carboxylesterases and epoxide hydrolases.

Phase I products do not characteristically eliminate quickly, but endure a subsequent reaction in which substrate such as glucuronic acid, sulfuric acid, acetic acid, or an amino acid combines with the existing or newly added or exposed functional group to form a highly polar conjugate to make them more easily excreted [39]. Proteins participating in pesticide disposition in the body are classified as phase I (oxidative), phase II (conjugative) metabolizing enzymes, or phase III transporters

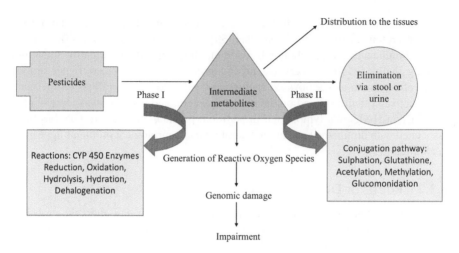

FIGURE 8.2 Metabolism of pesticides in Phase I and II.

involved in efflux mechanisms. The main enzymes of phase I metabolism are heme-thiolate proteins of the CYPs. Phase I enzymes generate functional groups that may consequently serve as a site for conjugation catalysed by phase II enzymes UDP-glucuronosyltransferases (UGT), sulfotransferases (SULT), glutathione s-transferase (GST), and n-acetyltransferase (NAT). These enzyme reactions are necessary for a lipophilic compound to biotransform into a water-soluble product that can be excreted in urine. Phase III transporters like p-glycoprotein (Pgp), multidrug resistance associated proteins (MRPs), and organic anion transporting polypeptide 2 (OATP2) are expressed in several tissues such as the liver, intestine, kidney, and brain, and play a significant role in pesticide absorption, distribution, and excretion. Alongside phase I and phase II enzyme induction/inhibition, pre-treatment with different inducers or inhibitors has been shown to vary the expression of phase III transporters, with the ultimate results of altered excretion of pesticides. Exposure to phase I, phase II, and phase III inducers might induce cellular stress response resulting in enhancement in gene expression, which eventually enhances the elimination and clearance of pesticides.

Biotransformation is a primary determinant of the toxicity of many pesticides and other chemicals. Biotransformation leads to detoxification and hastened removal of some pesticides. The parent compound, for instance, is responsible for the neurotoxic action of pyrethroids. These compounds are inactivated by the concerted actions of carboxylesterases and P450s-catalysed hydroxylation and subsequent conjugation [40]. OPs are also detoxified by esterase-catalyzed hydrolysis, even though desulfuration by P450s gives rise to oxons, the neurotoxic moieties that associate with and inhibit acetylcholinesterase [41]. The distinct acute toxicity of chlorpyrifos in immature rats is assigned to their deficiencies of chlorpyrifos-oxonase, which is the A-esterase that hydrolyzes the oxon [42], and of carboxylesterase [43]. Therefore, recognition of metabolic differences is indispensable to comprehension of variances in the toxicity of xenobiotics in different cells, tissues, species, strains, sexes,

TABLE 8.1
Several Common Pesticides Metabolised by a Particular CYP

Pesticide	Participating CYPs	Role
Diazion	CYP1A1, CYP2C9, CYP1A2.CYP2C19, YP2D6, CYP2E1 CYP3A4, CYP3A7, CYP2C8, CYP3A5	Desulfuration [47]
Carbofuran	CYP1A2, CYP2C19, CYP3A4	Ring oxidation [58]
Diuron	CYP1A1, CYP2C9, CYP1A2.CYP2C19, CYP2D6, CYP2E1, CYP3A4, CYP3A7, CYP2C8, CYP3A5	N-Demethylation [62]
Carbosulfan	CYP1A1, CYP2C9, CYP1A2.CYP2C19, CYP2D6, CYP2E1 CYP3A4, CYP3A7, CYP2C8	N-S Clevage Sulfoxidation [63]
DEET	CYP3A4, CYP2C19, CYP2D6, CYP2E1, CYP3A5	N-Deethylation [58]
Endosulfan	CYP2B6, CYP2C9, CYP3A4, CYP3A5, CYP3A7	Sulfoxidation [64]
Malathion	CYP1A2, CYP2B6, CYP3A5, CYP3A7	Desulfuration [65]
Parathion	CYP1A2, CYP2C9, CYP2D6, CYP3A5, CYP3A7	Desulfuration [47]
Atrazine	CYP1A1, CYP2C9, CYP1A2, CYP2C19, CYP2D6, CYP2E1, CYP3A4, CYP3A7	N-Deethylation N-Deisopropylation [66]
Carbaryl	CYP1A1, CYP2C9, CYP1A2.CYP2C19, CYP2D6, CYP2E1 CYP3A4, CYP3A7, CYP2C8	Aromatic hydroxylation Methyl Oxidation [67]

races, and age groups. Main phase I enzyme is the superfamily of CYP-dependent monooxygenases. P450s exist as a large superfamily of proteins and are the principal enzymes involved in the oxidation of pesticides [44]. The common chemical reactions involved in phase I are aromatic hydroxylations, aliphatic hydroxylations, oxidative N-dealkylations, oxidative O-dealkylations, S-oxidations, reductions, and hydrolysis. UGTs, SULTs, NAT, and GSTs of phase II enzymes help in detoxification of pesticides by catalyzing conjugation reactions [45]. Several common pesticides metabolised by a particular CYP are described in Table 8.1.

The metabolizing enzymes are responsible for protecting the organism by rapidly processing chemicals to inert derivatives that can be easily removed from the body through urine or bile [46]. Besides detoxification, most of the time these enzymes also mediate the toxicity of chemicals through metabolic activation of protoxins and pro-carcinogens, which is thought to play an essential role in individual susceptibility to chemical-induced diseases and cancer. The liver is the main organ for pesticide metabolism and transformation reactions. In addition to the use of OPs in agriculture, disease control, and as remedial agents, they are also used in industries as solvents, flame retardants, and in defence forces as nerve agents [28]. OPs inhibit acetylcholinesterase resulting in harmful effects on human health. Studies have been conducted in order to determine the relationship between OP exposure and cancers of different types. Experimental studies both in vitro and in vivo have reported that a number of OPs exert genotoxic action. OPs are first and primarily metabolized by numerous hepatic CYPs that turn into an active intermediate organophosphate-oxons (OP-oxons) [47]. These active intermediate OP-oxons are hydrolyzed by paraoxonase to 4-nitrophenol and diethyl phosphate. These oxons are known to be the mediator of sensitive OP toxicity, due to their capacity to unite and hinder acetylcholinesterase in the nervous system and at neuromuscular junctions [48]. The genotype of an individual can remarkably influence the disposition of a chemical and determine their susceptibility to its toxicity. Exposure to different chemicals can result in different gene expression, which in turn can lead to different pharmacodynamic effects. Along with genomics toxicological research can provide improved understanding of how various xenobiotics act in the human body.

8.4 METABOLISM OF ENDOCRINE DISRUPTORS

Endocrine-disrupting chemicals (EDCs) are natural or human-made compounds that tend to interfere with the endocrine system of an organism [49], [50]. EDCs can mimic or inhibit naturally arising hormones by binding classical nuclear hormone receptors or by disrupting other pathways regulating hormone synthesis or action, resulting in disruption of normal physiology and homeostatic processes of the organism [51]. In recent years, growing clinical, experimental, developmental, and physiological data show that EDCs disrupt cellular and whole-body metabolism [52]. Certainly, new research has recognized exposures to these metabolism-disrupting chemicals (MDCs) as playing a contributing role in a broad spectrum of metabolic disorders in humans, comprising obesity, diabetes, dyslipidemia, nonalcoholic fatty liver disease (NAFLD), and cardiovascular dysfunction [53].

Among various mechanisms of action, lipophilic EDCs compounds can bind specifically to nuclear receptors and can displace the corresponding endogenous ligands to modulate hormone-responsive pathways. Enduring organic pollutants such as OPs, dioxins, and polyfluoroalkyl compounds and nonpersistent pollutants such as bisphenol A and several phthalates are thought to be responsible for the metabolic disruption activity. Among different mechanisms of action, lipophilic EDC compounds can bind specifically to nuclear receptors and can relocate the equivalent endogenous ligands to modulate hormone-responsive pathways. Persistent organic pollutants such as OPs, dioxins, and polyfluoroalkyl compounds, and nonpersistent pollutants such as bisphenol A and several phthalates are thought of metabolic disruption activity [54]. A wide range of substances both natural like hormones and phytoestrogens such as genistein and man-made ones are believed to cause endocrine disruption, including pharmaceuticals, dioxin and dioxin-like compounds, polychlorinated biphenyls, DDT and other pesticides, and plasticizers such as bisphenol A. Endocrine disruptors may be found in many everyday products, including plastic bottles, metal food cans, detergents, flame retardants, food, toys, cosmetics, and pesticides [55]. The EDCs are bio-transformed by CYPs, human estrogen sulfotransferase 1E1 (SULT1E1), and other phase II enzymes, and can lead to the formation of bioactive metabolites.

8.5 MUTAGENIC POLLUTANTS

Mutagens are materials that change the genetic information of an organism, typically by changing DNA. Mutagens are generally also carcinogens as mutations often cause cancer. General mutagens are ethidium bromide, formaldehyde, dioxane, and nicotine. Mutation is the replacement of nitrogen base with another in one or both the strands or the addition or deletion of a base pair in a DNA molecule [56].

Xenobiotic metabolism takes place in the hepatocyte and the role of CYPs are pivotal in biotransformation of xenobiotics. Apart from hydroxylation, CYPs catalyse various chemical reactions like oxygenation and dealkylation. Various DNA adducts are fixed as mutations through replication of DNA. Reactive oxygen species generated by pollutants also induce the creation of DNA adducts. DNA adducts have been detected as a marker for the exposure of humans and wildlife to mutagens. Due to its high sensitivity the 32P-postlabel-thin layer chromatography (TLC) method is widely used for the analysis of DNA adducts formed by PAH and related bulky compounds. Conversely, new systems are required for detecting mutations induced in genomic DNA in vivo to monitor environmental mutagens [57].

The roles of human CYP2A6 and CYP2E1 in the mutagenic activation of NDMA, NDEA, NDPA, and NDBA have been shown by using the *S. typhimurium* YG7108 cells expressing CYP2A6 or CYP2E1. Pharmacokinetic studies of several environmental toxicants in *Cyp1* knockout mouse lines have shown that CYP1A1, CYP1A2, and CYP1B1 might be beneficial or damaging based on their time-specific, organ-specific, tissue-specific, and cell-type-specific expression [58]. These new discoveries imply that animal studies and human epidemiological studies might need to be reconsidered. Whereas oral drugs more commonly use the portal vein system (and first-pass elimination kinetics), we suggest that the lymphatic system might be more important in delivering the very hydrophobic oral polycyclic aromatic hydrocarbons (PAHs) and polyhalogenated aromatic hydrocarbons (PHAHs) to target tissues [59]. This is particularly pertinent to clinical medicine, because oral exposure to these procarcinogens is the most predominant route. More focus on pharmacokinetic and pharmacodynamic studies of hydrophobic procarcinogens in animal models and their extrapolation to humans is required. A two-tiered phenomenon occurs when it comes to the chance of developing cancer as a consequence of the environment. First, human inter-individual variation in the upfront hPpM characteristics represented by phase I genes will have a major effect on a small (5 – 15%) portion of any population that lacks large polymorphisms in the downstream target genes. Second, all downstream targets of phase-I-mediated reactive intermediates always contribute to the unimodal gradient response, and each of these targets may have a unique essential polymorphism. [60], [61].

8.6 CONCLUSION

Human CYP enzymes constitute a superfamily of membrane-bound hemoproteins that are responsible for the metabolism of a variety of clinically, physiologically, and toxicologically vital compounds. These heme-thiolate monooxygenases play a pivotal role in the detoxification of xenobiotics, taking part in the metabolism of several structurally diverge compounds. During the metabolism of certain pesticides, a variety of reactive and unstable intermediates can be formed, which attack DNA, causing cell toxicity and transformation. Individuals vary in the levels of expression and catalytic activities of metabolic enzymes that activate or detoxify pesticides in several organs. With the prevailing PCR-based techniques rapid development of molecular biology approaches for the identification of CYP gene polymorphisms and genotyping screening can be done and importance of different polymorphic variants can be discovered.

REFERENCES

[1] P. Chowdhary, V. Hare, and A. Raj, Book review: Environmental pollutants and their bioremediation approaches, *Frontiers in bioengineering and biotechnology* 6 (2018) 193.

[2] P.O. Ukaogo, U. Ewuzie, and C.V. Onwuka, Environmental pollution: causes, effects, and the remedies, in Pankaj Chowdhary, Abhay Raj, Digvijay Verma and Yusuf Akhter (eds.) *Microorganisms for Sustainable Environment and* Health, Elsevier 2020, pp. 419–429.

[3] L.G.A. Barboza, A. Cózar, B.C. Gimenez, T.L. Barros, P.J. Kershaw, and L. Guilhermino, *Macroplastics pollution in the marine environment, World seas: An environmental evaluation,* Elsevier 2019, pp. 305–328.

[4] F.A. da Silva, E. Feldberg, N.D.M. Carvalho, S.M.H. Rangel, C.H. Schneider, G.A. Carvalho-Zilse, V.F. da Silva, and M.C. Gross, Effects of environmental pollution on the rDNAomics of Amazonian fish, *Environmental pollution* 252 (2019) 180–187.

[5] M. Shafiq, S. Anjum, C. Hano, I. Anjum, and B.H. Abbasi, An overview of the applications of nanomaterials and nanodevices in the food industry, Foods 9(2) (2020) 148.

[6] D.E. Schraufnagel, J.R. Balmes, C.T. Cowl, S. De Matteis, S.-H. Jung, K. Mortimer, R. Perez-Padilla, M.B. Rice, H. Riojas-Rodriguez, and A. Sood, Air pollution and noncommunicable diseases: a review by the Forum of International Respiratory Societies' Environmental Committee, Part 2*:* Air pollution and organ systems, Chest 155(2) (2019) 417–426.

[7] L. Song, B. Zhang, B. Liu, M. Wu, L. Zhang, L. Wang, S. Xu, Z. Cao, and Y. Wang, Effects of maternal exposure to ambient air pollution on newborn telomere length, Environment international 128 (2019) 254–260.

[8] Herbstman, J. B., Tang, D., Zhu, D., Qu, L., Sjödin, A., Li, Z., ... & Perera, F. P. (2012). Prenatal exposure to polycyclic aromatic hydrocarbons, benzo [a] pyrene–DNA adducts, and genomic DNA methylation in cord blood. Environmental health perspectives, 120(5), 733–738.

[9] K. Kansara, P. Patel, R.K. Shukla, A. Pandya, R. Shanker, A. Kumar, and A. Dhawan, Synthesis of biocompatible iron oxide nanoparticles as a drug delivery vehicle, *International journal of nanomedicine* 13(T-NANO 2014 Abstracts) (2018) 79.

[10] G.S. Gupta, A. Kumar, V.A. Senapati, A.K. Pandey, R. Shanker, and A. Dhawan, Laboratory scale microbial food chain to study bioaccumulation, biomagnification, and ecotoxicity of cadmium telluride quantum dots, *Environmental science & technology* 51(3) (2017) 1695–1706.

[11] G.S. Gupta, A. Kumar, R. Shanker, and A. Dhawan, Assessment of agglomeration, co-sedimentation and trophic transfer of titanium dioxide nanoparticles in a laboratory-scale predator-prey model system, *Sci Rep* 6(1) (2016) 1–13.

[12] N. Dasgupta, S. Ranjan, D. Mundekkad, C. Ramalingam, R. Shanker, and A. Kumar, Nanotechnology in agro-food: from field to plate, *Food Research International* 69 (2015) 381–400.

[13] V. Sharma, A. Kumar, and A. Dhawan, Nanomaterials: exposure, effects and toxicity assessment, Proceedings of the National Academy of Sciences, *India Section B: Biological Sciences* 82(1) (2012) 3–11.

[14] A. Kumar, R. Shanker, and A. Dhawan, Nanotoxicity: aquatic organisms and ecosystems, aquananotechnology: *Global Prospects* (2014) 97.
[15] R.K. Shukla, A. Kumar, N.V.S. Vallabani, A.K. Pandey, and A. Dhawan, Titanium dioxide nanoparticle-induced oxidative stress triggers DNA damage and hepatic injury in mice, *Nanomedicine* 9(9) (2014) 1423–1434.
[16] V.A. Senapati, A. Kumar, G.S. Gupta, A.K. Pandey, and A. Dhawan, ZnO nanoparticles induced inflammatory response and genotoxicity in human blood cells: a mechanistic approach, *Food and chemical toxicology* 85 (2015) 61–70.
[17] E. Croom, Metabolism of xenobiotics of human environments, *Progress in molecular biology and translational science* 112 (2012) 31–88.
[18] M.R. Juchau, The role of the placenta in developmental toxicology, in Keith Snell (ed.) *Developmental Toxicology,* Springer 1982, pp. 187–209.
[19] C.H. Johnson, A.D. Patterson, J.R. Idle, and F.J. Gonzalez, Xenobiotic metabolomics: major impact on the metabolome, *Annual review of pharmacology and toxicology* 52 (2012) 37–56.
[20] O.A. Almazroo, M.K. Miah, and R. Venkataramanan, Drug metabolism in the liver, *Clinics in liver disease* 21(1) (2017) 1–20.
[21] K. Bachmann, Drug metabolism. In Miles Hacker, William Messer and Kenneth Bachmann (eds.) *Pharmacology* 2009. Academic Press, pp. 131–173. https://doi.org/10.1016/C2009-0-01474-3
[22] M. Krajinovic, H. Sinnett, C. Richer, D. Labuda, and D. Sinnett, Role of NQO1, MPO and CYP2E1 genetic polymorphisms in the susceptibility to childhood acute lymphoblastic leukemia, *International journal of cancer* 97(2) (2002) 230–236.
[23] H. Autrup, Genetic polymorphisms in human xenobiotica metabolizing enzymes as susceptibility factors in toxic response, *Mutation research/genetic toxicology and environmental mutagenesis* 464(1) (2000) 65–76.
[24] X. Yang, S. Solomon, L.R. Fraser, A.F. Trombino, D. Liu, G.E. Sonenshein, E.V. Hestermann, and D.H. Sherr, Constitutive regulation of CYP1B1 by the aryl hydrocarbon receptor (AhR) in pre-malignant and malignant mammary tissue, *Journal of cellular biochemistry* 104(2) (2008) 402–417.
[25] T. Oinonen, and O.K. Lindros, Zonation of hepatic cytochrome P-450 expression and regulation, *Biochemical journal* 329(1) (1998) 17–35.
[26] M.C. Alavanja, J.A. Hoppin, and F. Kamel, Health effects of chronic pesticide exposure: cancer and neurotoxicity, *Annu. rev. public health* 25 (2004) 155–197.
[27] N.R. Council, *Intentional human dosing studies for EPA regulatory purposes: Scientific and ethical issues* 2004. The National Academies Press.
[28] W. Hayes, Dosage and other factors influencing toxicity, *Handbook of pesticide toxicology* 1 (1991) 39–105.
[29] K. Jaga, and C. Dharmani, Sources of exposure to and public health implications of organophosphate pesticides, *Revista panamericana de salud pública* 14 (2003) 171–185.
[30] S. Karami-Mohajeri, and M. Abdollahi, Toxic influence of organophosphate, carbamate, and organochlorine pesticides on cellular metabolism of lipids, proteins, and carbohydrates: a systematic review, *Human & experimental toxicology* 30(9) (2011) 1119–1140.
[31] D. Li, Q. Huang, M. Lu, L. Zhang, Z. Yang, M. Zong, and L. Tao, The organophosphate insecticide chlorpyrifos confers its genotoxic effects by inducing DNA damage and cell apoptosis, Chemosphere 135 (2015) 387–393.
[32] M.C. Alavanja, M. Dosemeci, C. Samanic, J. Lubin, C.F. Lynch, C. Knott, J. Barker, J.A. Hoppin, D.P. Sandler, and J. Coble, Pesticides and lung cancer risk in the

agricultural health study cohort, *American journal of epidemiology* 160(9) (2004) 876–885.
[33] D.-Z. Hung, H.-J. Yang, Y.-F. Li, C.-L. Lin, S.-Y. Chang, F.-C. Sung, and S.C. Tai, The long-term effects of organophosphates poisoning as a risk factor of CVDs: a nationwide population-based cohort study, *PLoS one* 10(9) (2015) e0137632.
[34] F. Jamal, Q.S. Haque, S. Singh, and S. Rastogi, *Retracted: The influence of organophosphate and carbamate on sperm chromatin and reproductive hormones among pesticide sprayers*, SAGE Publications, 2016.
[35] V.A. Rauh, W.E. Garcia, R.M. Whyatt, M.K. Horton, D.B. Barr, and E.D. Louis, Prenatal exposure to the organophosphate pesticide chlorpyrifos and childhood tremor, *Neurotoxicology* 51 (2015) 80–86.
[36] A. Berthet, N.B. Hopf, A. Miles, P. Spring, N. Charrière, A. Garrigou, I. Baldi, and D. Vernez, Human skin in vitro permeation of bentazon and isoproturon formulations with or without protective clothing suit, *Archives of toxicology* 88(1) (2014) 77–88.
[37] E. MacFarlane, R. Carey, T. Keegel, S. El-Zaemay, and L. Fritschi, Dermal exposure associated with occupational end use of pesticides and the role of protective measures, *Safety and health at work* 4(3) (2013) 136–141.
[38] F. Oesch, M.E. Herrero, J.G. Hengstler, M. Lohmann, and M. Arand, Metabolic detoxification: implications for thresholds, *Toxicologic pathology* 28(3) (2000) 382–387.
[39] H. Joo, K. Choi, and E. Hodgson, Human metabolism of atrazine, *Pesticide biochemistry and physiology* 98(1) (2010) 73–79.
[40] D.M. Soderlund, J.M. Clark, L.P. Sheets, L.S. Mullin, V.J. Piccirillo, D. Sargent, J.T. Stevens, and M.L. Weiner, Mechanisms of pyrethroid neurotoxicity: implications for cumulative risk assessment, *Toxicology* 171(1) (2002) 3–59.
[41] L.G. Sultatos, Mammalian toxicology of organophosphorus pesticides, *Journal of toxicology and environmental health, Part A Current Issues* 43(3) (1994) 271–289.
[42] S. Mortensen, S. Chanda, M. Hooper, and S. Padilla, Maturational differences in chlorpyrifos-oxonase activity may contribute to age-related sensitivity to chlorpyrifos, *Journal of biochemical toxicology* 11(6) (1996) 279–287.
[43] V.C. Moser, S.M. Chanda, S.R. Mortensen, and S. Padilla, Age- and gender-related differences in sensitivity to chlorpyrifos in the rat reflect developmental profiles of esterase activities, *Toxicological sciences* 46(2) (1998) 211–222.
[44] K. Abass, M. Turpeinen, A. Rautio, J. Hakkola, and O. Pelkonen, Metabolism of pesticides by human cytochrome P450 enzymes in vitro—a survey, *Insecticides—advances in integrated pest management* (2012) 165–194. DOI:10.5772/28088
[45] C. Infante-Rivard, D. Labuda, M. Krajinovic, and D. Sinnett, Risk of childhood leukemia associated with exposure to pesticides and with gene polymorphisms, *Epidemiology* 10(5) (1999) 481–487.
[46] E. Arinç, J.B. Schenkman, and E. Hodgson, *Molecular aspects of monooxygenases and bioactivation of toxic compounds*, Springer Science & Business Media 2012.
[47] E. Mutch and F.M. Williams, Diazinon, chlorpyrifos and parathion are metabolised by multiple cytochromes P450 in human liver, *Toxicology* 224(1–2) (2006) 22–32.
[48] I. Cascorbi, J. Brockmöller, P.M. Mrozikiewicz, A. Müller, and I. Roots, Arylamine N-acetyltransferase activity in man, *Drug metabolism reviews* 31(2) (1999) 489–502.
[49] F.W. Bazer, G. Wu, G.A. Johnson, and X. Wang, Environmental factors affecting pregnancy: endocrine disrupters, nutrients and metabolic pathways, *Molecular and cellular endocrinology* 398(1–2) (2014) 53–68.
[50] Y. Gibert, R.M. Sargis, and A. Nadal, Endocrine disrupters and metabolism, *Frontiers in endocrinology* 10 (2019) 859.

[51] K.P. Phillips and W.G. Foster, Key developments in endocrine disrupter research and human health, *Journal of toxicology and environmental health, Part B* 11(3–4) (2008) 322–344.

[52] R.M. Sargis and R.A. Simmons, Environmental neglect: endocrine disruptors as underappreciated but potentially modifiable diabetes risk factors, *Diabetologia* 62(10) (2019) 1811–1822.

[53] U. Shahnazaryan, M. Wójcik, T. Bednarczuk, and A. Kuryłowicz, Role of obesogens in the pathogenesis of obesity, *Medicina* 55(9) (2019) 515.

[54] C. Casals-Casas, and B. Desvergne, Endocrine disruptors: from endocrine to metabolic disruption, *Annual review of physiology* 73 (2011) 135–162.

[55] C. Monneret, What is an endocrine disruptor?, *Comptes rendus biologies* 340(9–10) (2017) 403–405.

[56] P. Manikandan, and S. Nagini, Cytochrome P450 structure, function and clinical significance: a review, *Current drug targets* 19(1) (2018) 38–54.

[57] H. Sato, and Y. Aoki, Mutagenesis by environmental pollutants and bio-monitoring of environmental mutagens, *Current drug metabolism* 3(3) (2002) 311–319.

[58] K.A. Usmani, R.L. Rose, J.A. Goldstein, W.G. Taylor, A.A. Brimfield, and E. Hodgson, In vitro human metabolism and interactions of repellent N, N-diethyl-m-toluamide, *Drug metabolism and disposition* 30(3) (2002) 289–294.

[59] M. Ingelman-Sundberg, S.C. Sim, A. Gomez, and C. Rodriguez-Antona, Influence of cytochrome P450 polymorphisms on drug therapies: pharmacogenetic, pharmacoepigenetic and clinical aspects, *Pharmacology & therapeutics* 116(3) (2007) 496–526.

[60] S. Rendic, and F.P. Guengerich, Contributions of human enzymes in carcinogen metabolism, *Chemical research in toxicology* 25(7) (2012) 1316–1383.

[61] Y. Jin, M. Zollinger, H. Borell, A. Zimmerlin, and C.J. Patten, CYP4F enzymes are responsible for the elimination of fingolimod (FTY720), a novel treatment of relapsing multiple sclerosis, *Drug metabolism and disposition* 39(2) (2011) 191–198.

[62] K. Abass, P. Reponen, M. Turpeinen, J. Jalonen, O Pelkonen, Characterization of diuron N-demethylation by mammalian hepatic microsomes and cDNA-expressed human cytochrome P450 enzymes, *Drug Metabolism and Disposition* 35(9) (2007) 1634–1641.

[63] K. Abass, P. Reponen, S. Mattila, O. Pelkonen, Metabolism of carbosulfan II. Human interindividual variability in its in vitro hepatic biotransformation and the identification of the cytochrome P450 isoforms involved, *Chemico-biological interactions* 185(3) (2010) 163–173.

[64] H-K. Lee, J-K. Moon, C-H. Chang, H. Choi, H-W. Park, B-S. Park, et al. Stereoselective metabolism of endosulfan by human liver microsomes and human cytochrome P450 isoforms, *Drug Metabolism and Disposition* 34(7) (2006) 1090–1095.

[65] F.M. Buratti, A. D'aniello, M.T. Volpe, A. Meneguz, E. Testai. Malathion bioactivation in the human liver: the contribution of different cytochrome P450 isoforms, *Drug Metabolism and Disposition* 33(3) (2005) 295–302.

[66] D.H. Lang, A.E. Rettie, R.H. Böcker. Identification of enzymes involved in the metabolism of atrazine, terbuthylazine, ametryne, and terbutryne in human liver microsomes, *Chemical Research in Toxicology* 10(9) (1997) 1037–1044.

[67] J. Tang, Y. Cao, R.L. Rose, E. Hodgson. In vitro metabolism of carbaryl by human cytochrome P450 and its inhibition by chlorpyrifos, *Chemico-biological Interactions* 141(3) (2002) 229–241.

9 Bioavailability, Bioconcentration, and Biomagnification of Pollutants

Shrushti Shah and Ashutosh Kumar

CONTENTS

9.1	Introduction	180
9.2	Food Chain and Food Web	182
	9.2.1 Trophic Levels	184
	9.2.2 Bioavailability, Bioconcentration, and Biomagnification – A Comparison	185
9.3	Pollutants	186
	9.3.1 Heavy Metal Wastes	186
	9.3.1.1 Bioavailability of Heavy Metals	187
	9.3.1.2 Bioconcentration or Bioaccumulation of Heavy Metals	187
	9.3.1.3 Biomagnification of Heavy Metals	190
	9.3.1.4 Radioactive Wastes	190
	9.3.1.5 Bioavailability, Bioaccumulation, and Biomagnification of Radioactive Isotops of Metals	191
	9.3.2 Organic Pollutants	191
	9.3.2.1 Bioavaibility of Organic Pollutants	192
	9.3.2.2 Bioaccumulation of Organic Pollutants	192
	9.3.2.3 Biomagnification of Organic Pollutants	194
	9.3.3 Microplastic Wastes	195
	9.3.3.1 Bioavailability of Microplastics	196
	9.3.3.2 Bioaccumulation or Bioconcentration of Microplastics	197
	9.3.3.3 Biomagnification of the Microplastics	197
	9.3.4 Nanomaterial Pollutants	199
	9.3.4.1 Bioavailability of Nanomaterials	200
	9.3.4.2 Bioaccumulation or Bioconcentration of Nanomaterials	201
	9.3.4.3 Biomagnification of Nanomaterials	203
9.4	Conclusion	203
	References	204

DOI: 10.1201/9781003244349-9

9.1 INTRODUCTION

Environmental pollution is a global issue as it threatens not only human life but also the life of all other organisms of different ecosystems. Any adverse effect caused by the pollutants in the living organisms will directly affect the carrying capacity of that ecosystem [1–4]. Also, the interactions of pollutants in food webs at different trophic levels and possible bio-magnifications is a big concern. Anthropogenic activities have led to an increase in unwanted substances or components in the environment, which has led to pollution. This includes industrial waste, agricultural waste, sludge, sewage waste, etc. There are some natural occurrences of environmental pollution as well like volcano eruption, forest fires, and other natural disasters, but the pollution derived from natural occurrences is much less than that of the wastes obtained from human activities in the past few decades. These different wastes leave behind environmental toxins and end up harming ecological components. Now, one might ask what are environmental toxins? The toxin is a word derived from the Greek word *toxikon,* which means poison [5]. Toxins are substances or compounds that are harmful or *poisonous* to humans as well as other components of the biosphere. The next question that comes to our mind is from where do these toxins arrive in the environment? Human activities are mainly responsible for environmental pollution. Pollutants are toxic (poisonous) or harmful to the environment and its components, and thus are called environmental toxins. This gives rise to a series of questions like What exactly are these pollutants? How do they come into contact with organisms? What happens when these toxins enter organisms' bodies? What harm do they cause to living organisms? What is their probability to biomagnify in different organisms? This chapter aims to address these major questions related to environmental pollution.

Advancements in science, technology, and high throughput instruments and machines facilitate the detection of some of these toxins and their extent of toxicity. However, it is still important to know what toxins are harmful to biota (living organisms) and to what extent the organisms can withstand these toxins. Hence, scientists have experimented with various organisms starting from organisms of lower classifications to higher classifications [6, 7]. These experiments are fruitful in acknowledging the effects of various toxins. Most of the organisms of ecosystems come into contact with these toxins from the environment either directly or indirectly [8–12]. For example, if the toxins are present in water bodies the aquatic organisms come into contact with these toxins via their habitat and food. Similarly, humans also come into contact with various pollutants via different mediums. Environmental pollutions are also of different types like air pollution, water pollution, soil pollution, radioactive pollution, etc., and all these different pollutions give rise to different yet harmful toxins. Toxins like poly aromatic hydrocarbons (PAHs), metal toxins, nanoparticles, microplastics, phthalate esters, benzoyl compounds, etc., are present in these different pollutions, which can harm humans and cause headache and respiratory diseases like asthma, chronic obstructive pulmonary disease (COPD), emphysema, bronchitis, cystic fibrosis, pneumonia, and lung cancer [13–15]. Soil pollution along with water pollution can cause severe problems like cardiac illness, gastroenteritis (inflammation of the gut), liver disease, cancer, and skin allergies [15–19]. Figure 9.1 describes the ways in which the pollutants can enter ecosystems.

Bioavailability, Bioconcentration, and Biomagnification of Pollutants

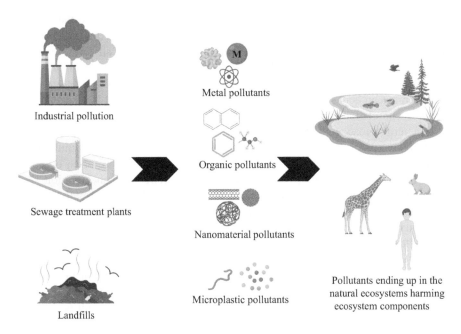

FIGURE 9.1 The different types of pollutions, their origins and effects.

The pollutants/chemicals can enter the body of organisms and cause adverse effects leading to toxicity. In scientific terms, the concentration of different chemicals or drugs (here toxins) that are found in the systemic circulation of organisms is called bioavailability. Bioavailability is the most commonly used term in the field of pharmacology as it deals with the drugs and medicines to be made available to humans. In this chapter, the term bioavailability has been used in relation to toxins. Humans come into contact with these toxins via food, water, and various consumer products; the toxins enter our systemic circulation and that is called bioavailability. When these toxins enter the body of an organism the body will try to reduce the concentration of these toxins by metabolism. However, some of these toxins accumulate in the body, which is called bioaccumulation or bioconcentration. The term bioconcentration is used with respect to aquatic ecosystems; it refers to the concentration of the toxin accumulated in the body. This accumulation when increases step by step in the food chain or trophic levels it increases gradually, and maximizes for the last level or last step of the chain and this phenomenon is known as biomagnification.

Bioavailability refers to the scale of active absorption of a component (i.e., a drug or a toxin) by the body that reaches the systemic circulatory system of the body. As the extent of pollution is increasing on a daily basis, organisms come into contact with different environmental toxins in various ways. This measurement is called bioavailability. Aquatic organisms are greatly affected by the waste disposed of in their habitat. Reports have demonstrated that due to increased demand for plastics, oceans and natural ecosystems are being increasingly polluted due to toxic plastic waste and marine organisms and marine botany are being affected [20–30]. Terrestrial organisms come into contact with these toxins by contaminated food, water, and the air they breathe, and humans have higher chances of exposure to these toxins as most consumer products contain toxins. Plants are an inseparable part of all ecosystems and environmental toxins cause equal harm to plants as well. Toxins present in the plant habitat, either soil or water, can be examined for the bioavailability measurement. However, researchers are interested in understanding the factors of bioconcentration and bioaccumulation of plants [31–35].

The term bioaccumulation suggests the accumulation of certain chemicals in the body (precisely in some of the tissues of the body). Bioaccumulation is the measure of accumulated chemicals in the tissues of the body of an organism and when this accumulation occurs via water it is called bioconcentration [36]. Bioconcentration is the term mostly used for aquatic toxicology. Usually, when a chemical or drug is taken by an organism from the outside, part of that chemical component is metabolized by the body while some of it gets stored or deposited in some tissues of the body. This is referred to as bioaccumulation. If the chemical enters the body directly by water absorption and is accumulated in some tissues of the body it is referred to as bioconcentration. A classical example of this is mercury contamination in aquatic animals [37]. Mercury is a metal waste (discarded by the industries) found in most natural aquatic habitats and as a result is absorbed from the water by aquatic organisms and is deposited in the tissues. This deposition is called bioconcentration. Bioconcentration can be measured with the formula of bioconcentration factor (BCF):

$$BCF = \frac{\text{Concentration of a compound in tissues of the body}}{\text{Concentration of compound in the water (i.e., in the surrounding habitat)}} \quad (9.1)$$

Similarly, the bioaccumulation factor can also be calculated. The deposited toxin when found in higher amount in the tissues of the body than the concentration present outside or the surrounding habitat bioconcentration factor helps to determine the extent of the bioconcentration. Recently published studies have described the bioconcentration factors for different aquatic species of marine, brackish, and freshwater ecosystems for bioaccumulative chemicals (e.g., per- and polyfluoroalkyl substances (PFAS), highly manufactured artificial chemicals used by the number of industries). Various studies have shown high occurrence of bioconcentration of compounds like pesticides, metals like zinc (Zn), cadmium (Cd), mercury (Hg), nickel (Ni), copper (Cu), molybdenum (Mo), nanomaterials, microplastics, etc., in the aquatic species of various geographic locations and have estimated the pervasive effects of these materials on ecosystems. Fish are the most commonly studied organism for examining bioconcentration [38–46]. To understand trophic transfer and biomagnification in detail, it is important to understand the intense phenomenon of the ecology (i.e., food chain).

9.2 FOOD CHAIN AND FOOD WEB

Ecology consists of different ecosystems like the terrestrial ecosystem, which includes forest ecosystem, grassland ecosystem, etc., and the aquatic ecosystem, which includes both marine and freshwater ecosystems. Each ecosystem also has its own food chain. The food chain is a straight connection of ecological components based on their dependence on food or energy. The food chain starts with the autotrophic components (i.e., plants) and ends with the consumer species. The food chain starts from the producers (i.e., plants), which then leads to primary consumers (the organisms that are dependent on plants), secondary consumers (the organisms that are dependent on the primary consumers for food), and tertiary consumers (the organisms that are dependent on the secondary consumers for food) and so on. This can be better understood from Figure 9.2.

The forest ecosystem has herbivorous animals as primary consumers that depend on plants and carnivorous animals that depend on primary consumers for their source of food and energy. The grassland ecosystem has four consumers: primary, secondary, tertiary, and quaternary consumers. While the aquatic ecosystem has three consumers as described. A much related term is the food web. The connection of all the food chains of different ecosystems in the biosphere gives rise to the food web. It connects the food chains by the relationship at various consumer levels in the food chain.

FIGURE 9.2 Three charts describing the food chain of (a) The forest ecosystem; (b) the grassland ecosystem (both (a) and (b) are terrestrial ecosystems); and (c) the aquatic ecosystem.

9.2.1 Trophic Levels

The individual level of the food chain is considered as a trophic level. The first trophic level is the autotrophs, then there are the primary consumers, secondary consumers, and tertiary consumers as second, third, and fourth trophic levels, respectively. The trophic levels are represented as the pyramid graph shown in Figure 9.3. The shape of the pyramid indicates that as one moves from lower to higher trophic levels the amount of energy decreases gradually, as postulated by thermodynamics laws; after each level significant energy is lost in the form of heat.

The potential energy by increasing a step in trophic levels decreases but the inverse occurs in the case of biomagnification. The amount of toxin increases as we move further in up in trophic levels.

Concerning biomagnification, a study done on the marine ecosystem describes the biomagnification of monomethyl mercury (MMHg), which is an abundantly found toxin in the marine ecosystem, and can be understood by referring to Figure 9.4.

The seven species at the bottom indicate the first trophic level and the species are of phytoplankton. The numbers indicate the size of the species in micrometer (μm). The small and large herbivorous zooplankton (written as HZ) depend on the phytoplankton for food while the carnivorous zooplankton (CZ) depend on herbivorous zooplankton for food so the concentration of MMHg present in the phytoplankton then reaches the herbivorous zooplankton with an increase. There is further increase of this toxin in carnivorous zooplankton, and in this way the concentration of a toxin is magnified in the above trophic levels. There are other larger organisms from higher trophic levels that feed on the carnivorous zooplankton for food, and this toxin concentration continues to increase with increasing trophic levels. This phenomenon of transfer of toxins in trophic levels is called trophic transfer. The study highlights the fact that larger zooplankton play an important role in biomagnification as the efficiency to excrete these toxins decreases with an increase in the size of the body, which explains the higher accumulation of these toxins and their gradual

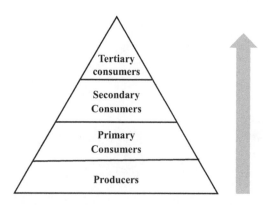

FIGURE 9.3 The trophic levels of the ecosystem. Arrow indicates the direction of increasing trophic levels.

Bioavailability, Bioconcentration, and Biomagnification of Pollutants

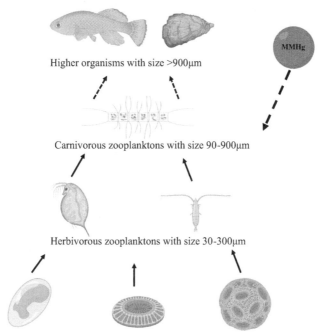

FIGURE 9.4 The biomagnification of monomethyl mercury in the marine ecosystem is depicted.

Source: recreated from Ref. [47].

increase with the increase in trophic levels [47]. Here, the toxin MMHg was studied but in the latter part of this chapter we will describe other toxins that invade the balance of ecosystems.

We can infer the relationship between biomagnification and trophic levels. Biomagnification is directly proportional to the trophic levels; with an increase in trophic levels the extent of biomagnification increases.

Biomagnification| α Trophic level (9.2)

9.2.2 Bioavailability, Bioconcentration, and Biomagnification – A Comparison

Bioavailability is the actual concentration of a toxin that is present in the environment with which organisms come into contact. Bioaccumulation or bioconcentration is the amount of toxin deposited in the body of an organism. Bioconcentration is dependent on time, as time increases the bioconcentration in an organism (directly proportional relationship (i.e., **|Bioconcentration| α Time**). These two factors (i.e., bioavailability and bioconcentration or bioaccumulation) are limited to a specific organism, but when it comes to biomagnification it occurs across the trophic levels and is not limited to a particular organism or species since it has involvement in the food chain levels. As bioconcentration is dependent on time, and biomagnification is dependent on the components of the food chain. Biomagnification is a phenomena used for prediction of species at a higher risk from certain toxins.

9.3 POLLUTANTS

At the start of this chapter several questions were asked: What exactly are these pollutants? How do they come into contact with the organisms? What happens when these toxins enter organisms' bodies? What harm do they cause to living organisms? This section of the chapter will address all these questions one by one. As mentioned above, pollutants are toxins that enter the environment by anthropogenic activities and hinder the balance of natural habitats or ecosystems. There are different types of pollutants, and they can be broadly classified into the following categories:

1. Inorganic pollutants –includes heavy metals and radioactive isotopes of the metals
2. Organic pollutants – includes agricultural and other wastes
3. Microplastics
4. Nano-material pollutants

Further classification of these toxins and detailed analysis of their bioavailability, bioaccumulation or bioconcentration, and biomagnification follows.

9.3.1 Heavy Metal Wastes

In general, there are two different types of metals: light metals with low molecular weight and density $< 5g/cm^3$ (e.g., beryllium, lithium, sodium, boron, etc.) and heavy metals with high molecular weight and density $> 5g/cm^3$ (e.g., zinc, manganese, iron, cadmium, etc.) [48]. The prevalence of heavy metals is higher in both terrestrial as well as aquatic ecosystems. This is a result of anthropogenic activities like excessive metal mining, extraction of metals from ores (i.e., smelting), factories and business units for casting metals, leaching of metals from landfills and waste dumping grounds, sewage, automobile pollution, the building of roads, and usage of heavy metals in agricultural products and pharmaceutical industries. There are several natural causes of heavy metal pollution like volcanic eruptions, corrosion of metals, erosion of soil, re-suspension of soil, and geological weathering, but it is reported that the metal pollution derived from natural occurrences is much lower than that of human activities [49].

Metals are not further broken down in nature as they are non-biodegradable. These metals are available in the surrounding habitat of the organisms either in soil or water, and are taken up by the organisms. Humans are exposed to heavy metals by ingestion of contaminated food and water, inhalation from the atmosphere, through use of certain pharmaceutical products, and contact via dermis. These metals are further stored in the soma, bioaccumulation or bioconcentration [50]. After getting bioaccumulated the heavy metals hinder the natural metabolism of the body, which leads to healthy conditions such as neurotransmitter inhibition and nerve cell damage leading to neurotoxicity, damage to lipid bilayer leading to cellular damage, misfolding, aggregation, structural and conformational change in proteins, loss in enzyme functions leading to loss of cellular function and metabolism, and DNA damage and hindrance in DNA repair mechanism leading to progression of cancer [51].

9.3.1.1 Bioavailability of Heavy Metals

Metal wastes are believed to be omnipresent in natural ecosystems. Measurement of bioavailability from waters is done by a technique called diffusive gradients in thin films (DGT). To measure bioavailability in soil-interacting processes like adsorption, desorption, complexation, dissociation, precipitation, dissolution, diffusion, and physiological parameters, which are strongly specific to the metals due to their different affinities and characteristics like pH, redox status, and conditions, and the content of minerals and oxides used [52, 53].

The bioavailability of metal in a particular resource, either soil or water, depends on various factors like its location, the amount of industrial waste, access by the human population, and the amount of rainfall and hence the metal compound that ends up in the resource. Table 9.1 lists the most common heavy metals, their forms, their industrial origin, and bioaccumulation evidence.

9.3.1.2 Bioconcentration or Bioaccumulation of Heavy Metals

Bioaccumulation of metals in organisms depends on various factors like the bioavailability of the metal in the habitats, behavior, niche, physiology, metabolism, and food consumption habits. Also, the accumulation concentration depends on the particular organism and its adaptations. However, studies done on the different habitats have shown that bioaccumulation occurs in all classes of organisms.

Reports derived from various parts of the globe indicate the accumulation of heavy metals in different organisms including humans. The metals arsenic (As), cadmium (Cd), chromium (Cr), copper (Cu), mercury (Hg), nickel (Ni), lead (Pb), and zinc (Zn) have been found to be accumulated in fish, crustaceans, mollusks, and echinoderms, where the concentration of mercury (Hg) and arsenic (As) was found to be relatively high [54]. Accumulation of these metals was also found to be fairly high in zooplanktons and microorganisms. The organisms of terrestrial ecosystems like arthropods and annelids also had accumulation of metals like lead (Pb), cadmium (Cd), copper (Cu), and zinc (Zn) while accumulation of Pb was found to be relatively high [55].

When these metals enter the human body via the oral route most of them are absorbed and enter blood circulation. Some of the metal amounts are excreted via urine, feces, and sweat while some are accumulated. The most common sites for metal accumulation are the liver, lungs, and kidneys. Table 9.1 describes the sites of accumulation in the body. Metals like chromium (Cr), arsenic (As), cadmium (Cd), and nickel (Ni) interfere with DNA repair, replication, and epigenetic mechanisms, which leads to DNA damage resulting in cancer.

TABLE 9.1
Different Metals, Their Uses and Accumulation Evidences

Sr no.	Name of the metal	Usage	Most occurring form	Accumulation evidences	Probable Accumulation sites in humans	Reference
1.	Arsenic (As)	Wood preservation, Glass manufacturing, insecticide production, Bronze production, Doping agent in semiconductors.	Arsenate (AsO_4^{-3}) Arsenite (AsO_3^{-3}) (more toxic than arsenate)	Bioaccumulation in Tilapia (freshwater fish) in intestine> liver> gill> muscles tissues. In carp (*Carassius auratus*) high accumulation in fish muscle was detected by dietary exposure.	Liver, kidneys, lungs and skin	[56–58]
2.	Cadmium (Cd)	Phosphate fertilizers, pesticides, Corrosion resistant plating, plastic production industries, nuclear reactors	Cadmium carbonate ($CdCO_3$) abundant in water and Cadmium sulphide (CDS) abundant in soil	Reported significant accumulation *Danio rerio* embryos resulting developmental toxicity. Bioaccumulation in the gills of carp (*Cyprinus carpio*).	Lungs, liver and kidneys	[56, 59, 60]
3.	Lead (Pb)	Hair dyes, insecticides, car batteries, computer screens, sports equipments, roofing in buildings, fossil fuels.	Sulfide, carbonate and sulfate forms abundant in soil and oxide in water with lesser pH	Bioaccumulation in carp(*Cyprinus carpio*) in Gill > Intestine > Muscle > Kidney > Liver (short time exposure) and in heart > Spleen > Ovary > Muscle after long time exposure.	Bones and erythrocytes results in placental transfer	[56, 61, 62]
4.	Mercury (Hg)	Barometers, thermometers, chlorine manufacture, fluorescent light bulbs, calomel electrodes, insecticides, rectifiers, catalyst, electrical switches	Methylated mercury (MHg), chloride and elemental form	Reported significant accumulation in carp (*Labeorohita*) kidneys. Amphibians for both adult and developmental stages and rats showed high accumulation in kidneys and brain.	Thyroid tissues, heart, muscles, adrenals, liver, skin, kidneys, breast, pancreas, prostate, testis and also taken up by CNS	[56, 63–65]
5.	Selenium (Se)	Electronics, colorant, glasswares, ceramics, paints, plastics, shampoos, animal feeds	Soluble form SeO_4^{-2} coupled with other oxide minerals	Bioaccumulation in a fish *Merluccius productus*in kidneys >liver >gonads>gills >muscle	Erythrocytes, liver and kidneys	[56, 66]

6.	Zinc (Zn)	Galvanization, paints, cosmetics, soaps, deodorants, weapons, electronics, batteries, plastic, ink, pharmaceuticals, rubber, textiles, X-ray screens, mixed in alloys, fluorescent lights	Zinc Oxide and sulphide (ZnO and ZnS)	Often studied in a mixture with other metals like copper and cadmium as accumulation is relatively less. Sites in carp are gills and kidneys.	Liver, bones, muscles, pancreas and kidneys	[56, 59]
7.	Chromium (Cr)	Metal ceramics, electroplating, leather tanning, Dye paints, production of synthetic rubies	Chromium (VI), chromium (III)	Bioaccumulation in *Cyprinus carpio* gills, intestine, muscles, skin and bones.	Kidneys and liver	[56, 67]
8.	Nickel (Ni)	Jewellery, Coins, Plating, Welding, alloys, welding, Rocket engines.	Nickel Oxide, Nickel Sulphide (NiO$_2$ and NiS)	Bioaccumulation in black fish (*Capoetafusca*) in gill > liver > muscle > skin. Accumulation in wood frog (*Lithobatessylvaticus*) leading to lethality.	Brain, lungs, adrenals, liver and kidneys	[56, 68, 69]

Heavy metals are also accumulated in plants. Aquatic plants are exposed to these metals in their very habitat so the nutrition they receive is contaminated similar to aquatic organisms (fishes, planktons, etc.), while terrestrial plants have been found to accumulate these metals in their root and shoot parts. These pollutants come into contact with plants via root (as a result of contaminated soil) and once they enter plant are circulated in the whole plant body by plant cargo (i.e., xylem and phloem) [70, 71].

9.3.1.3 Biomagnification of Heavy Metals

Biomagnification is the step after bioaccumulation. The metals that tend to accumulate more magnify further in the food chain. In other words, the metals that cannot be easily excreted have higher tendencies to biomagnify. Evidence of biomagnification of metals like arsenic (As), cadmium (Cd), mercury (Hg), and lead (Pb) have been found to higher as compared to other metals. The example described in Figure 9.5 describes mercury biomagnification in aquatic ecosystem. However, in the terrestrial ecosystem, metals like Pb and As have reported no biomagnification while Hg and Cd have shown biomagnification up to two trophic levels. In terrestrial habitats transfer of toxin from diet (i.e., plants and edible organisms and from the soil of agriculture) is calculated using the transfer factor (TF) [72].

9.3.1.4 Radioactive Wastes

The word radioactive often makes people think of Madam Marie Curie, the inventor of radioactivity, or the Chernobyl incident where there was a huge spill of nuclear radioactive wastes. But what exactly is radioactivity and what can be considered as a radioactive pollutant? The words radioactivity, nuclear decay, and radioactive decay are synonymous and refer to emission of an ionizing radiation, which is the result of continuous disintegration of atomic nuclei. Release of these kinds of particles in nature by anthropogenic activities leads to radioactive pollution. All the atoms found until now in chemistry have a radioactive isotope, which is one of the properties of the elements, but this does not mean that all of them have equal industrial importance and all of them cause pollution.

There are two different types of radiation: ionizing and non-ionizing. Ionizing is the main cause of radioactive contamination, which includes radiation like X-rays, beta rays, alpha rays, and gamma rays (i.e., the radiation with higher energy). Non-ionizing radiation includes UV rays, visible light, infrared, microwaves, and also cellular radiatios, which cause harm to certain species of birds.

Radioactive pollution in nature is mainly caused by the radioactive isotopes of metals used on large scale in industries. Radioactive pollutants are elements such as uranium (U), strontium (Sr), thorium (Th), technetium (Tc), cesium (Cs), cadmium (Cd), etc., and their compounds, which end up in the natural environment as a result of human activities like discharges from nuclear reactors used in power generation, and excess mining in addition to the use of the radioactive materials in medicine, agricultural, research, industrial manufacturing, etc. Though there are protocols for disposing of radioactive waste it ends up in nature, exposing humans and ecosystems.

9.3.1.5 Bioavailability, Bioaccumulation, and Biomagnification of Radioactive Isotops of Metals

Uranium is a widely used radioactive material in industries. In nature it consists of three different radioactive isotopes: ^{234}U, ^{235}U, and ^{238}U where the composition of ^{238}U is the highest [73]. Uranium is omnipresent in the earth's crust with concentrations of 3 mg kg^{-1} and 3 ug L^{-1} in sea water [74]. Since in nature uranium is found in mixtures of its isotopes humans have created a new version of uranium (i.e., the depleted uranium), which has much less uranium isotope content. ^{235}U, also known as fissile isotope, is less radioactive but has similar toxicity as uranium found in the nature. Depleted uranium has applications in both civil and military fields. It is used in counterweights for aircraft, as catalysts in oil and gas industries, and alloyed in special steels such as in military applications due to its high density, relatively low cost, and hardness. Uranium with higher ^{235}U isotope has the highest radioactivity and is used in nuclear power plants for generation of energy, for generating nuclear weapons, and as a fuel in nuclear fission reactors. These industries release wastes and aerosols that tend to be found in the soil and become embedded there and eventually with rain water spread in water resources. The other radioactive compounds found in nature in relatively higher amounts are radioactive cadmium ^{113}Cd and ^{116}Cd and radioactive cesium ^{137}Cs having applications in generation of nuclear power [75–78]. A study done in a natural environment found significant accumulation of arsenic, nickel, and uranium in a North American crustacean species (i.e., *Hyalellaazteca*) found in fresh and brackish waters. This organism feeds on algae and diatoms but serves as the food source for higher organisms. So, along with bioaccumulation the possibility of biomagnification is expected [79].

9.3.2 ORGANIC POLLUTANTS

Organic does not always mean completely biodegradable or amicable to biota and environment. This specific class of pollutants is called organic because of their physical and chemical properties. Organic pollutants consist of hydrocarbons linked in aromatic or aliphatic chains with covalent and hydrogen interactions. To name a few, phenols, azo dyes, polyaromatic hydrocarbons, polychlorinated biphenyls, dichlorodiphenyltrichloroethane (DDT), phthalate esters, etc., are persistent organic pollutants (POPs), which means that their presence is ubiquitous in the environment. But from where do they derive in nature? They are very derivative of other types of pollutants (i.e., via industrial and domestic waste effluents). In fact, natural resources like rivers, lakes, oceans, sediments, and soil are pools of organic pollutants [80].

POPs are global hazards and exposures to these pollutants have proved to be dangerous to all classes of organisms in both terrestrial and aquatic ecosystems. The effects of these POPs are decline in population for specific species, diseases, deformities, congenital abnormalities, and metabolic hindrance in the body of organisms. In addition, human exposure to POPs causes reproductive, developmental, endocrine, neurological, immunological, and behavioral disorders. As a result, various countries have issued guidelines for using these compounds and many of them are banned in several countries. For example, in 1972, the United States banned the use of DDT.

On the basis of one of chemical properties (i.e., volatility or boiling point of these compounds) the USEPA and WHO categorized them into three classes:1) very volatile organic compounds (VVOCs often gaseous); 2) volatile organic compounds (VOCs); and 3) semi-volatile organic compounds (SVOCs). In addition, the USEPA claims that this arbitrary classification was done to quantify indoor air.

Other than this, there is no other classification of organic pollutants. Also, most of the POPs fall under the third category of classification.

9.3.2.1 Bioavaibility of Organic Pollutants

POPs are broadly classified, and include: Polycyclic Aromatic Hydrocarbons (PAHs), which include compounds like naphthalene, phenanthreneneanthrecene, benzopyrene, coronene, etc.; Polychlorinated Biphenyls (PCBs), which has myriad structures with more than one chlorine attached to each of the phenyl rings; dioxins, which have two oxygen atoms; phthalate esters, which include diethyl phthalate (DEP), dibutylphthalate (DBP), dethylhexylphthalate (DEHP), etc.; and others include DDT, alderin, dielderins, furanes, phenols with benzene ring and -OH group, Azo compounds, chlordaene, mirex, etc.

POPs have a wider range of the applications as they are used in agricultural products such as insecticides, pesticides, fungicides, and fertilizers; in paper industries; in rubber industries to obtain synthetic rubber; in electronics industries; in paint industries; as fire retardants; and also produced during combustion of fossil fuels, etc. This results in their conspicuous existence in effluents and sludge leading to pollution of natural resources. In addition, the organic pollutants that fall under the category of volatile and very volatile organic compounds are the pollutants that can travel by the medium of air. Most of these air-traveling organic pollutants, which are also persistent in nature, are obtained by burning of fossil fuels (i.e., via vehicular or industrial air pollution). Certain PAHs, PCBs, and aliphatic hydrocarbons are found in contaminated air as well.

9.3.2.2 Bioaccumulation of Organic Pollutants

As organic pollutants are lipophilic (hydrophobic) in nature they can readily enter the body of an organism. As these compounds are hydrophobic, they tend to accumulate more in lipophilic regions of the body like liver and muscles though their accumulation in kidney is also reported often [81].

Interestingly, though organic pollutants can not be biodegraded, they enter the body of an organism and metabolize resulting in the derivative of that organic compound, which is then accumulated in the body unlike many other inorganic pollutants. For example, when humans were highly exposed to DDT instead of its derivative (i.e., DDE (dichlorodiphyldichloroethylene)) it was found in the blood, which signifies its processing in the body where loss of hydrogen chloride (HCl) takes place [82]. Bioaccumulation of organic pollutants and their derivatives is evident as seen in Table 9.2.

From these accumulation studies it can be inferred that in higher classes of organisms the major accumulation site for organic compounds is likely liver, muscles, and adipose tissues (fat storing tissues) followed by other major organs.

TABLE 9.2
Different Organic Pollutants, Their Uses and Accumulation Evidences

Sr. No	Name of the pollutant	Usage or occurance	Evidences of Bioaccumulation or bioconcentration	References
1.	Polyaromatic hydrocarbons (PAHs)	Burning of fossil fuels in vehicles, incomplete combustion of oil, crude and natural gas	In Mekong catfish and cyprinids when collected from the natural habitat Bioaccumulation in liver and muscles.	[83]
2.	Polychlorinated Biphenyls (PCBs)	Electronics; to produce heat resistant equipments, plasticizers, dyes, colors and pigments, and paper industry.	In *Hypostomuscommersoni* a freshwater fish bioaccumulation 27 ± 78.7 and 69.2 ± 18.1 ng/g were found in liver and muscles. Oligochaeta worms (*Lumbriculus variegatus*) when exposed to PAHs and PCBs from sediments for 28 days showed significant bioaccumulation In American bobcats (Lynx rufus) PCB bioaccumulation in liver was 562.97 ng/g on an average.	[84–86]
3.	Hexachlorobenzenes (HCB)	Agricultural products like fungicides and fertilizers, termite resistant wood, military products like tracer in bullets.	Trahira (*Hopliasmalabaricus*) HCB and DDT bioaccumulation in liver and muscles. Bioaccumulation of HCBs in rat adipose tissues has been reported long ago.	[87, 88]
4.	Azo compounds	Dyes, paints, polishes, fabrics and food colours.	In goldfish (*Carassicus auratus*) Accumulation of HC orange (an azo dye) in the liver.	[89]
5.	Dioxin and dioxin like PCBs (dlPCBs)	Paper industries for bleaching the paper pulp, agricultural products like herbicides and pesticides; smelting	In zebrafish (*Brachdanio rerio*) early stage Accumulation of octacholorobenzo-p-dioxin (OCDD) ranging from 61–94 mg/kg In fishes like Sabalo and Prochilodus accumulation of dlPCBs 6.8 ± 3.9 to 281 ± 155 ng/g dw	[90, 91]
6.	dichlorodiphenyltrichloroethane (DDT)	Agricultural products like insecticides, pesticides and fertilizers.	In yellowfin tuna (*Thunnus albacares*) from natural habitat; bioaccumulation of DDT in muscles and its metanbolites ranging 92.1–221.8 ng/g lw In carp (*Cyprinus carpio*) DDT metabolites were bioaccumulated more from sediment than water.	[92, 93]

(*continued*)

TABLE 9.2 (Continued)
Different Organic Pollutants, Their Uses and Accumulation Evidences

Sr. No	Name of the pollutant	Usage or occurance	Evidences of Bioaccumulation or bioconcentration	References
7.	Phenols	Dyes, pesticides, agricultural products, pharmaceutical industries, wood preservation, oil refinery wastes.	In Mozambique tilapia (*Oreochromis mossambicus*) bioaccumulation in kidney, liver, gills and muscles. In wild fish bioaccumulation of phenolic compounds in liver and muscles tissues. In common carp (*Cyprinus carpio*) bioaccumulation in kidney, liver and muscle tissues.	[94–96]
8.	Aldrin and dieldrin	Mostly used as agricultural products like insecticides and pesticides. Also, referred as organochlorinated compounds.	In Cyprinus carpio 85 mg/g ww in liver tissue (maximum) followed by gills, muscle, intestine and kidney. In *Hypostomuscommersoni* a freshwater fish bioaccumulation was found in liver and gills.	[85, 97]

In mammalian study models (rats and mice) degree of toxicity of these pollutants and effects on body mechanisms are measured to showcase the probable effects on humans [98–106]. For example, toxic effects of organic compounds in rats are found to cause endocrine disruption, neurobehavioral disorders, hepatotoxicity, and lung lesions when exposed to contaminated air, skin disease, etc. In severe cases of contamination cancers can also be predicted. In addition, evidence of fetal transmission and transfer through milk in mammals have also been found [107, 108].

9.3.2.3 Biomagnification of Organic Pollutants

As the metabolites of POPs tend to accumulate, occurrence of biomagnification is manifest. Many researchers have reported the biomagnification and trophic transfer of organic pollutants [109–111]. Biomagnification of DDT is the classical example to acknowledge this phenomenon in aquatic food chain, as can be seen in Figure 9.5.

In aquatic food chain organisms divided in three different trophic levels, microalgae (*Dunaliellatertiolecta*), mussel (*Mytilus galloprovincialis*), and fish (*Dicentrarchuslabrax*), were tested for biomagnification of benzo(a)pyrene and dimethylbenz(a)anthracene [112]. Significant accumulation at trophic level 2 (i.e., in mussels) was reported while fish tended to have coping mechanism in the contaminated environment. Zhang et al. reported the trophic transfer of dioxins from copepods (marine pytoplankton) to fish. The dietry exposure of dioxins tends to accumulate in phytoplankton, but tends to biomagnify in fish (i.e., successive food chain elements) [113].

Bioavailability, Bioconcentration, and Biomagnification of Pollutants

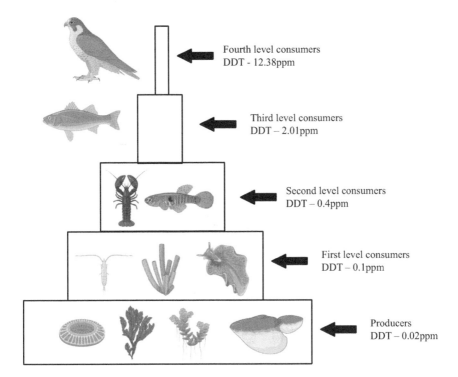

FIGURE 9.5 Biomagnification of DDT – an organic pollutant.

In addition, trophic transfer of polychlorinated biphenyls (PCBs) by aquatic and terrestrial insects to tree swallow *(Tachycinetabicolor)* nestlings 2,827.76 ±505.67 µg/kg was reported with lower concentrations in the insects it feeds on. Other than that, two terrestrial food chains, Eurasian tree sparrow (*Passer montanus*– lower trophic level) – common kestrel (*Falco tinnunculus*– higher trophic level) and brown rat (*Rattus norvegicus*– lower trophic level) – eagle owl *(Bubo Bubo)*, were selected to disclose the trophic transfer of organohalogen pollutants. The biomagnification of polybrominated diphenyl ethers (PBDEs) and polychlorinated biphenyls (PCBs) was found, followed by DDT [114, 115].

9.3.3 Microplastic Wastes

Plastics are considered amongst the major pollutants contaminating marine, freshwater, and terrestrial ecosystems mainly because of their resistance to degradation and persistence in the environment. The demand for plastics is increasing drastically. According to reports published by the IUCN (International Union for Conservation of nature) over 300 million tons of plastic is manufactured every year and 8 million tons of it ends up in oceans annually distorting water ecosystems. Based on size plastics can be classified into four different classes: i) Macroplastics with size > 25mm; ii) mesoplastics with size 25 to 5 mm;iii) microplastics with size 1 to 5mm;

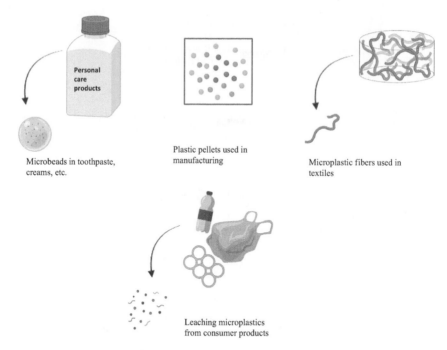

FIGURE 9.6 Types of microplastics and their origin.

and iv) nanoplastics with size in nm (nanometers), discovered recently. Why only microplastics? Compared to other plastics described above microplastics tend to aggregate more in ecosystems as well as in the bodies of organisms due to their smaller size and highest availability indicating the maximum threat to ecosystems. Microplastics are recognized as POPs as they can easily and rapidly migrate from the environment to organisms. Microplastics are further classified into two different categories: primary and secondary microplastics. When the microplastics enter the environment directly in micro-size range like in the form of microbeads used in the soaps and detergents they are said to be primary microplastics, while the ones microplastics derive from breakdown or leaching of bigger plastics, which results from weathering and exposure to UV radiation from sunlight, are called secondary microplastics [116, 117]. Classification of microplastics and their origin can be seen in Figure 9.6.

Microplastics have industrial and domestic applications. They are used in beauty products like scrubs, soaps, body washes, beauty creams, cosmetics; toys, paints, adhesives, and insect repellents; in housing to confer flexibility, strength, and durability; and to manufacture medical devices, packaging, vehicles, agriculture, textiles, and electronics products. Industrial waste drainage systems and wastewater treatment plants allow microplastics to enter water ecosystems.

9.3.3.1 Bioavailability of Microplastics

Around 80% of marine plastic debris comes from terrestrial sources *(IUCN)*. Microplastics are not only present in marine waters but their contamination is severe in freshwaters as well. Most are microplastics (fibers and fragments) but some of the secondary microplastics are also found in waters. These plastics as discussed above end up in the waters often by industrial waste and sewage treatment plants. Most microplastic studies focus on the occurrences in the aquatic environment because of the lack of promising analytical methods to identify microplastics in the complex organic matrix of soil. Though some studies focus on soil and show that microplastic pollution is a result of solid waste landfills, dumping grounds, and underground sludge and sewage treatment plants, which are the major sources of primary microplastics. The studies also claim that land would have 423-fold more contamination of soil compared to the aquatic ecosystems [118–122].

The secondary microplastics are often in a liquid state, derived from the products that use these microplastics in their manufacturing process. Once these products are discarded, they end up in waste or sewage and due to exposure to sunlight break down into liquid microplastics, also called phthalate esters.

9.3.3.2 Bioaccumulation or Bioconcentration of Microplastics

Aquatic organisms as well as terrestrial organisms tend to bioaccumulate these plastics. In aquatic ecosystems, the microplastics are present in the habitat of the organisms; hence, they tend to come into contact with the microplastics very easily. Ingestion of microplastics occurs easily in aquatic Ecosystems. The solid microplastics are non-biodegradable in nature and hence accumulate in the body of the organism, which may result in death. While the study of bioaccumulation of microplastics is newer than other areas of research, such as the study of metals and other organic pollutants, researchers have shown that bioaccumulation and magnification have potential human risks.

The phthalate esters, which fall under the category of liquid or leached microplastics, follow certain metabolic processes in the body of organisms and are converted into metabolites, but some of these metabolites are neither further broken down nor excreted (analogous to the metabolism of other POPs discussed above), staying in the body of organisms permanently (i.e., bioaccumulated). These accumulated microplastics end up harming organisms resulting in malformations, congenital defects, and diseases, which sometimes lead to death.

Terrestrial organisms tend to come into contact with microplastics mostly via food and air. Organisms living in contaminated soil eating contaminated vegetation ingest the microplastics. Humans come into contact with these microplastics mostly via ingestion of crops grown in contaminated agricultural areas (agricultural lands are found to be the most contaminated with microplastics) or by eating contaminated aquatic seafood [123]. Organisms belonging to the lower class of the kingdom Animalia tend to bioaccumulate as well as biomagnify these contaminants. Table 9.3 gives information about commonly occurring microplastics and their bioaccumulation

9.3.3.3 Biomagnification of the Microplastics

Biomagnification of microplastics in aquatic ecosystems is explained in Figure 9.7. Starting from the left side, the first image indicates the primary and secondary microplastics, which are directly ingested by small organisms like zooplankton,

TABLE 9.3
Different Microplastics, Their Uses and Accumulation Evidences

Sr no.	Microplastic compounds	Uses	Bioaccumulation evidences	References
1.	Polyamide Molecular formula $[C_6H_{11}ON]_n$	Textile industries, automobile industries, kitchen utilities, sports industry, etc.	Bioaccumulation of polyamide nylon fibers in the digestive system of the fishes found in natural habitat.	[124]
2.	Polyester Molecular formula $[C_2H_4O_2]_n$	Textile industries to manufacture fabrics and dyes, electronics items, musical instruments, vehicles, etc.	Bioaccumulation of polyesters in digestive glands of the edible oysters.	[125]
3.	Polypropylene Molecular formula $[C_3H_6]_n$	To manufacture heat resistant plastics, fibre plastics and has various commercial and industrial applications.	Bioaccumulation in cockles(*Anadaraantiquate*) from natural Tanzanian cost.	[126]
4.	Polystyrene (PS)	Rigid plastic polymer, commercial packaging, synthetic resin, temperature resistant plastic products, vessels, etc.	Predicted bioaccumulation in liver, kidney and gut of mice. Bioaccumulation and hepato-toxicity in fresh water fish red tilapis (Oreo*chromis niloticus*) in gut, gills, brain and liver. Bioaccumulation in corals can reach upto 150 ng/gd	[16, 127, 128]
5.	Polyethylene (PE) Molecular formula $[C_2H_2]_n$	Most used plastic for making the plastic bags, cosmetics, packaging materials, etc.	Bioaccumulation of PE fibers in marine fishes collected from the vicinity of effluent discharge. Bioaccumulation in zebra fish (*Danio rerio*) causing neurotoxicity.	[129, 130]
6.	Polyvinyl chloride (PVC) Molecular formula $[C_3H_6]_n$	Falls under the category of thermoplastics, highly durable, used in construction, pharmaceutical industries, house appliances, medical devices, packaging materials, electronics, textile industries, etc.	Combined bioaccumuoation with other pollutants in liver and gut (PVC, PE and PP with other pollutants)	[123]

Bioavailability, Bioconcentration, and Biomagnification of Pollutants 199

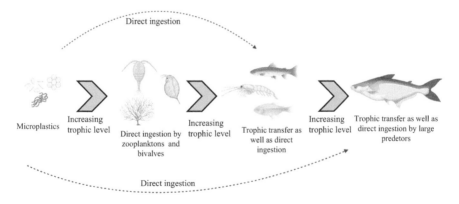

FIGURE 9.7 Biomagnification of microplastics.

Source: recreated from Ref. [123].

phytoplankton, and bivalves. These microplastics are accumulated in the body. The organisms from the other trophic levels feed on these small organisms and tend to accumulate a higher amount of microplastics in the body. This system is followed by the organisms from the higher trophic levels that tend to have more and more bioaccumulation of these contaminants leading to biomagnification. This figure also indicates the possibility of humans encountering microplastics via ingestion of these organisms. While no concrete evidence of biomagnified microplastics in the terrestrial food chain has been found, the transfer factor can indicate the transfer of these pollutants from food and other external resources.

9.3.4 Nanomaterial Pollutants

Today the growing field of nanotechnology has applications in industrial, technological, healthcare, electronics, and various other sectors. The technology that deals with nano-particles is called nanotechnology. Nanoparticles as the name itself suggests are fine particles with adiameter in nanometers (10^{-9}m). These particles are not visible under normal optical microscopes; electron microscopes are required for their visualization. Nanoparticles can be classified into different classes: metallic nanoparticles, non-metallic nanoparticles, which are mostly carbon-based nanoparticles, ceramic nanoparticles, semiconductor nanoparticles, polymeric nanoparticles, and lipid-based nanoparticles. In addition to nanoparticles, there are other nanomaterials like nanotubes, nanofibers, imaging nanoprobes, quantum dots, nanosheets, and nanowires [131–134].

The common metallic nanoparticles are Gold (Au), Silver (Ag), Iron (Fe_2O_3), Zinc Oxide (ZnO), Silica (SiO_2), Cadmium sulphide (CdS), Copper oxide (CuO), Cerium oxide (CeO), and Titanium dioxide (TiO_2) nanoparticles. These metallic nanoparticles are commonly used to produce consumer products and industrial goods and are engineered with very specific properties by physical and chemical processes. Nanotubes and nanofibers also have wide industrial applications. These nanomaterials

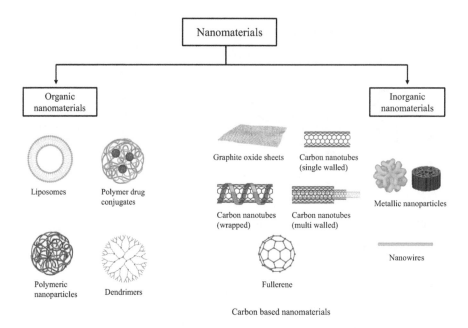

FIGURE 9.8 Types of nanomaterials and their classification.

may end up in natural environments through industrial waste and improper handling of various consumer and healthcare products. Today, the use of nanoparticles is also evident in the food industry. There has been an increase in the number of patents, publications, and intellectual property rights in the field of nano-agri-food as well as in recent research trends in food processing, packaging, nutraceutical delivery, quality control, and functional food [135, 136].

Nanomaterials are classified as organic nanomaterials or inorganic nanomaterials. Organic nanoparticles include liposomes, polymeric nanoparticles, polymer-drug conjugates, and dendrimers. Inorganic nanomaterials include metallic nanoparticles, carbon-based nanoparticles, carbon nanotubes, nanographene, and magnetic nanoparticles. This classification is described in Figure 9.8.

Both of these types of nanomaterials have their different applications in different fields. Organic nanomaterials can be further broken down or biodegraded in nature but inorganic nanoparticles are not biodegradable and amalgamation of such nanomaterials in nature causes interference in environmental processes.

9.3.4.1 Bioavailability of Nanomaterials

Nanomaterials are commonly used in pharmaceutical industries for drug development as well as diagnostic purposes, and in commercial products like soaps, detergents, cosmetics, paints, and in textiles, electronics, food and agriculture, catalysis and construction [137–139]. These nanomaterials are produced industrially for applications in consumer products and hence are called engineered nanomaterials (ENMs) [140]. Thusnanomaterials are found in industrial waste as well as waste from consumer products like paints, body wash, soaps, etc. This waste ends up in the environment, and causes harm to both terrestrial as well as aquatic ecosystems [141].

Moreover, nanomaterials present in wastewater interact with other components of wastewater, which results in change of overall properties of the wastewater. For example, it has been understudies that fullerenes can adsorb onto the natural organic matter existing in wastewater and can change the size and morphology of the matter, which results in magnification of fullerene concentration [142–145]. Moreover, the presence of these nanomaterials in the atmosphere resulting from aerosols, road traffic, and static combustion sources causes dispersion over wider distances (due to their smaller size they can travel fast and far from the origin), which can lead to their landing in soil and water resources contaminating them, the main cause of occurrence of nanoparticles in soil. Figure 9.8 shows the pathways through which nanomaterials can end up in nature through anthropogenic endeavors [146–150].

Silver, copper oxide, cerium oxide, iron oxide, titanium dioxide, zinc oxide, carbon nanotubes, and fullerenes are reported to be the most widely used and researched nanomaterials [151]. Table 9.4 lists the detailed uses and occurrence of the widely used nanomaterials.

Advancements and new technologies have allowed researchers to develop nanomaterials and also manufacture them on an industrial scale but reliable methodology and measurement techniques to analyze these nanomaterials in environmental resources are still not available. While probabilistic modeling studies are useful in estimating environmental concentrations along more precise tools are needed [152].

9.3.4.2 Bioaccumulation or Bioconcentration of Nanomaterials

Bioavailability of non-biodegradable nanomaterials tends to cause bioaccumulation in the biota. It has been reported that various factors play important roles in bioaccumulation of nanomaterials. These factors are size, composition of particles, and shape (rods, spheres, fibers, platelets). Several studies have reported that particles having smaller size and dimension tend to accumulate more in organisms as they are more bioavailable to them [153, 154]. Also, for metallic nanoparticles the uptake is faster in the case of ionic forms of metals. In fact, Au, TiO_2, and SiO_2 are less bioavailable and toxic than CuO, Ag, and ZnO nanoparticles [155].

Bioaccumulation of nanomaterials results by direct exposure via food and habitat. Because of this in aquatic animals the maximum accumulation is found in liver, intestines, and kidneys. In addition to this, humans are exposed to these nanoparticles via ingestion, inhalation of contaminated air, and use of common consumer products [156].

Table 9.4 lists the widely occurring nanomaterials and their bioaccumulation in various model organisms. There are myriad studies done on bioaccumulation of nanomaterials in plant species. The studies done on model organisms showcase the degree of bioaccumulation of certain nanomaterial pollutants. Apart from accumulation, nanomaterials cause toxicities in the body of organisms, which can lead to diseases and health conditions. For example, exposure of nanoparticles can cause malformations in zebrafish embryo and larvae [157, 158]. The negative effects of nanoparticles are evident in microorganisms as well. ZnO and TiO_2 and other

TABLE 9.4
Different Nanoparticles, Their Uses and Accumulation Evidences

Sr. No.	Name of nanomaterial	Type	Usage	Evident bioaccumulation or Bioconcentration	Reference
1.	ZnO nanoparticles	Metallic nanoparticles (inorganic)	Rubber, paints, coatings, cosmetics and biomedicine	In blackfish (*Capoetafusca*) Trend of accumulation intestine > gill > kidney > liver with concentrations 1.47–1.71 mg/kg and 093–1.23 mg/kg in intestines and gills.	[167, 168]
2.	CuO nanoparticles	Metallic nanoparticles (inorganic)	Catalyst, superconducting and heat resistant materials, ceramic resistors, semi and superconductors	In gold fish (*Carassius auratus*) Liver and intestine-major sites of accumulation of both CuO and ZnO concentrations ranging from 53–153 and 63–199 µg/kg respectively.	[169]
3.	TiO_2 nanoparticles	Metallic nanoparticles (inorganic)	paints, ceramics, paper, plastic, rubber, catalysts, textiles, cosmetics, food packaging,, electronics, medicine	In African clewed frog (*Xenopus laevis*) Accumulation in gut In Zebrafish (*Danio rerio*) Accumulation in Gut, gills, liver and kidneys	[170–173]
4.	Fe_3O_4 nanoparticles	Metallic nanoparticles (inorganic)	Catalysis, magnetic resosnance imaging, electronics	In blackfish (*Capoetafusca*) Accumulation found highest in liver followed by gills.	[174]
5.	Fullerenes and its derivatives	Carbon-based NM (inorganic)	Electrocatalysts, coating agents, electronics, lubricants, cosmetics,	In Zebrafish (*Danio rerio*) Maximum accumulation in intestines, liver followed by gills. Bioaccumulation in *Daphia magna* 31.20 ± 1.59 mg/g dry weight	[175]
6.	Graphene and its derivatives	Carbon-based NM (inorganic)	Electronics, sensors, lubricants, medicine, photodetectors,.	In *common carp* (*Cyprinus carpio*) Accumulation in intestine, liver, gills and kidneys.	[176]

nanoparticles have been found to cause DNA damage, mutations, and oxidative stress in bacterial cells, which leads to decreased cell viability [159, 160]. Human lung cells when exposed to ZnO nanoparticles cause genetic defects, which can lead to cancer. When human and mice liver cells are exposed to TiO_2 nanoparticles apoptosis, DNA damage, and oxidative stress occurs [161–165].

Nanomaterials like liposomes, polymer drug conjugates, and dendrimers are confined to specific studies related to biological and drug discoveries, while metallic nanoparticles, carbon-based nanomaterials, and nanofibers have a wider range of application industrially and hence are abundant in nature as pollutants. Thus, further studies are needed to understand the fate of these nanomaterials once released into nature. While nanomaterials are have proved to be efficient drug delivery vehicles in the body for most types of cancers, they can also be toxic due to their tendency to accumulate [166].

9.3.4.3 Biomagnification of Nanomaterials

A few studies have been done to investigate the trophic transfer of nanomaterials. Bouldin et.al. (2008) disclosed the trophic transfer of CdS nanoparticles (10–25nm) from freshwater algae (*Pseudokirchneriellasubcapitata*) to zooplanktons (*eriadaphniadubia*) [177]. Zhu et al. (2010) examined the transport of TiO_2 nanoparticles (21nm) from crustaceans (*Daphnia magna*) to zebrafish (*Danio rerio*). Wang et al. (2008) investigated the transfer of ZnO, Al_2O_3, and TiO_2 nanoparticlesfrom bacteria (*Escherichia coli*)to worms (*Caenorhabditis elegans*) through dietary uptake. In addition to that a study done on model organisms of terrestrial ecosystem (i.e., plant and primary consumer) showcased trophic transfer. Tobacco (*Nicotiana tabacum L*) was the producer and tobacco hornworm (*Munducasexta*) was primary consumer and the trophic transfer of Au NP (5, 10, and 15 nm)along with more accumulation of Au NP in tobacco hornworms than in tobacco plants were apparent, which clearly explains the phenomena of biomagnification [178].

Apart from trophic transfer of metallic nanoparticles the ^{13}C-labeled fullerenols, which are carbon-based nanomaterials, were investigated for three trophic levels in aquatic food chain starting from phytoplankton (*Scenedesmus obliquus*) (i.e., first trophic level), then crustaceans (*Daphia magna*) (i.e., second trophic level), and zebrafish (*Danio rerio*) (i.e., third trophic level), as seen in Figure 9.9. The results showed the systemic biomagnified transfer of these nanomaterials leading to higher accumulation of ^{13}C fullerenols in liver, intestine, and brain tissues of zebrafishes [175]. However, today scientists are predicting the trophic transfer of nanoparticles with the help of predator-prey model system established in the laboratory [179].

9.4 CONCLUSION

As environmental pollution is increasing on a daily basis, the concentrations of some omnipresent pollutants are also increasing in a commensurate manner. Hence, it is obvious we need to study the effects of these pollutants or toxins on humans as well as ecosystems. The equilibrium of nature is constantly being disturbed, and it is important to understand the effects of toxins. This chapter aimed to answer questions related to the noxious effects of pollutants like metals (especially heavy metals),

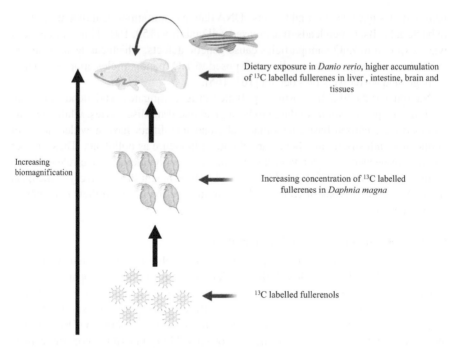

FIGURE 9.9 Biomagnification of ^{13}C fullerenols.

Source: Recreated from Ref. [175].

organic pollutants, microplastics, and nanomaterials by discussing bioavailability, bioconcentration or bioaccumulation, and biomagnification. These three phenomena play a significant role in understanding the status and extent of the damage caused by contaminants in a specific ecosystem. Bioavailability describes the amount of a toxin present in an ecosystem and that is taken up by an organism, bioconcentration and bioaccumulation provide the information of these toxins retained by the bodies of organisms, and biomagnification elucidates the further action of these adulterants in the progressing food chain and food webs. This chapter cited studies and evidence by various researches to elaborate on the negative effects of these pollutants.

REFERENCES

[1] Arnold, K.E., et al., *Medicating the Environment: Assessing Risks of Pharmaceuticals to Wildlife and Ecosystems*. 2014.The Royal Society.
[2] Chagnon, M., et al., Risks of large-scale use of systemic insecticides to ecosystem functioning and services. *Environmental Science and Pollution Research*, 2015. 22(1): pp. 119–134.
[3] Chen, M., et al., Development of a comprehensive assessment model for coral reef island carrying capacity (CORE-CC).*Scientific Reports,* 2021. 11(1): pp. 1–12.
[4] Leung, K.M., et al., Toward sustainable environmental quality: Priority research questions for Asia. *Environmental Toxicology and Chemistry,* 2020. 39(8): pp. 1485–1505.

[5] Magowska, A., *The Natural History of the Concept of Antidote.* Toxicology Reports, 2021.
[6] Ching, T., et al., Opportunities and obstacles for deep learning in biology and medicine. *Journal of the Royal Society Interface,* 2018. 15(141):p. 20170387.
[7] Fanelli, D., A theory and methodology to quantify knowledge. *Royal Society Open Science,* 2019. 6 (4): p. 181055.
[8] Haigh, R., et al., Effects of ocean acidification on temperate coastal marine ecosystems and fisheries in the Northeast Pacific. *PloS One,* 2015. 10(2): p. e0117533.
[9] Lara, R., et al., Aquatic ecosystems, human health, and ecohydrology. *Treatise on Estuarine and Coastal Science,* 2011. p. 263.
[10] Lyon-Colbert, A., S. Su, and C. Cude, A systematic literature review for evidence of Aphanizomenon flos-aquae toxigenicity in recreational waters and toxicity of dietary supplements: 2000–2017.*Toxins,* 2018. 10(7):p. 254.
[11] Roué, M., H.T. Darius, and M. Chinain, Solid phase adsorption toxin tracking (SPATT) technology for the monitoring of aquatic toxins: A review.*Toxins,* 2018. 10(4): p. 167.
[12] Shennan, C., Biotic interactions, ecological knowledge and agriculture. *Philosophical Transactions of the Royal Society B: Biological Sciences,* 2008. 363(1492): pp. 717–739.
[13] Jacquemin, B., et al., Ambient air pollution and adult asthma incidence in six European cohorts (ESCAPE).*Environmental Health Perspectives,* 2015. 123(6): pp. 613–621.
[14] Kim, D., et al., Air pollutants and early origins of respiratory diseases. *Chronic Diseases and Translational Medicine,* 2018. 4(2): pp. 75–94.
[15] Li, J., et al., Major air pollutants and risk of COPD exacerbations: a systematic review and meta-analysis. *International Journal of Chronic Obstructive Pulmonary Disease,* 2016. 11: p. 3079.
[16] Yang, Y.F., et al., Toxicity-based toxicokinetic/toxicodynamic assessment for bioaccumulation of polystyrene microplastics in mice *Hazard Mater,* 2019. 366 : pp. 703–713.
[17] Fernández-Llamazares, Á., et al., A state-of-the-art review of indigenous peoples and environmental pollution. *Integrated Environmental Assessment and Management,* 2020. 16(3): pp. 324–341.
[18] Odiete, W.E., Novel metric for managing the protection of humanity and the environment against pollution and its adverse effects.*Heliyon,* 2020. 6(11): p. e05555.
[19] Tudi, M., et al., Agriculture development, pesticide application and its impact on the environment. *International Journal of Environmental Research and Public Health,* 2021. 18(3): p. 1112.
[20] Andrade, M.C., et al., First account of plastic pollution impacting freshwater fishes in the Amazon: ingestion of plastic debris by piranhas and other serrasalmids with diverse feeding habits. *Environmental Pollution,* 2019. 244: pp. 766–773.
[21] Andrady, A.L., Microplastics in the marine environment. *Mar Pollut Bull,* 2011. 62(8): pp. 1596–1605.
[22] Duan, W., et al., Environmental behavior and eco-toxicity of xylene in aquatic environments: a review. *Ecotoxicology and Environmental Safety,* 2017. 145 : pp. 324–332.
[23] Huang, D., et al., Microplastics and nanoplastics in the environment: macroscopic transport and effects on creatures. *Journal of Hazardous Materials,* 2020: volume 407 p. 124399.

[24] Hund-Rinke, K., and M. Simon, Ecotoxic effect of photocatalytic active nanoparticles (TiO$_2$) on algae and daphnids (8 pp). *Environmental Science and Pollution Research*, 2006. 13(4): pp. 225–232.
[25] Karbalaei, S., et al., Occurrence, sources, human health impacts and mitigation of microplastic pollution. *Environmental Science and Pollution Research*, 2018. 25(36): pp. 36046–36063.
[26] Kögel, T., et al., Micro-and nanoplastic toxicity on aquatic life: determining factors. *Science of the Total Environment*, 2020. 709: p. 136050.
[27] Sharma, S., and S. Chatterjee, Microplastic pollution, a threat to marine ecosystem and human health: a short review. *Environmental Science and Pollution Research*, 2017. 24(27): pp. 21530–21547.
[28] Wan, J.-K., et al., Distribution of microplastics and nanoplastics in aquatic ecosystems and their impacts on aquatic organisms, with emphasis on microalgae. *Reviews of Environmental Contamination and Toxicology*, 2018. 246: pp. 133–158.
[29] Windsor, F.M., et al., Environment and food web structure interact to alter the trophic magnification of persistent chemicals across river ecosystems. *Science of the Total Environment*, 2020. 717:p. 137271.
[30] Xu, S., et al., Microplastics in aquatic environments: occurrence, accumulation, and biological effects. *Science of the Total Environment*, 2020. 703 : p. 134699.
[31] Barkay, T., M. Gillman, and R.R. Turner, Effects of dissolved organic carbon and salinity on bioavailability of mercury. *Applied and Environmental Microbiology*, 1997. 63(11): pp. 4267–4271.
[32] Brix, K.V., et al., Development of empirical bioavailability models for metals. *Environmental Toxicology and Chemistry*, 2020. 39(1): pp. 85–100.
[33] Liu, F., et al., Unravelling metal speciation in the microenvironment surrounding phytoplankton cells to improve predictions of metal bioavailability. *Environmental Science &Technology*, 2020. 54(13): pp. 8177–8185.
[34] Schlekat, C., W. Stubblefield, and K. Gallagher, State of the science on metal bioavailability modeling: introduction to the outcome of a Society of Environmental Toxicology and Chemistry technical workshop. *Environmental Toxicology and Chemistry*, 2020. 39(1): pp. 42–47.
[35] Shaked, Y., et al., Insights into the bioavailability of oceanic dissolved Fe from phytoplankton uptake kinetics. *The ISME Journal*, 2020. 14(5): pp. 1182–1193.
[36] Schmitz, K.S., Chapter 4 – Life Science, in *Physical Chemistry*, K.S. Schmitz, Editor. 2018, Elsevier. pp. 755–832.
[37] Bernhoft, R.A., Mercury toxicity and treatment: a review of the literature. *J Environ Public Health*, 2012. 2012: p. 460508.
[38] Ajima, M.N., et al., Bioaccumulation of heavy metals in Mbaa River and the impact on aquatic ecosystem. *Environ Monit Assess*, 2015. 187(12): p. 768.
[39] Burkhard, L.P., Evaluation of Published Bioconcentration Factor (BCF) and Bioaccumulation Factor (BAF) data for per- and polyfluoroalkyl substances across aquatic species. *Environ Toxicol Chem*, 2021. 40(6): pp. 1530–1543.
[40] Cerveny, D., et al., Bioconcentration and behavioral effects of four benzodiazepines and their environmentally relevant mixture in wild fish. *Sci Total Environ*, 2020. 702: p. 134780.
[41] Chen, X., et al., Bioconcentration and developmental neurotoxicity of novel brominated flame retardants, hexabromobenzene and pentabromobenzene in zebrafish. *Environ Pollut*, 2021. 268(Pt B): p. 115895.
[42] Chopra, A.K., M.K. Sharma, and S.Chamoli, Bioaccumulation of organochlorine pesticides in aquatic system – an overview. *Environ Monit Assess*, 2011. 173 (1–4): pp. 905–916.

[43] McGeer, J.C., et al., Inverse relationship between bioconcentration factor and exposure concentration for metals: implications for hazard assessment of metals in the aquatic environment. *Environ Toxicol Chem*, 2003. 22(5): pp. 1017–1037.

[44] Mendoza-Carranza, M., et al., Distribution and bioconcentration of heavy metals in a tropical aquatic food web: a case study of a tropical estuarine lagoon in SE Mexico. Environ Pollut, 2016. 210: pp. 155–165.

[45] Zhang, P., et al., Distribution and bioavailability of ceria nanoparticles in an aquatic ecosystem model.*Chemosphere*, 2012. 89(5): pp. 530–535.

[46] Zhang, Y., et al., Heavy metals in aquatic organisms of different trophic levels and their potential human health risk in Bohai Bay, China. *Environ Sci Pollut Res Int*, 2016. 23(17): pp. 17801–17810.

[47] Wu, P., et al., Biomagnification of methylmercury in a marine plankton ecosystem. *Environ Sci Technol*, 2020. 54(9): pp. 5446–5455.

[48] Walker, C.H., R. Sibly, and D.B. Peakall, *Principles of Ecotoxicology*. 2005. CRC Press.

[49] Gautam, P.K., et al., *Heavy Metals in the Environment: Fate, Transport, Toxicity and Remediation Technologies*. Nava Science publishers, Inc 2016. pp. 101–130.

[50] Engwa, G.A., et al., Edited by Ozgur Karcioglu and Banu Arslan Mechanism and health effects of heavy metal toxicity in humans, in *Poisoning in the Modern World – New Tricks for an Old Dog*, IntechOpen.2019. 10.

[51] Masindi, V., and K. Muedi, Edited by Hosam El-Din M. Saleh and Refaat F. Aglan *Environmental Contamination by Heavy Metals*. IntechOpen 2018.

[52] Gao, L., et al., DGT: A promising technology for in-situ measurement of metal speciation in the environment. *Sci Total Environ*, 2020. 715 : p. 136810.

[53] Blume, H.-P., et al., *Scheffer/schachtschabel: Lehrbuch der bodenkunde*. 2016. Springer-Verlag.

[54] Bonsignore, M., et al., Bioaccumulation of heavy metals in fish, crustaceans, molluscs and echinoderms from the Tuscany coast. *Ecotoxicol Environ Saf*, 2018. 162 : pp. 554–562.

[55] Jaiswal, A., A. Verma, and P. Jaiswal, Detrimental effects of heavy metals in soil, plants, and aquatic ecosystems and in humans. *J Environ Pathol Toxicol Oncol*, 2018. 37(3): pp. 183–197.

[56] Haynes, W.M., *CRC Handbook of Chemistry and Physics*. 2014.CRC Press.

[57] Pei, J., et al., The bioaccumulation and tissue distribution of arsenic species in Tilapia. *Int J Environ Res Public Health*, 2019. 16(5) 757 .

[58] Cui, D., et al., The dynamic changes of arsenic biotransformation and bioaccumulation in muscle of freshwater food fish crucian carp during chronic diet borne exposure. *J Environ Sci (China)*, 2021. 100: pp. 74–81.

[59] Delahaut, V., et al., Toxicity and bioaccumulation of Cadmium, Copper and Zinc in a direct comparison at equitoxic concentrations in common carp (Cyprinus carpio) juveniles. *PLoS One*, 2020. 15(4): p. e0220485.

[60] Redelstein, R., et al., Bioaccumulation and molecular effects of sediment-bound metals in zebrafish embryos. *Environ Sci Pollut Res Int*, 2015. 22(21): pp. 16290–16304.

[61] Łuszczek-Trojnar, E., et al., Copper and lead accumulation in common carp females during long-term dietary exposure to these metals in pond conditions. *Aquaculture Research*, 2016. 47(7): pp. 2334–2348.

[62] Rajeshkumar, S., et al., Effects of exposure to multiple heavy metals on biochemical and histopathological alterations in common carp, Cyprinus carpio L. *Fish Shellfish Immunol*, 2017.70: pp. 461–472.

[63] Bergeron, C.M., et al., Mercury accumulation along a contamination gradient and nondestructive indices of bioaccumulation in amphibians. *Environ Toxicol Chem*, 2010. 29(4): pp. 980–988.

[64] Ghosh, D., and D.K. Mandal, Histopathological effects and bioaccumulation of mercury in the kidney of an Indian major carp, Labeo rohita (Hamilton). *Bull Environ Contam Toxicol*, 2012. 89(3): pp. 479–483.

[65] Li, P., et al., Mercury bioaccumulation and its toxic effects in rats fed with methylmercury polluted rice. *Sci Total Environ*, 2018. 633: pp. 93–99.

[66] Acosta-Lizárraga, L.G., et al., Bioaccumulation of mercury and selenium in tissues of the mesopelagic fish Pacific hake (Merluccius productus) from the northern Gulf of California and the risk assessment on human health. *Chemosphere*, 2020. 255: p. 126941.

[67] Ali, Z., et al., Toxicity and bioaccumulation of manganese and chromium in different organs of common carp (Cyprinus carpio) fish. *Toxicol Rep*, 2021. 8: pp. 343–348.

[68] Klemish, J.L., et al., Nickel toxicity in wood frog tadpoles: bioaccumulation and sublethal effects on body condition, food consumption, activity, and chemosensory function. *Environmental Toxicology and Chemistry*, 2018. 37(9): pp. 2458–2466.

[69] Mansouri, B., M. Ebrahimpour, and H. Babaei, Bioaccumulation and elimination of nickel in the organs of black fish (Capoeta fusca). *Toxicol Ind Health*, 2012. 28(4): pp. 361–8.

[70] Gall, J.E., R.S. Boyd, and N. Rajakaruna, Transfer of heavy metals through terrestrial food webs: a review. *Environ Monit Assess*, 2015. 187(4): p. 201.

[71] Raskin, I., et al., Bioconcentration of heavy metals by plants. *Current Opinion in Biotechnology*, 1994. 5: pp. 285–290.

[72] Ali, H., and E. Khan, Trophic transfer, bioaccumulation, and biomagnification of non-essential hazardous heavy metals and metalloids in food chains/webs – Concepts and implications for wildlife and human health. *Human and Ecological Risk Assessment: An International Journal*, 2019. 25(6): pp. 1353–1376.

[73] Will Horwath (Ed.) *CRC Handbook of Chemistry and Physics,* 57th edition. *Soil Science Society of America Journal*, 1977. 41(4): pp. 665–830

[74] Ku, T.-l., K.G. Knauss, and G.G. Mathieu, Uranium in open ocean: concentration and isotopic composition. *Deep Sea Research*, 1977. 24: pp. 1005–1017.

[75] Betti, M., Civil use of depleted uranium. *J Environ Radioact*, 2003. 64(2–3): pp. 113–119.

[76] Bleise, A., P.R. Danesi, and W. Burkart, Properties, use and health effects of depleted uranium (DU): a general overview. *J Environ Radioact*, 2003. 64(2–3): pp. 93–112.

[77] Scoullos, M.J., et al., *Mercury – Cadmium – Lead Handbook for Sustainable Heavy Metals Policy and Regulation*. Vol. 31. 2012. Springer Science & Business Media.

[78] Ragnarsdottir, K., *Depleted Uranium in Kosovo*. Post-conflict assessment report. 2001.

[79] Goulet, R.R., and P. Thompson, Bioaccumulation and toxicity of uranium, arsenic, and nickel to juvenile and adult Hyalella azteca in spiked sediment bioassays. *Environ Toxicol Chem*, 2018. 37(9): pp. 2340–2349.

[80] Sulzberger, B., et al., Solar UV radiation in a changing world: roles of cryosphere-land-water-atmosphere interfaces in global biogeochemical cycles. *Photochem Photobiol Sci*, 2019. 18(3): pp. 747–774.

[81] Li, Q.Q., et al., Persistent organic pollutants and adverse health effects in humans. *J Toxicol Environ Health A*, 2006. 69(21): pp. 1987–2005.

[82] Kezios, K.L., et al., Dichlorodiphenyltrichloroethane (DDT), DDT metabolites and pregnancy outcomes. *Reprod Toxicol*, 2013. 35: pp. 156–164.

[83] Phanwichien, K., et al., The ecological complexity of the Thai-Laos Mekong River: III. Health status of Mekong catfish and cyprinids, evidence of bioaccumulative effects. *J Environ Sci Health A Tox Hazard Subst Environ Eng*, 2010. 45(13): pp. 1681–1688.
[84] Boyles, E., and C.K. Nielsen, Bioaccumulation of PCBs in a Wild North American Felid. *Bull Environ Contam Toxicol*, 2017. 98(1): pp. 71–75.
[85] Bussolaro, D., et al., Bioaccumulation and related effects of PCBs and organochlorinated pesticides in freshwater fish Hypostomus commersoni. *J Environ Monit*, 2012. 14(8): pp. 2154–2163.
[86] Tuikka, A.I., et al., Predicting the bioaccumulation of polyaromatic hydrocarbons and polychlorinated biphenyls in benthic animals in sediments. *Sci Total Environ*, 2016. 563–564: pp. 396–404.
[87] Linder, R.E., et al., Long-term accumulation of hexachlorobenzene in adipose tissue of parent and filial rats. *Toxicol Lett*, 1983. 15(2–3): pp. 237–243.
[88] Miranda, A.L., et al., Bioaccumulation of chlorinated pesticides and PCBs in the tropical freshwater fish Hoplias malabaricus: histopathological, physiological, and immunological findings. *Environ Int*, 2008. 34(7): pp. 939–949.
[89] Sun, Y., et al., Bioaccumulation and ROS generation in liver of freshwater fish, goldfish Carassius auratus under HC Orange No 1 exposure. *Environ Toxicol*, 2007. 22(3): pp. 256–263.
[90] Berends, A.G., et al., Bioaccumulation and lack of toxicity of octachlorodibenzofuran (OCDF) and octachlorodibenzo-p-dioxin (OCDD) to early-life stages of zebra fish (Brachydanio rerio). *Chemosphere*, 1997. 35(4): pp. 853–865.
[91] Cappelletti, N., et al., Bioaccumulation of dioxin-like PCBs and PBDEs by detritus-feeding fish in the Rio de la Plata estuary, Argentina. *Environ Sci Pollut Res Int*, 2015. 22(9): pp. 7093–7100.
[92] Di, S., et al., Bioaccumulation of dichlorodiphenyltrichloroethanes (DDTs) in carp in a water/sediment microcosm: important role of sediment particulate matter and bioturbation. *Environ Sci Pollut Res Int*, 2019. 26(10): pp. 9500–9507.
[93] Sun, R.X., et al., Bioaccumulation and human health risk assessment of DDT and its metabolites (DDTs) in yellowfin tuna (Thunnus albacares) and their prey from the South China Sea. *Mar Pollut Bull*, 2020. 158 : p. 111396.
[94] Lv, Y.Z., et al., Bioaccumulation, metabolism, and risk assessment of phenolic endocrine disrupting chemicals in specific tissues of wild fish. *Chemosphere*, 2019. 226: pp. 607–615.
[95] Sannadurgappa, D., N.H. Ravindranath, and R.H. Aladakatti, Toxicity, bioaccumulation and metabolism of phenol in the freshwater fish Oreochromis mossambicus. *J Basic Clin Physiol Pharmacol*, 2007. 18(1): pp. 65–77.
[96] Wang, Q., et al., Toxicokinetics and bioaccumulation characteristics of bisphenol analogues in common carp (Cyprinus carpio). *Ecotoxicol Environ Saf*, 2020. 191: p. 110183.
[97] Satyanarayan, S., Ramakant, and A. Satyanarayan, Bioaccumulation studies of organochlorinated pesticides in tissues of Cyprinus carpio. *J Environ Sci Health B*, 2005. 40(3): pp. 397–412.
[98] Al-Griw, M.A., et al., Cellular and molecular etiology of hepatocyte injury in a murine model of environmentally induced liver abnormality. *Open Vet J*, 2016. 6(3): pp. 150–157.
[99] Al-Griw, M.A., et al., Environmentally toxicant exposures induced intragenerational transmission of liver abnormalities in mice. *Open Vet J*, 2017. 7(3): pp. 244–253.

[100] Bruckner, J.V., et al., Acute, short-term, and subchronic oral toxicity of 1,1,1-trichloroethane in rats. *Toxicol Sci*, 2001. 60(2): pp. 363–372.

[101] Damstra, T., Potential effects of certain persistent organic pollutants and endocrine disrupting chemicals on the health of children. *J Toxicol Clin Toxicol*, 2002. 40(4): pp. 457–65.

[102] Guo, W., et al., Persistent organic pollutants in food: contamination sources, health effects and detection methods. *Int J Environ Res Public Health*, 2019. 16(22) 4361.

[103] Hennig, B., et al., Nutrition can modulate the toxicity of environmental pollutants: implications in risk assessment and human health. *Environ Health Perspect*, 2012. 120(6): pp. 771–774.

[104] Ostrem Loss, E.M., and J.H. Yu, Bioremediation and microbial metabolism of benzo(a)pyrene. *Mol Microbiol*, 2018. 109(4): pp. 433–444.

[105] Tsatsakis, A.M., M. Christakis-Hampsas, and J. Liesivuori, The increasing significance of biomonitoring for pesticides and organic pollutants. *Toxicol Lett*, 2012. 210(2): pp. 107–109.

[106] Zuloaga, O., et al., Overview of extraction, clean-up and detection techniques for the determination of organic pollutants in sewage sludge: a review. *Anal Chim Acta*, 2012. 736 : pp. 7–29.

[107] Kilanowicz, A., et al., Prenatal toxicity and maternal-fetal distribution of 1,3,5,8-tetrachloronaphthalene (1,3,5,8-TeCN) in Wistar rats. *Chemosphere*, 2019. 226: pp. 75–84.

[108] Muralidhara, S., et al., Acute, subacute, and subchronic oral toxicity studies of 1,1-dichloroethane in rats: application to risk evaluation. *Toxicol Sci*, 2001. 64(1): pp. 135–145.

[109] Corsolini, S., and G. Sarà, The trophic transfer of persistent pollutants (HCB, DDTs, PCBs) within polar marine food webs. *Chemosphere*, 2017. 177 : pp. 189–199.

[110] Volschenk, C.M., et al., Bioaccumulation of persistent organic pollutants and their trophic transfer through the food web: human health risks to the rural communities reliant on fish from South Africa's largest floodplain. *Sci Total Environ*, 2019. 685 : pp. 1116–1126.

[111] Windsor, F.M., et al., Biological traits and the transfer of persistent organic pollutants through river food webs. *Environ Sci Technol*, 2019. 53(22): pp. 13246–13256.

[112] D'adamo, R., et al., Bioaccumulation and biomagnification of polycyclic aromatic hydrocarbons in aquatic organisms. *Marine Chemistry*, 1997. 56(1–2): pp. 45–49.

[113] Zhang, Q., L. Yang, and W.X. Wang, Bioaccumulation and trophic transfer of dioxins in marine copepods and fish. *Environ Pollut*, 2011. 159(12): pp. 3390–3397.

[114] Maul, J.D., et al., Bioaccumulation and trophic transfer of polychlorinated biphenyls by aquatic and terrestrial insects to tree swallows (Tachycineta bicolor). *Environ Toxicol Chem*, 2006. 25(4):pp. 1017–1025.

[115] Yu, L., et al., Occurrence and biomagnification of organohalogen pollutants in two terrestrial predatory food chains. *Chemosphere*, 2013. 93(3): pp. 506–511.

[116] Lakshmi Kavya, A.N.V., S. Sundarrajan, and S. Ramakrishna, Identification and characterization of micro-plastics in the marine environment: amini review. *Mar Pollut Bull*, 2020. 160 : p. 111704.

[117] Auta, H.S., C.U. Emenike, and S.H. Fauziah, Distribution and importance of microplastics in the marine environment: a review of the sources, fate, effects, and potential solutions. *Environ Int*, 2017. 102: pp. 165–176.

[118] Desforges, J.P., et al., Widespread distribution of microplastics in subsurface seawater in the NE Pacific Ocean. *Mar Pollut Bull*, 2014. 79(1–2): pp. 94–99.

[119] Horton, A.A., et al., Microplastics in freshwater and terrestrial environments: evaluating the current understanding to identify the knowledge gaps and future research priorities. *Sci Total Environ*, 2017. 586 : pp. 127–141.
[120] Jambeck, J.R., et al., Marine pollution. Plastic waste inputs from land into the ocean. *Science*, 2015. 347(6223): pp. 768–771.
[121] Nizzetto, L., M. Futter, and S. Langaas, Are agricultural soils dumps for microplastics of urban origin? *Environ Sci Technol*, 2016. 50(20): pp. 10777–10779.
[122] Rillig, M.C., Microplastic in terrestrial ecosystems and the soil? *Environ Sci Technol*, 2012. 46(12):pp. 6453–4.
[123] Sheng, C., S. Zhang, and Y. Zhang, The influence of different polymer types of microplastics on adsorption, accumulation, and toxicity of triclosan in zebrafish. *J Hazard Mater*, 2021. 402 : p. 123733.
[124] Lusher, A.L., M. McHugh, and R.C. Thompson, Occurrence of microplastics in the gastrointestinal tract of pelagic and demersal fish from the English Channel. *Mar Pollut Bull*, 2013. 67(1–2): pp. 94–99.
[125] Zhu, X., et al., Bioaccumulation of microplastics and its in vivo interactions with trace metals in edible oysters. *Mar Pollut Bull*, 2020. 154: p. 111079.
[126] Mayoma, B.S., et al., Microplastics in beach sediments and cockles (Anadara antiquata) along the Tanzanian coastline. *Bull Environ Contam Toxicol*, 2020. 105(4): pp. 513–521.
[127] Aminot, Y., et al., Leaching of flame-retardants from polystyrene debris: bioaccumulation and potential effects on coral. *Mar Pollut Bull*, 2020. 151: p. 110862.
[128] Zhang, S., et al., Interactive effects of polystyrene microplastics and roxithromycin on bioaccumulation and biochemical status in the freshwater fish red tilapia (Oreochromis niloticus). *Sci Total Environ*, 2019. 648: pp. 1431–1439.
[129] Mak, C.W., et al., Microplastics from effluents of sewage treatment works and stormwater discharging into the Victoria Harbor, Hong Kong. *Mar Pollut Bull*, 2020. 157:p. 111181.
[130] Zhang, J., et al., Combined effects of polyethylene and organic contaminant on zebrafish (Danio rerio): accumulation of 9-Nitroanthracene, biomarkers and intestinal microbiota. *Environ Pollut*, 2021. 277: p. 116767.
[131] Astruc, D., *Nanoparticles and Catalysis*. 2008. John Wiley & Sons.
[132] Campelo, J.M., et al., Sustainable preparation of supported metal nanoparticles and their applications in catalysis. *ChemSusChem*, 2009. 2(1): pp. 18–45.
[133] Nam, N.H., and N.H. Luong, Chapter 7 – Nanoparticles: synthesis and applications, in *Materials for Biomedical Engineering*, V. Grumezescu and A.M. Grumezescu, Editors. 2019, Elsevier. pp. 211–240.
[134] Mullin, J.W., *Encyclopedia of Chemical Technology*, Vol. 2, 4th edition. Jacqueline I. Kroschwitz & Mary Howe-Grant, Editors. 1992, John Wiley &Sons. ISBN 0 471 52669 X. *Journal of Chemical Technology & Biotechnology*, 1993. 56(4): pp. 421–422.
[135] Dasgupta, N., et al., Nanotechnology in agro-food: from field to plate. *Food Research International*, 2015.69: pp. 381–400.
[136] Ranjan, S., et al., Nanoscience and nanotechnologies in food industries: opportunities and research trends. *Journal of Nanoparticle Research*, 2014. 16(6): p. 2464.
[137] Shafiq, M., et al., An overview of the applications of nanomaterials and nanodevices in the food industry. *Foods (Basel, Switzerland)*, 2020. 9(2): p. 148.
[138] Kansara, K., et al., Synthesis of biocompatible iron oxide nanoparticles as a drug delivery vehicle. *International Journal of Nanomedicine*, 2018. 13(T-NANO 2014 Abstracts): pp. 79–82.

[139] Savaliya, R., et al., Nanotechnology in disease diagnostic techniques. *Curr Drug Metab,* 2015. 16(8): pp. 645–661.

[140] Sharma, V., A. Kumar, and A. Dhawan, Nanomaterials: exposure, effects and toxicity assessment. *Proceedings of the National Academy of Sciences, India Section B: Biological Sciences,* 2012. 82(1): pp. 3–11.

[141] Kumar, A., and R. Shanker, *Nanotoxicity: Aquatic Organisms and Ecosystems.* 2014. Taylor and Francis.

[142] Duncan, L.K., J.R. Jinschek, and P.J. Vikesland, C60 colloid formation in aqueous systems: effects of preparation method on size, structure, and surface charge. *Environ Sci Technol,* 2008. 42(1):pp. 173–178.

[143] Brar, S.K., et al., Engineered nanoparticles in wastewater and wastewater sludge--evidence and impacts. *Waste Manag,* 2010. 30(3): pp. 504–520.

[144] Markiewicz, M., et al., Changing environments and biomolecule coronas: consequences and challenges for the design of environmentally acceptable engineered nanoparticles. *Green Chemistry,* 2018. 20(18): pp. 4133–4168.

[145] Wang, L., et al., Nanocarbon materials in water disinfection: state-of-the-art and future directions. *Nanoscale,* 2019. 11(20): pp. 9819–9839.

[146] Biswas, P., and C.Y. Wu, Nanoparticles and the environment. *J Air Waste Manag Assoc,* 2005. 55(6): pp. 708–746.

[147] Bishoge, O.K., et al., Remediation of water and wastewater by using engineered nanomaterials: a review. *Journal of Environmental Science and Health, Part A,* 2018. 53(6): pp. 537–554.

[148] Hadef, F., An introduction to nanomaterials, in Dasgupta, N., Ranjan, S., Lichtfouse, E (eds.) *Environmental Nanotechnology.* 2018, Springer. pp. 1–58.

[149] Saleem, H., and S.J. Zaidi, Sustainable use of nanomaterials in textiles and their environmental impact. *Materials,* 2020. 13(22): p. 5134.

[150] Shi, J., et al., Sources and concentration of nanoparticles (< 10 nm diameter) in the urban atmosphere. *Atmospheric Environment,* 2001. 35: pp. 1193–1202.

[151] Wang, R., et al., A general strategy for nanohybrids synthesis via coupled competitive reactions controlled in a hybrid process. *Scientific Reports,* 2015. 5(1):pp. 1–14.

[152] Baalousha, M., and J. Lead, *Characterization of Nanomaterials in Complex Environmental and Biological Media.* 2015.Elsevier.

[153] Cornelis, G., et al., Transport of silver nanoparticles in saturated columns of natural soils. *Science of the Total Environment,* 2013. 463 : pp. 120–130.

[154] Cozzari, M., et al., Bioaccumulation and oxidative stress responses measured in the estuarine ragworm (Nereis diversicolor) exposed to dissolved, nano-and bulk-sized silver. *Environmental Pollution,* 2015. 198: pp. 32–40.

[155] Dai, L., et al., Influence of copper oxide nanoparticle form and shape on toxicity and bioaccumulation in the deposit feeder, Capitella teleta. *Marine Environmental Research,* 2015. 111 : pp. 99–106.

[156] Kuehr, S., V. Kosfeld, and C. Schlechtriem, Bioaccumulation assessment of nanomaterials using freshwater invertebrate species. *Environmental Sciences Europe,* 2021. 33(1): pp. 1–36.

[157] Kansara, K., et al., Montmorillonite clay and humic acid modulate the behavior of copper oxide nanoparticles in aqueous environment and induces developmental defects in zebrafish embryo. *Environmental Pollution,* 2019. 255: p. 113313.

[158] Kansara, K., etal., Assessment of the impact of abiotic factors on the stability of engineered nanomaterials in fish embryo media. *Emergent Materials,* 2021. 4(5): pp. 1339–1350.

[159] Kumar, A., et al., Engineered ZnO and TiO2 nanoparticles induce oxidative stress and DNA damage leading to reduced viability of Escherichia coli. *Free Radical Biology and Medicine*, 2011. 51(10): pp. 1872–1881.

[160] Kumar, A., et al., Cellular uptake and mutagenic potential of metal oxide nanoparticles in bacterial cells. *Chemosphere*, 2011. 83(8): pp. 1124–1132.

[161] Kumar, A., et al., Zinc oxide nanoparticles affect the expression of p53, Ras p21 and JNKs: an ex vivo/in vitro exposure study in respiratory disease patients. *Mutagenesis*, 2018. 33(3): pp.237–245.

[162] Shukla, R.K., et al., TiO$_2$ nanoparticles induce oxidative DNA damage and apoptosis in human liver cells. *Nanotoxicology*, 2013. 7(1): pp. 48–60.

[163] Kumar, A. and A. Dhawan, Genotoxic and carcinogenic potential of engineered nanoparticles: an update. *Archives of Toxicology*, 2013. 87(11): pp. 1883–1900.

[164] Kansara, K., et al., TiO$_2$ nanoparticles induce DNA double strand breaks and cell cycle arrest in human alveolar cells. *Environmental and Molecular Mutagenesis*, 2015. 56(2): pp. 204–217.

[165] Shukla, R.K., et al., Titanium dioxide nanoparticle-induced oxidative stress triggers DNA damage and hepatic injury in mice. *Nanomedicine*, 2014. 9(9): pp. 1423–1434.

[166] Xie, J., et al., Nanomaterial-based blood-brain-barrier (BBB) crossing strategies. *Biomaterials*, 2019. 224: p. 119491.

[167] Jiang, J., J. Pi, and J. Cai, The advancing of zinc oxide nanoparticles for biomedical applications. *Bioinorganic Chemistry and Applications*, 2018. Hindawi, pp 1062562-1062562.

[168] Sayadi, M.H., et al., Bioaccumulation and toxicokinetics of zinc oxide nanoparticles (ZnO NPs) co-exposed with graphene nanosheets (GNs) in the blackfish (Capoeta fusca). *Chemosphere*, 2021. 269: p. 128689.

[169] Ates, M., et al., Accumulation and toxicity of CuO and ZnO nanoparticles through waterborne and dietary exposure of goldfish (Carassius auratus). *Environmental Toxicology*, 2015. 30(1): pp. 119–128.

[170] *Hawley's Condensed Chemical Dictionary*, 15th edition. Richard J. Lewis, Editor. 2007, Sr. John Wiley & Sons, Inc. ISBN 978-0-471-76865-4. *Journal of the American Chemical Society*, 2007. 129(16): pp. 5296–5296.

[171] Bacchetta, R., et al., Nano-sized CuO, TiO$_2$ and ZnO affect Xenopus laevis development. *Nanotoxicology*, 2012. 6(4): pp. 381–398.

[172] Chen, J., et al., Effects of titanium dioxide nano-particles on growth and some histological parameters of zebrafish (Danio rerio) after a long-term exposure. *Aquatic Toxicology*, 2011. 101(3–4): pp. 493–499.

[173] Griffitt, R.J., et al., Comparison of molecular and histological changes in zebrafish gills exposed to metallic nanoparticles. *Toxicological Sciences*, 2009. 107(2): pp. 404–415.

[174] Sayadi, M.H., et al., Exposure effects of iron oxide nanoparticles and iron salts in blackfish (Capoeta fusca): acute toxicity, bioaccumulation, depuration, and tissue histopathology. *Chemosphere*, 2020. 247: p. 125900.

[175] Shi, Q., et al., Bioaccumulation, biodistribution, and depuration of 13C-labelled fullerenols in zebrafish through dietary exposure. *Ecotoxicology and environmental safety*, 2020. 191: p. 110173.

[176] Qiang, L., et al., Facilitated bioaccumulation of perfluorooctanesulfonate in common carp (Cyprinus carpio) by graphene oxide and remission mechanism of fulvic acid. *Environmental Science &Technology*, 2016. 50(21): pp. 11627–11636.

[177] Zhu, X., et al., Trophic transfer of TiO$_2$ nanoparticles from daphnia to zebrafish in a simplified freshwater food chain. *Chemosphere*, 2010. 79(9): pp. 928–933.

[178] Judy, J.D., J.M. Unrine, and P.M. Bertsch, Evidence for biomagnification of gold nanoparticles within a terrestrial food chain. *Environmental Science & Technology,* 2011. 45(2): pp. 776–781.

[179] Gupta, G.S., et al., Assessment of agglomeration, co-sedimentation and trophic transfer of titanium dioxide nanoparticles in a laboratory-scale predator-prey model system. *Scientific Reports,* 2016. 6(1): p. 31422.

10 High Throughput Approaches for Engineered Nanomaterial-Induced Ecotoxicity Assessment

Krupa Kansara and Dhiraj Devidas Bhatia

CONTENTS

10.1 Introduction 215
10.2 HTS: Experimental Design and Imaging Techniques 217
      10.2.1 Cell-based and Biochemical Assays for HTS 217
      10.2.2 Imaging Techniques 220
10.3 HTS Contribution in Ecotoxicity 221
Acknowledgments 226
References 226

10.1 INTRODUCTION

Nanomaterials (NMs) and nanoparticles (NPs) are very different and distinct from their bulk counterparts, especially in chemical composition, surface charge, size, agglomeration, aggregation, and solubility. NMs and NPs have at least one dimension with less than 100 nm and thus possess extraordinary physicochemical properties [1–7]. The increasing trend of nano-based research papers in PubMed in the last decade alone is proof of their cross-cutting potential in different disciplines and vast number of applications in multiple sectors [8–16]. NMs have been defined on the basis of size, length scale, differential properties, and novel functionalities. The extraordinary physicochemical properties of NMs have boosted their uses in diversified industrial and consumer-based products (Figure 10.1) [17–23]. However, increased usage and their presence in consumer product sectors raise major concerns for humans and especially aquatic environments as the potential effects of these NMs have been poorly characterized [24,25]. Currently, the limitation of the published data or conflicting published data on the adverse effects of NMs make it difficult to understand or generalize about the risk associated with NMs to humans and aquatic environment. Apart from NMs, new novel chemical products also cause risk to ecological

DOI: 10.1201/9781003244349-10

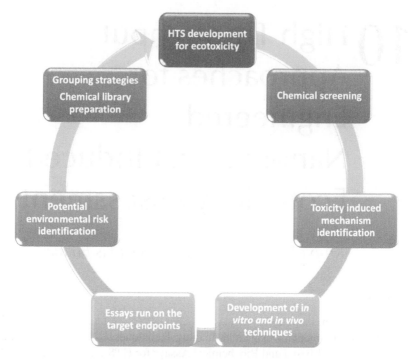

FIGURE 10.1 Nanomaterial-based products available indifferent sectors in global market (Products image courtesy: ©2006 David Hawxhurst, Woodrow Wilson International Center for Scholars).

environments. Hence, there is an urgent need to understand and clarify the toxic effects and mechanisms involved in the toxicity of NMs and new novel chemical compounds.

The environment is at risk as its contamination comes from human activities, expansion of industries, advancement of agriculture activities, and improper channels of waste removal. The diverse mixture of these different inputs makes it difficult to identify potentially adverse interactions. Precautionary measures come into focus as prevention policies help in identifying the potential risk of new products prior to their release for public consumption. However, with rapid development in fields such as nanobiotechnology and materials sciences, it is still nearly impracticable to screen all new products in a cost-effective and realistic timeframe.

In this era, due to the large number of NMs and other chemical products currently in use high throughput screening (HTS) techniques are clearly needed as the traditional *in vivo* set up is laborious and time consuming [26,27]. Globally, over 100 million animals are experimented on annually and it takes two to three years for one complete chemical toxicity evaluation [28]. HTS is an emerging tool for biochemical screening that can analyse cellular and biochemical responses in a timely, cost-effective, and less laborious manner through digital mechanisms and miniaturization [29,30]. Through HTS it is possible to assess cellular pathways and signalling

High Throughput Approaches

pathways to measure target endpoints, which allows predictive profiles for new target developments [31]. HTS was first introduced in the biopharmaceutical and chemical industries for evaluating the effects of novel chemical compounds in a rapid way. The best examples of HTS usage for ecotoxicity assessment are the ToxCast and Tox21 programs, which were established by the US Environmental Protection Agency (US EPA). These programmes successfully screened more than 10,000 chemicals that are directly associated with environmental toxicity and covered more than 700 HTS assays including 300 cell signalling pathways [32–38]. The produced data by various HTS platforms are available in public domains such as PubChem, ACToR, and ChEMBL [28,36,37]. These big HTS datasets are large in volume and have complex and more linkable information compared to conventional *in vivo* and *in vitro* datasets [39,40]. The HTS assays are very helpful to develop chemical libraries that give detailed information on potentially toxic compounds and their dose-dependant profiles to establish hazardous ranking of NMs by generating biological and environmental data [41]. While there are many advantages of HTS, we have to consider how to determine the suitability and reliability of these datasets for regulatory and safety purposes [28]. In this respect, the advantages and challenges of HTS, the HTS role, and different techniques in addressing emergent contaminations such as ENMs for ecotoxicity are discussed in detail in this chapter.

10.2 HTS: EXPERIMENTAL DESIGN AND IMAGING TECHNIQUES

HTS approaches are associated majorly with cells or whole organism and HTS readouts for different biochemical experiments are fluorescence, luminescence, scintillation, absorbance, and optical based. The overall cost for each setup and accuracy of data reproduction are major factors to consider when choosing the most appropriate readouts; however, the primary detection for biochemical assays are fluorescence-based only [42,43]. The major advantage of using fluorescence-based readouts is assay stability, ease of handling, and multiple endpoint inclusion in real-time scenario [44]. The use of fluorescence-based readouts for shorter wavelength excitation such as less than 400 nm is discouraged as it might interfere with engineered nanomaterials (ENMs) or other chemical compounds [45–47].

10.2.1 CELL-BASED AND BIOCHEMICAL ASSAYS FOR HTS

Cell-based assays can be incorporated to investigate whole cellular pathways with different time points using HTS. Cell-based assays can provide the insights of intracellular pathway disruption caused by ENMs or chemical compounds that cannot be obtained through biochemical assays [48,49]. HTS platforms are commonly used for cell-based studies such as cell survival, internal pathways, proliferation, and differentiation. HTS gives a detailed understanding of different cell functions and physiology. It can be used to scale down the tedious cell-based methods, and these platforms allow to include ENMs and chemical libraries that exhibit some hazardous effects or interruption in biological activities. Major bio-pharmaceuticals use two-dimensional cellular microarray assay platforms such as 96- and 384-well plates [50]. The advantage of a cellular microarray platform is that the minimum volume of diversified biomolecules and cells can be used with multiple parameters and endpoints to assess cellular physiology [48,51]. Variegated molecules such as ENMs, new chemical compounds, antibodies, and polymers can be analysed using fully automated spotting technology [52]. The cellular microarray of 2D cell lines can be used to screen for small molecules. A suitable example for the screening is CHO cell line as it could be considered for such system. There is flexibility in the HTS platform for cellular assays focusing on signalling pathway for choosing the readout. A key feature of cell-based screening is that multiple targets are screened at once, with the readout being the outcome of a cellular pathway or network [53].

Biochemical screening using HTS platforms generally utilizes a targeted purified protein and optimizes the ligand binding or inhibits enzymatic activity *in vitro* [54]. These optimizing assays are performed in a competition format, wherein the known ligand or substrate replaces the compound under study. These HTS assays are typically performed in 384-well plate formats, which can provide a good amount of experiment volume from 20–50 μL, which is screening volume. Optical-based readouts such as absorbance, fluorescence. or luminescence are used in the platform [55].

The HTS platform is famous for assessing the toxicity of ENMs and accurate design and planning is required. A few key points need to be considered before starting the toxicity assessment via

HTS. For example, inter-laboratory comparison is required for routine assessments help reduce the variability and identify the source of variance [56]. To reduce the human errors adoption of fully automated and robotic liquid and sample handling is needed, and the randomness in experimental setup is required to reduce biasness. HTS approaches is the clear choice for ENMs and cell interactions as we can identify the ENM-induced changes in cellular surfaces, proliferation, membrane permeability, genotoxicity, reactive oxygen species generation, and other molecular modifications. mass/pH, DNA and chromosome damage, activation of transcription factors, mitochondrial membrane potential changes, oxidative stress monitoring, and post-translational modification can also be detailed [57]. A major technical challenge of ENM toxicity assessment via HTS can arise, as ENM toxicology screening depends on the characterization of ENMs in exposure media, which is time consuming and cannot be performed via HTS platform. This limitation can be partially addressed if the complete characterization and stability of an ENM is done in the exposure media and then the HTS platform be used on various cell lines using different concentrations and time points. All experiments should be performed in triplicate with replicate samples to achieve statistical significance. The basic objective for HTS setup is to predictthe toxicity; ENM-specific positive and negative controls, finalised endpoints, relevant concentrations of ENMs, various time points, significant statistical assessment, validated assays, and graphical representation of data. The graphical representation is useful for multi-parametric data as the ENM concentration, exposure time points, and relevant concentrations can be easily visualized. Various methods and instrumentation have been used to study of toxicity of ENMs using HTS platforms, and they will now be described in detail.

TABLE 10.1
Advantages and Disadvantages of HTS Techniques to Assess the ENMs Toxicity

Technique	Advantage	Disadvantage
Flow cytometry	• Relatively rapid, inexpensive, and quantitative • Multiple parameters can be analysed • Internalization can be accounted via side scattering and forward scattering followed by uptake analysis • Localize and identify cells with and without NMs to study heterogeneity • Highly recognised for rapid quantification of cell-based assays in which ENMs do not interact with cells	• Fluorophores-based quantification in which ENMs may interfere with fluorophores and give false-positive results • Absorbance and fluorescence signals by ENMs may results by its own optical properties • Reagents and markers can be adsorbed by different chemical compositions of ENMs • Cellular surfaces and receptors can be masked by aggregated ENMs • Currently it is developed and used for 96-well plate systems only • Different control systems should be incorporated to identify the interference of NMs with fluorophores
HTS flow cytometry	• ENM uptake by different cell lines can be analysed specifically • Internalization of ENMs in various cells can detect via side-scattering parameter • Size-based internalization of NMs uptake can be determine	• In apoptotic scenario, the cellular granularity is increased and it may increase side scattering; such cases can be validated through different methods
Confocal laser scanning spectroscopy	• Inexpensive and highly validated method • Z-stacks, multiplexing options with combined imaging option, and combined imaging options	• Reduce performance or potential if it's not automated
Atomic emission spectroscopy	• Highly useful for uptake and to identify the particle localization • Quantifies specific chemical elements in ppb range of ENMs at cellular or tissue level	• Very expensive • Technical knowledge is needed
ICP-MS	• Recommended for low detection limit such as ppb • Battery elements can be selective for screening	• Only inorganic ENMs can be quantified • ENM uptake inside the cells or ENMs masked on the cell surfaces or ENMs present between the two cells cannot determine and quantified
Single particle ICP-MS	• Individual ENMs can generate pulse and can be recorded • It can differentiate the dissolved and particulate forms of ENMs	• Only inorganic ENMs can be quantified • ENMs uptake inside the cells or ENMs masked on the cell surfaces or ENMs present between the two cells cannot be determined and quantified

(continued)

TABLE 10.1 (Continued)
Advantages and Disadvantages of HTS Techniques to Assess the ENMs Toxicity

Technique	Advantage	Disadvantage
IBM techniques (μPIXE and/or μRBS)	• Spatially resolved elemental imaging can be possible • Parts per million range sensitivity can be acquired • ENM uptake inside the cells or ENMs masked on the cell surfaces or ENMs present between the two cells can determine and quantified • Final concentration of ENMs can be quantified as well	• Cannot determine the dissolved and particulate forms of ENMs
TEM and ToF-SIMS	• Accurate and continuation visualization of ENM uptake at cellular level	• Techniques are costly, time, and maintenance cost of the instruments are relatively high
CRM	• Very popular non-invasive technique • Relatively fast and quantify the genotoxic and cytotoxic potential of ENMs via different assays such as apoptosis, ROS generation, DNA fragmentation in the different model systems (*in vivo* and *in vitro*)	• It cannot quantify the dissolved matter released by various ENMs in systems

10.2.2 Imaging Techniques

In the investigation of toxicological responses of ENMs in various cell lines and *in vivo* models, the intracellular distribution is crucial factor as it is difficult to predict the responses in various models because of several transformation events. Flow cytometry and confocal microscopy are popular tools to assess and quantify the cellular uptake of various ENMs in both in vitro and in vivo systems. These techniques have major limitations as they required fluorescently labelled ENMs or fluorescently labelled molecules for analysis and fluorescent dye may interrupt the chemical properties of ENMs, which ultimately produces a false-positive dataset. Fluorescent-labelled free imagine techniques are advantageous as they can provide authentic ENM assessment for HTS platforms. A few of these techniques are described as follows:

> A. Semi-quantitative imagine techniques for ENM internalization include transmission electron microscopy (TEM), time-of-flight secondary ion mass spectrometry (ToF-SIMS), and confocal Raman micro spectroscopy (CRM) [58]. These methods allow the internalization of ENMs at cellular level in their biological environment. TEM is capable of visualizing the internalized ENMs in ultra-thin sections such as 50–100 nm as well [59,60]. The chemical

composition identification of ENMs can be analysed through X-ray energy-dispersive spectrometry (EDS) coupled with TEM. High-depth resolution and contrast 3D reconstruction of ENMs in cells/tissues can be visualized by focused ion beam-scanning electron microscopy (FIB-SEM) [61]. Moreover, 3D distribution of ENMs in different cells and tissues can be analysed with ToF-SIMS with high energy beam raster. Physiological changes in cell-based lipids induced by ENMs with resolution of 300 nm can be detected [62,63]. The major drawbacks of these techniques are cost and time required. CRM is the most suitable example for HTS, which could be adopted in the near future as it is rapid and less expensive than TEM and ToF-SIMS. CRM has a resolution around 200 nm and can provide 3D images with ENM localization at cellular levels. CRM is most suitable for manufactured ENMs as these nanoconstructs have specific wavelengths that are different than their bulk counterparts. ENM- or chemical compound-induced toxicity can be easily quantified by CRM as it is used to analyse cell cycle stages, genotoxicity, and cell survival [64–68]. However, the sensitivity and resolutions of CRM is limited compared to TEM.
B. Atom emission spectroscopy (AES) is another powerful tool for the quantification of a vast range of elements. AES works on the radiation emission in which characterization of every element is possible. AES can quantify within ppb range of accuracy for specific ENMs or chemical compounds inside cellular systems [69,70]. Mass spectroscopy (MS) is another technique to quantify the internalization of ENMs in a cellular context. It is a very sensitive technique and provides information based on mass-to-charge ratio of particular particles. Additionally, MS with inductively coupled plasma (ICP-MS) is widely used to analyse ENM uptake in cells [71]. A suspension of large numbers cells/tissue sample has to undergo dissolution prior to analysis in ICP-MS. This technique can differentiate the dissolved and particulate form of ENMs or chemical compounds whether organic or inorganic. This technique cannot differentiate or identify the localization of ENMs inside the cells. It cannot distinguish if the ENMs are internalised or attached on the cell surface or located in between the cells [72]. It can show the difference between the control cells without ENMs and uptake of ENMs in cells based on pulses recorded [73].

10.3 HTS CONTRIBUTION IN ECOTOXICITY

Of the large volume of ENMs and other chemicals produced (~ 100,000) annually in the global market, almost 30% contribute to neural toxicity [74]. This exponentially increased market needs a tool HTS to address the toxicological fate of ENMs. However, the HTS platform is quite new for assessing environmental toxicology compared to its use in drug development applications [31]. Currently, only a few publications are available that reference HTS platforms, thus the flexibility and adaption of these platforms need to be addressed in detail. To date, the application of HTS for environmental toxicology is restricted to cell-based and mechanism-based pathways on model organisms such as zebrafish, *C. elegance,* and *S. cerevisiae*

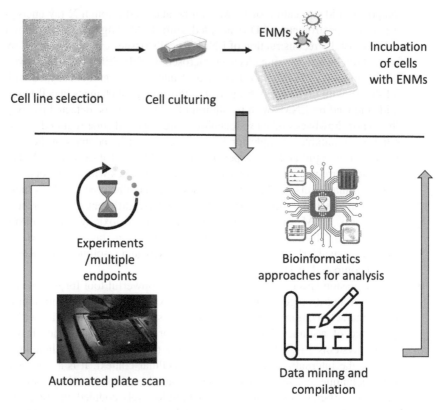

FIGURE 10.2 Experimental workflow (*in vitro* approach) for effective high-throughput screening.

[75]. These model organisms are widely used due to low cost and high throughput compared to large organisms for toxicity assessment. In the past, zebrafish and *C. elegance* were used as a *in vivo* endpoint, which linked higher levels of chemical and biological activities [76,77]. The main objectives of these studies was to check suitability of the HTS platform for the toxicity screening. To assess the ENM toxicity and establish a database, multiparametric endpoints can be used on whole organisms using 24-, 96-, and 384-well plates for rapid and accurate dose-dependant format. The strategies and workflow of HTS are described in Figures 10.2 and 10.3.

The conventional testing organism for HTS is zebrafish as it is a famous model organism and the toxicity protocols on zebrafish have been already standardized. The famous use of HTS strategies on zebrafish was in the ToxCast program, where researchers analysed 310 chemicals [78]. The experimental setup was on developing embryos in which they assessed the viability in 96-well plate using two assays format: one concentration (80 chemicals/plate) and ~ 8 chemicals for dose-dependant response up to 144 h. This approach was used to determine whether one concentration study is sufficient to determine the chemical toxicity compared to a dose-dependant response. This study validated that a single concentration parameter could pre-screen

High Throughput Approaches

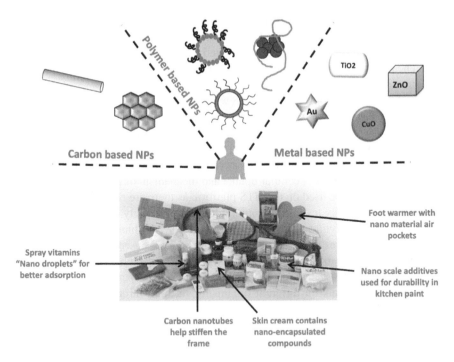

FIGURE 10.3 High-throughput screening (HTS) strategy to identify potential hazardous chemicals and preparing library.

the toxicity with an accuracy of 84%, after performing and reproducing interplate and intraplate setup. Moreover, this study identified four chemical classes that showed adverse effects and moderate toxicity and provided an excellent example of how the HTS platform could increase the throughput of chemical screening and mechanism of action for toxicity. In another study, researchers demonstrated the implementation of a HTS platform with automated image analysis to assess the adverse effects of dissolved metal oxide NPs on zebrafish embryo hatching [79].

C. elegance is another conventional model organism that has been commonly used for the development of HTS platforms. One study was conducted on *C. elegance* to assess the toxicological potential of 20 different metal ENMs on HTS. The study was carried out on whole organism culture with liquid (24-well plates) and solid (6-well plate) media. *C. elegance* were exposed to four different concentrations/each ENMs from 3 days to 4 weeks' time interval. Physiological measurements such as body size, locomotion, survival, and fitness at different time points were assessed, and HTS strategy was used with microplate reader (QuantWorm) to streamline the testing. The results indicated that the dissolution rate, size, and chemical composition of metal ENMs determined the toxicity fate, but carbon nanotubes did not exhibit size- and shape-based toxicity.

HTS is recognised as a suitable platform for model organisms to screen hazardous chemicals, but the admission of additional species is required to be more environmentally relevant. This goal can be achieved through incorporating invertebrate organism via *in vitro* methods [80]. Oysters and sentinel are ideal organisms for the HTS platform as they have been developed in biomonitoring assays for environmental health assessment and the development of these unconventional organisms for HTS strategy is encouraged [81].

One of the unconventional uses of HTS was of RT-W1 cells (*Oncorhynchus mykiss* gill cells) for Ag NPs (silver nanoparticles) having different shapes and sizes. The aim of the study was to identify the toxicity based on the different shapes of AgNPs in the form of nanowires, nanoplates, and nanospheres on cell survival and reactive oxygen species generation. The study observed that the size, shape, and dissolution of Ag NPs play crucial roles in cellular uptake and mechanism of toxicity [82]. Primary cell culture can also be used on the HTS platform as there are only a few publication available on ecotoxicity testing. Quantum dot-induced toxicity was assessed on gills and haemocytes of *Mytilus galloprovincialis*; cell survival, biomarker response, and reactive oxygen species were measured. Another study focused on *Mytilus* embryos using 96-well plate at different time points to pre-screen potential contaminants [83]. The strategy used a larger range of variations with dose-dependant suitability. The ecotoxicity assessments on different organisms via high throughput screening are described in Table 10.2.

There is an urgent need for innovative techniques like HTS as the conventional methodologies for ecotoxicity assessments for ENMs or chemical compounds are time consuming and are not able to maintain pace with current innovations. The complete evaluation of the currently available ENMs and chemical compounds would take more than 35 years to complete [102]. The HTS approaches for ecotoxicity assessment with respect to ENMs will permit reduced time and better cost effectiveness with increased data outcomes. Benefit-over-cost approaches have been used to modify existing technologies and HTS will provide the solution for hazardous identification in a time-constrained manner with cost effectiveness. Ecotoxicological assessments using HTS approaches can be expensive as fully automated, highly sensitive, and sophisticated equipment are used on various platforms. Greater investment in HTS platforms can be justified depending on how many ENMs and different samples can be analysed compared to traditional laboratory-based experiments in a time-efficient manner. Additionally, robustness and adaptive HTS platforms for ecotoxicology assessments can provide early and accurate results, which will help to find the potential of ENMs to cause damage to different ecosystems, environments, and ultimately human beings, thus maximizing the benefits of innovations.

Due to the increased use of ENMs in various consumer-based and healthcare sectors, accurate and fast test systems for ENM screening are needed to ensure the safety of living beings. HTS platforms have the potential to assess a large number of materials with different *in vivo* and *in vitro* test systems at the same time in a short period of time in a cost-effective manner. These approaches reduce experimental repetition and save manual labour. However, validation of HTS platforms is required as their relevance to *in vivo* and *in vitro* conditions may differ. A validated HTS platform is required to assess the dose- and time-dependant predicted ecotoxicity and more datasets for uptake and internalization of ENMs at cellular and subcellular level should be developed. The HTS endpoints should cover crucial and important toxicity

TABLE 10.2
Ecotoxicity Assessments on Different Organisms via High Throughput Screening Approaches

Organism	Species	Assessment/Assays	Number of Toxicant Screenings	References
Fish	Zebrafish	Developmental malformities, Survival rate, biological activities	9	[77]
	Zebrafish	Cardiac activity with different endpoints	10	[84]
	Zebrafish	Larval malformities	-	[85]
	Zebrafish	Hatching success and chorion enzyme (ZHE1) activity	24	[79]
	Zebrafish	Hatching success and omics profile	3	[86]
	Zebrafish	Hatching success and chorion enzyme (ZHE1) activity, fluorescent imaging	1	[87]
	Zebrafish	Survival rate and malformities	309	[78]
	Zebrafish	Survival rate and developmental defects	5	[88]
	Oncorhynchus mykiss (gill cells)	Cellular viability, SOD (superoxide dismutase) activity	5	[82]
Algae	Platymonas subcordiformis and Platymonas helgolandica var. tsingtaoensis	Cell viability	3	[89]
Bacteria	E.coli	Cell viability	6	[90]
	V. fischeri	Cell viability	8	[91]
	V. fischeri	Cell viability	11	[92]
	E. coli and P. aeruginosa	Reactive generation species (ROS and SOD)	2	[93]
	E. coli and S.cerevisiae	Cell viability and omics approaches	5	[94]
Invertebrates	C. elegans	Fitness, life span, body size, locomotion	20	[76]
	C. elegans	Feeding behaviour	4	[95]
	C. elegans	fluorescent-based imaging	7	[96]
	Dugesia japonica	Enzymatic activity for phase 1 and 2	5	[97]
	Mytilus galloprovincialis	Developmental changes	9	[83]
	Calanus finmarchicus	Realtime PCR analysis	1	[98]
	Echinogammarus marinus	Genomic expression profiling	-	[99]
	D. magna	FT-ICR analysis	1	[100]
	D. magna	nucleotide sequencing	1	[101]

parameters such as cytotoxicity, oxidative stress generated by ROS, genotoxicity, carcinogenic potential, and cellular uptake route. ENMs, based on different chemical compositions and sizes, may display the same physiochemical properties and can give the false-positive results of traditional toxicity assays [103–105]. Systematic positive and negative controls should be incorporated to avoid false-positive results and to confirm the sensitivity of the experimental setup as various ENMs may interfere with experimental media.

HTS is a promising platform for researchers to generate large datasets in short time frames and thus streamline the validation process along with standard operating procedures. Additionally, these parameters can reduce the variability of results and increase the independence of users. Reliable information regarding ENM-induced toxicity via HTS approaches can help the research community identify potential dangerous materials and allow complete automation and desired output in a more efficient manner.

ACKNOWLEDGMENTS

Funding received from GSBTM—Gujarat State Biotechnology Mission, Government of Gujarat. The Indian Institute of Technology Gandhinagar for all the facilities is also acknowledged.

REFERENCES

[1] Kansara K, Patel P, Shah D, et al. TiO$_2$ nanoparticles induce DNA double strand breaks and cell cycle arrest in human alveolar cells. *Environmental and Molecular Mutagenesis*. 2015;56(2):204–217.

[2] Patel P, Kansara K, Senapati VA, et al. Cell cycle dependent cellular uptake of zinc oxide nanoparticles in human epidermal cells. *Mutagenesis*. 2016;31(4): 481–490.

[3] Patel P, Kansara K, Singh R, et al. Cellular internalization and antioxidant activity of cerium oxide nanoparticles in human monocytic leukemia cells. *International Journal of Nanomedicine*. 2018;13(T-NANO 2014 Abstracts):39.

[4] Kansara K, Sathish C, Vinu A, et al. Assessment of the impact of abiotic factors on the stability of engineered nanomaterials in fish embryo media. *Emergent Materials*. 2021; 4(5):1–12.

[5] Chaudhari R, Patel P, Meghani N, et al. Fabrication of methotrexate-loaded gold nanoconjugates and its enhanced anticancer activity in breast cancer. *3 Biotech*. 2021;11(4):1–13.

[6] Kansara K, and Kumar A. In vitro methods to assess the cellular toxicity of nanoparticles. In: *Nanotoxicity*, Edited by Susai Rajendran, Tuan Anh Nguyen, Ritesh K. Shukla, Anita Mukherjee and Chandraiah Godugu. Elsevier; 2020. pp. 21–40.

[7] Vallabani NS, Sengupta S, Shukla RK, et al. ZnO nanoparticles-associated mitochondrial stress-induced apoptosis and G2/M arrest in HaCaT cells: a mechanistic approach. *Mutagenesis*. 2019;34(3):265–277.

[8] Dhawan A, Singh S, Kumar A, et al. *Nanobiotechnology: Human Health and the Environment*. CRC Press; 2018.

[9] Senapati VA, and Kumar A. ZnO nanoparticles dissolution, penetration and toxicity in human epidermal cells. Influence of pH. *Environmental Chemistry Letters*. 2018;16(3):1129–1135.

[10] Kumar A, Senapati VA, and Dhawan A. Protocols for in vitro and in vivo toxicity assessment of engineered nanoparticles. In: *Nanotoxicology*, Edited by Alok Dhawan, Diana Anderson, Rishi Shanker. 2017. pp. 9–132.

[11] Senapati VA, Gupta GS, Pandey AK, et al. Zinc oxide nanoparticle induced age dependent immunotoxicity in BALB/c mice. *Toxicology Research*. 2017;6(3):342–352.

[12] Shanker R, Dobrovolsky VN, Dhawan A, et al. *Mutagenicity: Part of the Methods in Toxicology and Pharmacology Series*. Elsevier Science & Technology. 2017.

[13] Kumar A, Singh S, Shanker R, et al. *Nanotoxicology: Challenges for Biologists*. Edited by Alok Dhawan, Diana Anderson, Rishi Shanker. 2017.

[14] Gupta GS, Kumar A, Shanker R, et al. Assessment of agglomeration, co-sedimentation and trophic transfer of titanium dioxide nanoparticles in a laboratory-scale predator-prey model system. *Scientific Reports*. 2016;6(1):1–13.

[15] Kumar A, Khan S, and Dhawan A. Comprehensive molecular analysis of the responses induced by titanium dioxide nanoparticles in human keratinocyte cells. *Journal of Translational Toxicology*. 2014;1(1):28–39.

[16] Kumar A, Khan S, and Dhawan A. Metal oxide nanoparticles elicit genotoxic responses in mammalian cells: a critical review. nanoscience and technology for mankind Allahabad: *National Academy of Sciences*. 2014:160–194.

[17] Kansara K, Kumar A, and Karakoti AS. Combination of humic acid and clay reduce the ecotoxic effect of TiO_2 NPs: A combined physico-chemical and genetic study using zebrafish embryo. *Science of the Total Environment*. 2020;698:134133.

[18] Gupta GS, Kansara K, Shah H, et al. Impact of humic acid on the fate and toxicity of titanium dioxide nanoparticles in Tetrahymena pyriformis and zebrafish embryos. *Nanoscale Advances*. 2019;1(1):219–227.

[19] Senapati VA, Kansara K, Shanker R, et al. Monitoring characteristics and genotoxic effects of engineered nanoparticle–protein corona. *Mutagenesis*. 2017;32(5):479–490.

[20] Kansara K, Paruthi A, Misra SK, et al. Montmorillonite clay and humic acid modulate the behavior of copper oxide nanoparticles in aqueous environment and induces developmental defects in zebrafish embryo. *Environmental Pollution*. 2019;255:113313.

[21] Patel P, Meghani N, Kansara K, et al. Nanotherapeutics for the treatment of cancer and arthritis. *Current Drug Metabolism*. 2019;20(6):430–445.

[22] Kumar A, Najafzadeh M, Jacob BK, et al. Zinc oxide nanoparticles affect the expression of p53, Ras p21 and JNKs: an ex vivo/in vitro exposure study in respiratory disease patients. *Mutagenesis*. 2015;30(2):237–245.

[23] Senapati VA, and Kumar A. Nanoparticles and the aquatic environment: application, impact and fate. In: *Nanobiotechnology*, Edited by Alok Dhawan, Sanjay Singh, Ashutosh Kumar, Rishi Shanker. CRC Press. 2018. pp. 299–322.

[24] Gupta GS, Shanker R, Dhawan A, et al. Impact of nanomaterials on the aquatic food chain. In: *Nanoscience in Food and Agriculture*, Edited by Shivendu Ranjan, Nandita Dasgupta, Eric Lichtfouse. 5: Springer. 2017. pp. 309–333.

[25] Kumar A, Shanker R, and Alok D. *Nanotoxicity: Aquatic Organisms and Ecosystems*. CRC Press. 2015. pp. 97–106.

[26] Cohen Y, Rallo R, Liu R, et al. In silico analysis of nanomaterials hazard and risk. *Accounts of Chemical Research*. 2013;46(3):802–812.

[27] Nel A, Xia T, Meng H, et al. Nanomaterial toxicity testing in the 21st century: use of a predictive toxicological approach and high-throughput screening. *Accounts of Chemical Research*. 2013;46(3):607–621.

[28] Judson R, Kavlock R, Martin M, et al. Perspectives on validation of high-throughput assays supporting 21st century toxicity testing. *Altex*. 2013;30(1):51.

[29] Wilson A, Reif DM, and Reich BJ. Hierarchical dose–response modeling for high-throughput toxicity screening of environmental chemicals. *Biometrics*. 2014;70(1):237–246.

[30] Patel T, Telesca D, George S, et al. Toxicity profiling of engineered nanomaterials via multivariate dose-response surface modeling. *The Annals of Applied Statistics*. 2012;6(4):1707.

[31] Dix DJ, Houck KA, Martin MT, et al. The ToxCast program for prioritizing toxicity testing of environmental chemicals. *Toxicological Sciences*. 2007;95(1):5–12.

[32] Kavlock RJ, Austin CP, and Tice R. Toxicity testing in the 21st century: implications for human health risk assessment. *Risk Analysis: An Official Publication of the Society for Risk Analysis*. 2009;29(4):485.

[33] Attene-Ramos MS, Huang R, Michael S, et al. Profiling of the Tox21 chemical collection for mitochondrial function to identify compounds that acutely decrease mitochondrial membrane potential. *Environmental Health Perspectives*. 2015;123(1):49–56.

[34] Tice RR, Austin CP, Kavlock RJ, et al. Improving the human hazard characterization of chemicals: a Tox21 update. *Environmental Health Perspectives*. 2013;121(7):756–765.

[35] Huang R, Xia M, Sakamuru S, et al. Modelling the Tox21 10 K chemical profiles for in vivo toxicity prediction and mechanism characterization. *Nature Communications*. 2016;7(1):1–10.

[36] Wang Y, Bryant SH, Cheng T, et al. Pubchem bioassay: 2017 update. *Nucleic Acids Research*. 2017;45(D1):D955–D963.

[37] Leach A. The ChEMBL database in 2017. *Nucleic Acids Research*. 2017;45:D945–D954.

[38] Judson R, Richard A, Dix D, et al. ACToR—aggregated computational toxicology resource. *Toxicology and Applied Pharmacology*. 2008;233(1):7–13.

[39] Zhu H, Zhang J, Kim MT, et al. Big data in chemical toxicity research: the use of high-throughput screening assays to identify potential toxicants. *Chemical Research in Toxicology*. 2014;27(10):1643–1651.

[40] Khoury MJ, and Ioannidis JP. Big data meets public health. *Science*. 2014;346(6213):1054–1055.

[41] Judson R, Houck K, Martin M, et al. In vitro and modelling approaches to risk assessment from the US Environmental Protection Agency ToxCast programme. *Basic & Clinical Pharmacology & Toxicology*. 2014;115(1):69–76.

[42] Zang R, Li D, Tang I-C, et al. Cell-based assays in high-throughput screening for drug discovery. *International Journal of Biotechnology for Wellness Industries*. 2012;1(1):31–51.

[43] Eggeling C, Brand L, Ullmann D, et al. Highly sensitive fluorescence detection technology currently available for HTS. *Drug Discovery Today*. 2003;8(14):632–641.

[44] An WF, and Tolliday NJ. Introduction: cell-based assays for high-throughput screening. *Cell-Based Assays for High-Throughput Screening*. 2009; 486:1–12.

[45] Kaminski T, and Geschwindner S. Perspectives on optical biosensor utility in small-molecule screening. *Expert Opinion on Drug Discovery*. 2017;12(11):1083–1086.

[46] Kaminski T, Gunnarsson A, and Geschwindner S. Harnessing the versatility of optical biosensors for target-based small-molecule drug discovery. *ACS Sensors*. 2017;2(1):10–15.

[47] Macarrón R, and Hertzberg RP. Design and implementation of high throughput screening assays. *Molecular Biotechnology*. 2011;47(3):270–285.

[48] Cader Z, Graf M, Burcin M, et al. Cell-based assays using differentiated human induced pluripotent cells. In: *Cell-Based Assays Using iPSCs for Drug Development and Testing*, Edited by Carl-Fredrik Mandenius, James A. Ross. Springer; 2019. pp. 1–14.

[49] Mandenius C-F, and Ross JA. *Cell-Based Assays Using IPSCs for Drug Development and Testing*. Springer; 2019.

[50] Lee DW, Doh I, and Nam D-H. Unified 2D and 3D cell-based high-throughput screening platform using a micropillar/microwell chip. *Sensors and Actuators B: Chemical*. 2016;228:523–528.

[51] Kelm JM, Lal-Nag M, Sittampalam GS, et al. Translational in vitro research: Integrating 3D drug discovery and development processes into the drug development pipeline. *Drug Discovery Today*. 2019;24(1):26–30.

[52] Aldewachi H, Al-Zidan RN, Conner MT, et al. High-throughput screening platforms in the discovery of novel drugs for neurodegenerative diseases. *Bioengineering*. 2021;8(2):30.

[53] Fröhlich F, Kessler T, Weindl D, et al. Efficient parameter estimation enables the prediction of drug response using a mechanistic pan-cancer pathway model. *Cell Systems*. 2018;7(6):567–579.e6.

[54] Fang Y. Ligand–receptor interaction platforms and their applications for drug discovery. *Expert Opinion on Drug Discovery*. 2012;7(10):969–988.

[55] Stoddart LA, White CW, Nguyen K, et al. Fluorescence- and bioluminescence-based approaches to study GPCR ligand binding. *British Journal of Pharmacology*. 2016;173(20):3028–3037.

[56] Prina-Mello A, Mohamed BM, Verma NK, et al. Advanced methodologies and techniques for assessing nanomaterial toxicity. *Nanotoxicology: Progress toward Nanomedicine*. 2014;155.

[57] Prina-Mello A, Crosbie-Staunton K, Salas G, et al. Multiparametric toxicity evaluation of SPIONs by high content screening technique: identification of biocompatible multifunctional nanoparticles for nanomedicine. *IEEE Transactions on Magnetics*. 2012;49(1):377–382.

[58] Monteiro-Riviere NA, Wiench K, Landsiedel R, et al. Safety evaluation of sunscreen formulations containing titanium dioxide and zinc oxide nanoparticles in UVB sunburned skin: an in vitro and in vivo study. *Toxicological Sciences*. 2011;123(1):264–280.

[59] Nemmar A, Al-Maskari S, Ali BH, et al. Cardiovascular and lung inflammatory effects induced by systemically administered diesel exhaust particles in rats. *American Journal of Physiology—Lung Cellular and Molecular Physiology*. 2007;292(3):L664–L670.

[60] Takenaka S, Karg E, Kreyling W, et al. Distribution pattern of inhaled ultrafine gold particles in the rat lung. *Inhalation Toxicology*. 2006;18(10):733–740.

[61] Heymann JA, Hayles M, Gestmann I, et al. Site-specific 3D imaging of cells and tissues with a dual beam microscope. *Journal of Structural Biology*. 2006;155(1):63–73.

[62] Haase A, Arlinghaus HF, Tentschert J, et al. Application of laser postionization secondary neutral mass spectrometry/time-of-flight secondary ion mass spectrometry in nanotoxicology: visualization of nanosilver in human macrophages and cellular responses. *ACS Nano*. 2011;5(4):3059–3068.

[63] Lee P-L, Chen B-C, Gollavelli G, et al. Development and validation of TOF-SIMS and CLSM imaging method for cytotoxicity study of ZnO nanoparticles in HaCaT cells. *Journal of Hazardous Materials*. 2014;277:3–12.

[64] Perna G, Lastella M, Lasalvia M, et al. Raman spectroscopy and atomic force microscopy study of cellular damage in human keratinocytes treated with HgCl$_2$. *Journal of Molecular Structure*. 2007;834:182–187.

[65] Pyrgiotakis G, Kundakcioglu OE, Pardalos PM, et al. Raman spectroscopy and support vector machines for quick toxicological evaluation of titania nanoparticles. *Journal of Raman Spectroscopy*. 2011;42(6):1222–1231.

[66] Zoladek A, Pascut FC, Patel P, et al. Non-invasive time-course imaging of apoptotic cells by confocal Raman micro-spectroscopy. *Journal of Raman Spectroscopy*. 2011;42(3):251–258.

[67] Uzunbajakava N, Lenferink A, Kraan Y, et al. Nonresonant confocal Raman imaging of DNA and protein distribution in apoptotic cells. *Biophysical Journal*. 2003;84(6):3968–3981.

[68] Krafft C, Knetschke T, Funk RH, et al. Studies on stress-induced changes at the subcellular level by Raman microspectroscopic mapping. *Analytical Chemistry*. 2006;78(13):4424–4429.

[69] Fischer HC, Fournier-Bidoz S, Chan W, et al. Quantitative detection of engineered nanoparticles in tissues and organs: an investigation of efficacy and linear dynamic ranges using ICP-AES. *Nanobiotechnology*. 2007;3(1):46–54.

[70] Albanese A, Tsoi KM, and Chan WC. Simultaneous quantification of cells and nanomaterials by inductive-coupled plasma techniques. *Journal of Laboratory Automation*. 2013;18(1):99–104.

[71] Malugin A, and Ghandehari H. Cellular uptake and toxicity of gold nanoparticles in prostate cancer cells: a comparative study of rods and spheres. *Journal of Applied Toxicology: An International Journal*. 2010;30(3):212–217.

[72] Allouni ZE, Gjerdet NR, Cimpan MR, et al. The effect of blood protein adsorption on cellular uptake of anatase TiO$_2$ nanoparticles. *International Journal of Nanomedicine*. 2015;10:687.

[73] Zhou X, Dorn M, Vogt J, et al. A quantitative study of the intracellular concentration of graphene/noble metal nanoparticle composites and their cytotoxicity. *Nanoscale*. 2014;6(15):8535–8542.

[74] Amiard-Triquet C, Amiard J-C, and Mouneyrac C. *Aquatic Ecotoxicology: Advancing Tools for Dealing with Emerging Risks*. Academic press; 2015.

[75] Szymański P, Markowicz M, and Mikiciuk-Olasik E. Adaptation of high-throughput screening in drug discovery—toxicological screening tests. *International Journal of Molecular Sciences*. 2012;13(1):427–452.

[76] Jung S-K, Qu X, Aleman-Meza B, et al. Multi-endpoint, high-throughput study of nanomaterial toxicity in Caenorhabditis elegans. *Environmental Science & Technology*. 2015;49(4):2477–2485.

[77] George S, Xia T, Rallo R, et al. Use of a high-throughput screening approach coupled with in vivo zebrafish embryo screening to develop hazard ranking for engineered nanomaterials. *ACS Nano*. 2011;5(3):1805–1817.

[78] Padilla S, Corum D, Padnos B, et al. Zebrafish developmental screening of the ToxCast™ Phase I chemical library. *Reproductive Toxicology*. 2012;33(2):174–187.

[79] Lin S, Zhao Y, Ji Z, et al. Zebrafish high-throughput screening to study the impact of dissolvable metal oxide nanoparticles on the hatching enzyme, ZHE1. *Small*. 2013;9(9–10):1776–1785.

[80] Elston R. Molluscan diseases: a tissue-culture perspective. In: *Aquatic Invertebrate Cell Culture*, Edited by Carmel Mothersil, Brian Austin. Springer-Praxis; 2000. pp. 183–203.

[81] Legras S, Mouneyrac C, Amiard J, et al. Changes in metallothionein concentrations in response to variation in natural factors (salinity, sex, weight) and metal contamination in crabs from a metal-rich estuary. *Journal of Experimental Marine Biology and Ecology*. 2000;246(2):259–279.

[82] George S, Lin S, Ji Z, et al. Surface defects on plate-shaped silver nanoparticles contribute to its hazard potential in a fish gill cell line and zebrafish embryos. *ACS Nano*. 2012;6(5):3745–3759.

[83] Fabbri R, Montagna M, Balbi T, et al. Adaptation of the bivalve embryotoxicity assay for the high throughput screening of emerging contaminants in Mytilus galloprovincialis. *Marine Environmental Research*. 2014;99:1–8.

[84] Raftery TD, Isales GM, Yozzo KL, et al. High-content screening assay for identification of chemicals impacting spontaneous activity in zebrafish embryos. *Environmental Science & Technology*. 2014;48(1):804–810.

[85] Pardo-Martin C, Chang T-Y, Koo BK, et al. High-throughput in vivo vertebrate screening. *Nature Methods*. 2010;7(8):634–636.

[86] Lin S, Zhao Y, Xia T, et al. High content screening in zebrafish speeds up hazard ranking of transition metal oxide nanoparticles. *ACS Nano*. 2011;5(9):7284–7295.

[87] Muller EB, Lin S, and Nisbet RM. Quantitative adverse outcome pathway analysis of hatching in zebrafish with CuO nanoparticles. *Environmental Science & Technology*. 2015;49(19):11817–11824.

[88] Olivares CI, Field JA, Simonich M, et al. Arsenic (III, V), indium (III), and gallium (III) toxicity to zebrafish embryos using a high-throughput multi-endpoint in vivo developmental and behavioral assay. *Chemosphere*. 2016;148:361–368.

[89] Zheng G-x, Li Y-j, Qi L-l, et al. Marine phytoplankton motility sensor integrated into a microfluidic chip for high-throughput pollutant toxicity assessment. *Marine Pollution Bulletin*. 2014;84(1–2):147–154.

[90] Ozaki AS. *Assessing the Effects of Titanium Dioxide Nanoparticles on Microbial Communities in Stream Sediment Using Artificial Streams and High Throughput Screening*. Loyola University Chicago; 2013.

[91] Pinto PC, Costa SP, Lima JL, et al. Automated high-throughput Vibrio fischeri assay for (eco) toxicity screening: application to ionic liquids. *Ecotoxicology and Environmental Safety*. 2012;80:97–102.

[92] Mortimer M, Kasemets K, Heinlaan M, et al. High throughput kinetic Vibrio fischeri bioluminescence inhibition assay for study of toxic effects of nanoparticles. *Toxicology In Vitro*. 2008;22(5):1412–1417.

[93] Priester JH, Singhal A, Wu B, et al. Integrated approach to evaluating the toxicity of novel cysteine-capped silver nanoparticles to Escherichia coli and Pseudomonas aeruginosa. *Analyst*. 2014;139(5):954–963.

[94] Ivask A, ElBadawy A, Kaweeteerawat C, et al. Toxicity mechanisms in Escherichia coli vary for silver nanoparticles and differ from ionic silver. *ACS Nano*. 2014;8(1):374–386.

[95] Boyd WA, McBride SJ, and Freedman JH. Effects of genetic mutations and chemical exposures on Caenorhabditis elegans feeding: evaluation of a novel, high-throughput screening assay. *PLoS One*. 2007;2(12):e1259.

[96] Pluskota A, Horzowski E, Bossinger O, et al. In Caenorhabditis elegans nanoparticle-bio-interactions become transparent: silica-nanoparticles induce reproductive senescence. *PloS One*. 2009;4(8):e6622.

[97] Li M-H. Development of in vivo biotransformation enzyme assays for ecotoxicity screening: in vivo measurement of phases I and II enzyme activities in freshwater planarians. *Ecotoxicology and Environmental Safety*. 2016;130:19–28.

[98] Hansen BH, Altin D, Booth A, et al. Molecular effects of diethanolamine exposure on Calanus finmarchicus (Crustacea: Copepoda). *Aquatic Toxicology*. 2010;99(2):212–222.

[99] Short S, Yang G, Kille P, et al. Vitellogenin is not an appropriate biomarker of feminisation in a crustacean. *Aquatic Toxicology*. 2014;153:89–97.

[100] Taylor NS, Weber RJ, Southam AD, et al. A new approach to toxicity testing in Daphnia magna: application of high throughput FT-ICR mass spectrometry metabolomics. *Metabolomics*. 2009;5(1):44–58.

[101] Watanabe H, Kobayashi K, Kato Y, et al. Transcriptome profiling in crustaceans as a tool for ecotoxicogenomics. *Cell Biology and Toxicology*. 2008;24(6):641–647.

[102] Choi J-Y, Ramachandran G, and Kandlikar M. *The Impact of Toxicity Testing Costs on Nanomaterial Regulation*. ACS Publications; 2009.

[103] Guadagnini R, Moreau K, Hussain S, et al. Toxicity evaluation of engineered nanoparticles for medical applications using pulmonary epithelial cells. *Nanotoxicology*. 2015;9(sup1):25–32.

[104] Stone V, Nowack B, Baun A, et al. Nanomaterials for environmental studies: classification, reference material issues, and strategies for physico-chemical characterisation. *Science of the Total Environment*. 2010;408(7):1745–1754.

[105] Kroll A, Pillukat MH, Hahn D, et al. Interference of engineered nanoparticles with in vitro toxicity assays. *Archives of Toxicology*. 2012;86(7):1123–1136.

11 Impact of Heavy Metals on Gut Microbial Ecosystem

Implications in Health and Disease

Dhirendra Pratap Singh, Ravinder Naik Dharavath, Raghunath Singh, Keya Patel, Shirali Patel, Vandana Bijalwan, Praveen Kolimi, Mahendra Bishnoi, and Santasabuj Das

CONTENTS

11.1	Introduction	233
11.2	Heavy Metals	235
11.3	Gut-microbiome	239
	11.3.1 GMB: Modulation of Xenobiotics Exposure	239
	11.3.2 GMB: Biosynthesis of Important Secondary Metabolites	240
11.4	The Bidirectional Relationship in HMs and GMB	242
	11.4.1 Lead (Pb)	242
	11.4.1.1 Pb and GMB	243
	11.4.2 Chromium (Cr)	243
	11.4.2.1 Cr and GMB	244
	11.4.3 Cadmium (Cd)	245
	11.4.3.1 Cd and GMB	246
	11.4.4 Arsenic (As)	248
	11.4.4.1 As and GMB	248
	11.4.5 Mercury (Hg)	249
11.5	Future Prospectives and Conclusion	250
References		260

11.1 INTRODUCTION

Advancements in technologies and industrial development have increased the heavy metal (HM) contamination in the food, water, and air consumed by living organisms. Environmental contamination by HMs has been of great concern for ecological and public health globally. HMs mostly occur naturally but are also derived from anthropogenic activities such as mining, agriculture, and industrialization. These are the metallic elements having densities higher than water (~5 times higher) (Duffus 2002; Tchounwou et al. 2012). In the context of human health, HMs are categorized as essential, such as cobalt (Co), copper (Cu), chromium (Cr), iron (Fe), magnesium (Mg), manganese (Mn), molybdenum (Mo), nickel (Ni), selenium (Se) and zinc (Zn)), and non-essential metals (such as aluminum (Al), antimony (Sb), arsenic (As), barium (Ba), beryllium (Be), bismuth (Bi), cadmium (Cd), gallium (Ga), germanium (Ge), gold (Au), indium (In), lead (Pb), lithium (Li), mercury (Hg), nickel (Ni), platinum (Pt), silver (Ag), strontium (Sr), tellurium (Te), thallium (Tl), tin (Sn), titanium (Ti), vanadium (V), and uranium (U)) (Chang et al., 1996; Tchounwou et al., 2012). Essential HMs have nutritional, physiological, and biochemical functions in humans and plants, whereas non-essential HMs have no beneficial biological effects but are toxic (Tchounwou et al., 2012). HMs produce their effects via interactions with major cellular organelle such as plasma membrane, endoplasmic reticulum, mitochondria, lysosomes, nuclei, and key enzymes involved in biotransformation, elimination, and damage repair (Wang and Shi, 2001). Interaction of HM ions with DNA and nuclear protein cause DNA damage leading to alteration in cell cycle, apoptosis and carcinogenesis (Beyersmann and Hartwig 2008). HMs also produce reactive oxygen species (ROS), and cause oxidative stress leading to cell death, which is the common mechanism for As, Cd, Cr, Hg, and Pb toxicity (Tchounwou et al., 2012). Occupational or accidental exposure to these HMs (majorly non-essential) can result in severe health issues, mainly neurological disorders, cardiovascular disorders, hepatic disorders, renal injuries, cancers, reproductive and developmental defects, autoimmune diseases, and metabolic disorders such as type-2 diabetes mellitus and obesity (Rehman et al., 2018; Duan et al., 2020; Khan and Wang, 2020).

Gut microbiota (GMB) is the collective term used for the trillions of microorganisms colonized in the human gastrointestinal tract (GIT). These microorganisms regulate several major physiological aspects of host immune system development and function. Evidence from animal and human studies suggest the pivotal role of GMB in brain development, function, and behavior (Fung et al., 2017). GMB has been regarded as a key player in the development and maintenance of function and homeostasis of GIT, which is in integration with the host's nervous and immune systems. On the basis of functioning under these complex integrations, GMB has been regarded as an "organ" or "superorganism" or "forgotten organ" (Assefa and Köhler, 2020; Duan et al., 2020). Besides these severe health conditions, HM toxicities have been greatly associated with alteration in the composition of intestinal flora or GMB. In recent years, the bidirectional relationship between GMB and HM toxicity has been established in such a way that we now know that HMs modulate the composition of GMB and/or GMB modulate the bioavailability, metabolism, and toxicity of HMs (Duan et al. 2020).

Environmental exposure to HMs affects various physiological functions and may have pathological consequences. GMB can serve as the first line of defense upon such exposures as it can directly interact with the various HM toxicants present in food and water to be consumed. However, the effect of these HMs on GMB or the ability of these gut flora in handling such xenobiotics is still not fully understood.

Impact of Heavy Metals on Gut Microbial Ecosystem   235

HM exposure can directly alter the GMB composition and these perturbations may promote various disorders. Similarly, the GMB composition could act as important determinant in the biotransformation and toxicities of HMs (Tinkov et al., 2018; Duan et al., 2020). Experiments from germ-free mice suggest that the intestinal microbiome limits the body burden of some HMs such as Pb and Cd (Breton et al., 2013b). A complex host-GMB relationship has been suggested in some of the studies where the microbiota affects the gut-barrier functions as well as the host response by expression of various metal ion transporters (Breton et al., 2013a), and co-selection of antibiotic resistance genes (Guo et al., 2014; Ding et al., 2019). Using various model organisms such as rodents, chicken, crayfish, and *Bufo gargarizans*, attempts have been made to understand these complex host-microbial relationships (Zheng et al., 2020; Zhou et al., 2020). Efforts have also been made for the remediation of HMs and other environmental toxicants affecting the host by utilizing the ability of some gut microbial species, such as *Lactobacilli* for Cd toxicity (Zhai et al., 2014). The ability of such probiotic strains has been linked to their ability to directly assimilate such HMs or by improving the gut barrier functions in the host. Their supplementation has also shown a reduction in oxidative stress and inflammation caused by HM exposures (Ojekunle et al., 2017). While most of the studies are preclinical, efforts are also being made toward their use in clinical settings. One such randomized, open-label pilot study assessed the effects of *L. rhamnosus* GR-1 (LGR-1)-supplemented yogurt in pregnant women and children. The results suggested that such supplementation protects the host from absorption of Hg and As (Bisanz et al., 2014). This chapter covers the impact of five major HMs/metalloids (As, Cd, Cr, Pb, and Hg) on this complex host-microbial relationship.

11.2 HEAVY METALS

HMs are generally defined as metals with relatively high densities, atomic weights, or atomic numbers. HMs are naturally occurring elements, present in low concentrations, and can be found in our planet's crust. The most basic properties of HMs are their high atomic number (>23) and high density (>5g/cm^3). They are some of the known toxicants that can affect biological systems at low concentrations (Mahurpawar, 2015). They are difficult to biotransform and thus can easily bioaccumulate (Beyersmann and Hartwig, 2008). Due to their widespread use in industrial and other settings, their untoward health effects cannot be ruled out. Many of them have no physiological roles and can cause various toxicities in biological systems. Toxic implications of those HMs on human health are some of the most sought-after toxicity studies.

Some HMs such as Cu, Se, Zn, and Cr^{3+} are required for human metabolism as trace elements and are categorized as essential but can be toxic at higher concentrations. However, many of these have no known physiological functions and appear toxic even at low levels (Martin, 2009). Biologically relevant toxicological studies are mostly based on metals and metalloids like As, Cr, Cd, Hg, and Pb because these toxicants can precipitate more pathologies than others due to having no known physiological functions. Pb, Cd, hexavalent chromium (Cr^{6+}), and Hg are toxic HMs that are distributed in the atmosphere to a significant degree by manufacturing plant's effluents, organic wastes, refuse burning, as well as transportation and power generation. They can be found at a great distance from the source due to the wind, depending on whether they are in gaseous or particulate form (Masindi and Muedi, 2018). Particulate matter/suspended metallic particles, when deposited on the surface of water or precipitate on the soil surface, can later reach the food chain when plants grow in those contaminated areas or by use of water from those sources (Gergen and Harmanescu, 2012). They enter our bodies in minute volumes through diet, potable water, and while breathing (Chary et al., 2008). Their presence is unusual because it is impossible to fully exclude them from the environment or even degrade them after they have entered it.

The level of danger posed by a metal is determined by numerous factors, including the dosage and method of exposure. Metals have varying effects on different animals. Age, gender, and genetic predisposition all play a role in toxicity within a single species. Tailings, toxic waste, mining, occupational exposure, paints, agricultural runoff, and treated wood are all potential causes of HM toxicities. Certain HMs, on the other hand, are of grave concern because they can damage multiple organ systems even at low levels of exposure. As, Cd, Cr, PB, and Hg are among the priority metals that are of public health concern due to their high toxicity. These elements are proven or suspected to be carcinogens in addition to being toxic and are found in abundance in the world, in air, food, and water at the same time. The dosage, route of exposure, and chemical species, as well as the age, gender, genetics, and nutritional status of those exposed, all influence their toxicity (Table 11.1). Metal ion toxicity in mammalian environments is caused by the metal ion-chemical reactivity with cellular structural proteins, enzymes, and the membrane structure. The organs that absorb the largest amounts of metals in vivo are normally the target organs of various metal toxicities. This is frequently determined by the route of exposure as well as the chemical properties of the metal, such as its valiancy state, volatility, lipid solubility, and so on.

HMs can affect hormone metabolism and may result in severe endocrine disruption. Endocrine-disrupting chemicals (EDCs) are one such class of external agents that impede the synthesis, secretion, transport, metabolism, action, or expulsion of natural bloodborne hormones, thereby resulting in negative health outcomes and an imbalance in toxicokinetics (Yilmaz et al., 2020). When xenobiotic routes and related genes are exposed to EDCs, enzymes and metabolites induce the biotransformation of these chemicals. The microbial output and consequences of metabolism of EDCs are then picked up by the host, which in turn affects glucose homeostasis, predominantly by influencing hepatic glycogenesis (Nadal et al., 2017; Lind and Lind, 2018). The only way for the EDCs to enter the human body is through the food chain, which is further metabolized by microbial communities present in the gut. Thus, we can conclude that these chemicals have chronic implications on human health, thereby making chemical-gut interaction studies even more imperative.

A wide range of different metal(loid) biotransformation processes have been reported, including reduction and oxidation, methylation and demethylation, hydrogenation, the addition of Sulpher (thiolation), as well as incorporation of the metal(loid) into more complex metal(loid) organic compounds (Table 11.1).

TABLE 11.1
General Physicochemical Properties of Selected Heavy Metals/Metalloids, Their Bioassimilation and Known Human Health Hazards

Parameter(s)	Arsenic (As)	Cadmium (Cd)	Chromium (Cr)	Lead (Pb)	Mercury (Hg)
Atomic number (Z)	33	48	24	82	80
Atomic mass (g/mole)	74.921	112.414	51.996	207.2	200.592
Density (g/cm^3)	5.727	8.65	7.19	11.34	13.534
Isotopes	^{73}As, ^{74}As, ^{75}As	^{106}Cd–^{116}Cd (11 Isotopes)	^{50}Cr–^{54}Cr (5 Isotopes)	^{204}Pb, ^{206}Pb, ^{207}Pb, ^{208}Pb	^{194}Hg–^{204}Hg (11 Isotopes)
Physical state/ Morphology	Gray/Yellow/Black solid metal	Silvery bluish-gray soft metal	Silver solid metal	Gray/Yellow/Black soft metal	Silver liquid
Melting point (K)	No melting temp.	594.22	2180	600.61	234.32
Boiling point (K)	887 K (Sublimation temperature)	1040	2944	2022	629.88
Oxidation states	−3, −2, −1, 0, +1, +2, +3, +4, +5	−2, +1, +2	−4, −2, −1, 0, +1, +2, + 3, +4, +5, +6	−4, −2, −1, +1, +2, +3, +4	−2, +1, +2
Permissible exposure limits (TWA)	0.01 ppb (OSHA)	0.005 ppm (OSHA)	1000 ppm (OSHA)	50 ppm (OSHA)	10 ppm (OSHA)
Industrial applications & Potential sources	Preservation of wood, Manufacturing of specific kinds of glass, Insecticides formulations, Doping agent in semiconductors, e.g. Gallium arsenide, used to change electric current into laser light, Pyrotechnics, Bronze production	Phosphate fertilizer, Pesticides, Nickel–Cadmium batteries, Glassware pigmentation, Corrosion-resistant plating, Stabilizer in plastic production, Nuclear reactors	Alloys, Metal ceramics, Electroplating, Leather tanning, Manufacturing of synthetic rubies, Dye paints, Chromium salts are used to color glass green	Hair dyes, Pottery lead glazes, Insecticides, Lead-acid batteries in cars, Computer screen sheets to safeguard from radiation, Ammunition and projectiles, Lead crystal glass, Cable sheeting, Weight belts for divers, Canister for corrosive liquids, In buildings for roofing, Stained glass windows, Lead piping	Barometers, Thermometers, Manufacturing chlorine, Gold recovery, Tooth fillings, Compact fluorescent lightbulbs, Photochemistry, Calomel electrodes, Insecticide, Rat poison, paints, Catalyst, Rectifiers, Electrical switches

(continued)

TABLE 11.1 (Continued)
General Physicochemical Properties of Selected Heavy Metals/Metalloids, Their Bioassimilation and Known Human Health Hazards

Parameter(s)	Arsenic (As)	Cadmium (Cd)	Chromium (Cr)	Lead (Pb)	Mercury (Hg)
Bioaccumulation	Fish, Gastropods, Crustacean, Carnivores, Herbivores, Saltmarsh plants, Marine algae, Diatom, seaweed, Groundwater (Rahman et al., 2012)	Marine flora and fauna, freshwater animals, feeds, fodders, and tissues of livestock. Plants, soil animals (earthworms, isopods, and gastropods) (Mortensen et al., 2018; Kar and Patra 2021)	Trees, food, and vegetable plants on ground. Microbes in air, water, and soil. Marine and freshwater animal and plant species. (Sharma et al., 2020)	Aquatic (Ocean and river) animal and plant species. Water bodies, soil, and vsoil-based living organisms. Edible crop and vegetable plants. Terrestrial birds. (Lee et al., 2019)	Aqua-marine invertebrates, vertebrates, plant species. Aquatic food chain, soil, water bodies. (Yan et al., 2019)
Bacteria mediated biotransformation	*Bacillus subtilis*, *B. medenterious vulgatus* *B. medenterious ruber* *Brevibacillus brevis* *Escherichia coli* (Mitra et al., 2017)	*Desulfovibrio desulfuricans* *Enterobacter cloacae* *Ochrobactrum anthropi* *Sphingomonas paucimobilis* *Stenotrophomonas sp.* *Pseudomonas aeruginosa* (Chellaiah 2018)	*Desulfovibrio desulfuricans* *Zoogloea ramigera* (Choudhary et al., 2017; Yin et al., 2019)	*Bacillus sp.* *Bacillus firmus* *Corynebacterium glutamicum* *Enterobacter cloacae* *Pseudomonas aeruginosa* *Pseudomonas putida* (Chatterjee et al., 2012; Pan et al., 2017)	*Pseudomonas aeruginosa* Sulfate-reducing bacteria (Wagner-Döbler 2013; Choudhary et al. 2017)
Human health issues	Carcinogenic Reproductive toxicity Infertility Genotoxic Neurotoxic Cardiotoxic Gastric irritation	Carcinogenic Nephrotoxic Neurotoxic Reproductive toxicity Genotoxic Osteoporosis	Cr (VI): Carcinogenic Hepatotoxic Nephrotoxic Gastric ulceration Pulmonary problems (Bronchitis, Pneumonia, Lung cancer)	Possible carcinogen Neurotoxic Nephrotoxic Reproductive toxicity Hypertension Iron deficiency	Unclassified carcinogen Reproductive toxicity Genotoxic Neurotoxic Down's Syndrome Minamata disease Teratogenic

11.3 GUT-MICROBIOME

The human GIT is colonized by more than 100 trillion individual microorganisms such as bacteria, viruses, yeast, and fungi, which is termed as GMB. The GMB itself encodes for more than 3 million genes, which is collectively ~150 times more than the human genome (~23,000 genes) (Rosenfeld, 2017; Rinninella et al., 2019). Altogether GMB can be considered as a separate organ weighing ~1–2 kg (Forsythe and Kunze, 2013). GMB is majorly comprised of phyla *Firmicutes* (genera *Clostridium, Lactobacillus, Bacillus, Enterococcus, and Ruminococcus*), *Bacteroidetes* (genera *Bacteroides* and *Prevotella*), *Actinobacteria* (genus *Bifidobacterium*), *Proteobacteria, Fusobacteria,* and *Verrucomicrobia*. Gut bacteria play important roles in regulating digestion, immunity, and metabolic homeostasis and are also involved in the synthesis, extraction, and absorption of several nutrients and metabolites (e.g., carbohydrates, lipids, short-chain fatty acids (SCFAs)), bile salts, amino acids, vitamins, and several neurotransmitters (Thursby and Juge, 2017; Rinninella et al., 2019). GMB plays a pivotal role in maintaining intestinal and blood-brain barrier (BBB) integrity and permeability (Kelly et al., 2015). In recent years altered GMB composition has been associated with severe mental illnesses (Kim and Shin, 2018; Mörkl et al., 2020), neurological disorders (Cryan et al., 2020), and metabolic disorders (Fan and Pedersen, 2021).

11.3.1 GMB: Modulation of Xenobiotics Exposures

In adults, GMB composition remains mostly stable, with slight fluctuation in individuals. It alters at an older age due to digestive tract physiological changes and diet (Faith et al., 2013). Nevertheless, antibiotic treatments or dietary changes alter this composition transiently. Studies suggest that short-term treatment with single-dose oral antibiotic alters the GMB for as long as four weeks for most of the species before re-establishing their original composition (Dethlefsen et al., 2008), but some species take several months or may not recover (Dethlefsen and Relman, 2011). Dietary factor plays a major role in the modulation of gut microbiota, as evident from the results from one study suggesting mice microbiome's 57% variability can be explained by dietary change as compared to <12% by a genetic mutation (Zhang et al., 2010). This microbial diversity can also be affected by various factors such as GIT factors (pH, bile acid, mucus layer composition, and transit time); microbial factors (metabolites, bacteriocins, enzymes and adhesion capabilities); nutrient factors (High fat diets, dietary fibers); and medications (Prakash et al., 2011; Turroni et al., 2014). However, perturbed GMB upon such exposure may affect various metabolic and physiological functions.

Assimilation and biotransformation of HMs in the gut take place in a bidirectional interaction with GMB. The direct effect of toxicants on the microbial ecosystem and remediation and detoxification of those toxicants by GMB provides an interesting research opportunity. GMB acts as the first line of defense by creating a physical barrier by forming biofilms, altering the pH and oxidative balance in the intestine to regulate absorption in systemic circulation of HMs. GMB also affects gut-barrier integrity, thus making the penetration of these toxicants difficult (Zhai et al., 2014).

TABLE 11.2
Studies Investigated the Interaction of Various Bacterial Species with Heavy Metals

Bacteria	Heavy Metals Interaction	Mode of Action	References
Sulphate reducing bacteria	Cd, Mn	Formation of precipitates of metal sulphides, phosphates, and bicarbonates	(Cruz Viggi et al., 2009)
Oxalobacter	Pb	Formation of oxalate precipitate of metals	(Eggers et al., 2019)
Lactobacillus	Pb, Cd, Mn	Inhibition of transport proteins	(Lin and Pan, 2019)
Lactobacillus and *Psudomonas sp.*	Pb, Cd, Cu	Binding of exopolysaccharides or binding protein to heavy metals	(Eggers et al., 2019)
Sulphate and iron-reducing bacteria, *Bacteroides, Clostridium sp., Alistipes, Bilophila, Methanoogens, Desulfovibrio, Deltaproteobacteria*	Hg, As, Bi, Fe	Methylation-demethylation and thiolation	(Bridou et al., 2011; Gilmour et al., 2013; DC. Rubin et al., 2014)
Bacteroides, Fecalibacterium, Pseudomonas	As	Oxido-reduction of metal ions and methylation of inorganic arsenic to organic species	(Cai et al., 2009; Alava et al., 2015; Yin et al., 2019; McDermott et al., 2020)
Streptomyces tendae	Cd	Sequestration by siderophores	(Dimkpa et al., 2009)
Psudomonos and *Bacillus*	Pb, Cd, Hg, Cr, Cu, MN	Formation of insoluble complexes by siderophores	(Zawadzka et al., 2009; Bridou et al., 2011; Ahmed and Holmström, 2014)

Some of the mechanisms for metabolism and absorption to host and involvement of specific bacterial populations are listed in Table 11.2. Studies have shown the interaction of various bacterial species with HMs.

11.3.2 GMB: Biosynthesis of Important Secondary Metabolites

The symbiotic microbial population has essential functions in the human body. Symbionts are also helpful in the metabolism of various xenobiotics and prevention of colonization of pathobionts (Nicholson et al., 2012). Some important functional outcomes of the residing microbial populations are the fermentation of indigestible carbohydrates (dietary fibers) and the utilization of glycans (e.g., mucins). Signaling molecules from commensal bacteria are essential for the gut-barrier functioning and gut-tissue repair. Some microbe-associated molecular patterns (MAMP) modulate mucosal immune responses (Rakoff-Nahoum et al., 2004; O'Hara and Shanahan, 2006; Kondepudi et al., 2015). It has also been suggested that the human genome

Impact of Heavy Metals on Gut Microbial Ecosystem

does not have the capabilities to encode all the necessities for improved immunological development, rather it depends on the gut-microbial products for this (Mazmanian et al., 2008). An example is a bacterial capsular polysaccharide, which maintains and supports the immune function of the host. Additionally, it helps in synthesizing vitamins (LeBlanc et al., 2013), bile acid transformation (Yatsunenko

TABLE 11.3
Gut Microbial Species Involved in the Synthesis of Important Metabolites

Metabolites	Related Bacterial Groups	Potential biological Function	References
SCFAs	*Roseburia, Eubacterium, Faecalibacterium, Coprococcus*	Reduced colonic pH, pathogen growth inhibitor, energy source to colonocytes, implicated in obesity, insulin resistance, etc.	(Wong et al., 2006; Zhao et al., 2019)
Choline metabolites	*F. prausnitzii, Bifidobacterium*	Lipid and glucose metabolism modulation, implicated in NAFLD, and diet-induced obesity.	(Samuel et al., 2008; Wang et al., 2011; Martin and Johnson, 2012; Heianza et al., 2019)
Bile acids	*Lactobacillus, Bifidobacteria, Clostridium, Enterobacter, Bacteroides*	Lipid assimilation, maintenance of intestinal barrier function, cholesterol biosynthesis, and energy homeostasis	(Ridlon et al., 2006; Swann et al., 2011; Parkar et al., 2013)
Phenolics	*Lactobacillus, Bifidobacterium, F. prausnitzii, C. difficile*	Metabolism of xenobiotics and their detoxification, excretion through urine	(Zheng et al., 2011)
Vitamins (K, B Complex)	*Bifidobacterium*	Complementary source, immunomodulation	(Koenig et al., 2011; LeBlanc et al., 2013; Stacchiotti et al., 2020)
Polyamines	*Campylobacter, Clostridium*	Anti-inflammatory, antitumoral effect, genotoxic effect	(Matsumoto et al., 2012)
Lipids	*Bifidobacterium, Roseburia, Lactobacillus, Klebsiella, Enterobacter, Citrobacter, Clostridium*	Affects intestinal barrier function, LPS-induced systemic inflammation, metabolic endotoxemia linked insulin resistance, activate immune system	(Cani et al., 2007; Serino et al., 2012)
Others: methanol, ethanol, glucose, urea, creatine, creatinine, LPS, endocannabinoids etc.	*Pseudobutyrivibrio, Ruminococcus, Faecalibacterium, Subdoligranulum, Bifidobacterium, Atopobium, Lactobacillus*	Direct or indirect synthesis and utilization	(Muccioli et al., 2010)

et al., 2012), dietary toxin and carcinogen metabolization, and intestinal angiogenesis regulation (Duca et al., 2014). Major short-chain fatty acids (SCFAs) (e.g., acetic, propionic, butyric) with other minor microbial metabolites (e.g., succinate, valerate, caproate, lactate, ethanol, etc.) are produced by gut microbes upon fermentation of non-digestible carbohydrates (Byrne et al., 2015). Some of the important metabolites and their synthesis by specific gut-microbial species are listed in Table11.3.

11.4 THE BIDIRECTIONAL RELATIONSHIP IN HMs AND GMB

11.4.1 Lead (Pb)

Lead is present in a wide variety of products and has previously been used as tetraethyl lead in gasoline. Potential causes of exposure include water, air, diet, and old paint cans via ingestion, inhalation, and transdermal exposure. The Centers for Disease Control and Prevention has established that there is no acceptable level of blood for Pb. The blood lead level of concern has been reduced from 10 to 5 µg/dL by this organization, but even lower levels can cause disease.

Tiny inorganic Pb particles can be absorbed through the respiratory tract, while large particles enter the body via oral route. Absorption of Pb is mostly done through the duodenum. However, the dermal route is not efficient for the absorption of Pb as compared to the former ones. Pb is distributed throughout the body independent of the route of ingestion. However, it mainly accumulates in the bones and red blood cells. Conditions such as pregnancy, menopause, lactation, and osteoporosis increase the bone resorption of Pb, thereby causing an elevation in its systemic levels. Pb can be transferred to a fetus/newborn via the blood-placental barrier and breastfeeding.

Organic and alkyl Pb compounds are actively metabolized through an oxidative dealkylation in the liver and catalyzed by the cytochrome P450 enzymes. Organic forms of Pb are mostly excreted through the urinary route, whereas the inorganic forms are eliminated via urine and fecal routes. However, very small amounts of Pb can be eliminated via other routes such as sweat, hair, nails, saliva, seminal fluid, and breast milk.

Chronic Pb exposure affects critical hormonal and neuronal systems, which regulate heart rate, peripheral vascular resistance, and cardiac output, thereby increasing blood pressure. Pb-induced hypertension is associated with adrenergic system abnormalities, including increased central sympathetic nervous system activity, decreased density of the vascular β-adrenergic receptors, and elevated plasma norepinephrine. Chronic Pb exposure induces nephrotoxicity by altering the physiological functions of several renal cell proteins, such as 2-microglobulin, δ-aminolevulinic dehydratase, and metallothionein. Being a competitive ligand of mitochondrial calcium transporter, Pb was found to enter the mitochondria and lead to cellular degeneration by disrupting the oxidative metabolism. Pb has also been observed to affect the hematopoietic system in both animals and humans and alters the haem biosynthesis by interfering with three enzyme activities, Aminolevulinic acid synthase (ALAS), ferrochelatase, and delta-aminolevulinate dehydratase (ALAD), leading to hypochromic and normocytic anemia, linked with reticulocytosis. Pb also alters the neural functions (long-term potentiation, memory processes, and spatial learning) by disrupting the neurotransmitter synthesis, conductance of ionic channels, ligand-receptor interactions, and dendritic branching.

11.4.1.1 Pb and GMB

Toxicity and exposure to Pb may also occur because of ingesting polluted food or water or inhaling lead particles. Lead has the potential to bioaccumulate in both the body and the environment. Considering its direct effect on gut microbial ecosystem, studies are suggestive of potential untoward health effects. Pb exposure in drinking water affected mouse GMB in a sex-specific manner (Wu et al., 2016). Exposure of Pb (50 or 100 ppm) and Cd (20, 100 ppm) in drinking water of 6-week old Balb/C female mice decreased the *Lachnospiraceae* abundances, but the relative abundances of *Lactobacillaeceae* and *Erysipelotrichaeceacae* increased, and there were also alterations in *Turicibacter* spp. abundances (Breton et al., 2013a). The metabolic disturbances of the gut were also determined when fecal samples were tested at 4 and 13 weeks after exposure, respectively. With age, the phylogenetic diversity of the GMB increased in control subjects, but this response was blunted in subjects treated with PbCl2. Some key alterations observed in this study were altered *Ruminococcus* spp., in with *Clostridiales*, *Oscillospira* spp. abundances (Gao et al., 2017). In addition, elevated liver triglyceride (TG) and pyruvate levels and modifications in various intestinal metabolites (including glutamate, isobutyrate, alanine and glycine, etc.) indicate a hepatic and intestinal metabolic disorder caused by the Pb, which were associated with altered GMB composition (Xia et al., 2018). Despite ceasing Pb exposure at 3 weeks of age, it was found that perinatal Pb exposure changed GMB composition in adult (age 40 weeks) offspring. Adult mice exposed perinatally to Pb had higher levels of bacteria from the genera *Pseudomonas*, *Enterobacter*, and *Desulfovibrio* than the controls (Sobolev and Begonia, 2008).

11.4.2 CHROMIUM (Cr)

Cr is a natural mineral and can be released from anthropogenic sources into the world. It is a prominent water pollutant, usually from industrial waste from tanning plants, steelworks, chromium plating or dyeing, timber conservation, chemical fertilizer, etc. (Pechova and Pavlata, 2007). It is a metallic element, with II, III, and VI being the most common oxidation states. Trivalent (III) forms are more stable than elemental forms (i.e., Hexavalent (VI)). Cr (III) in small amounts is considered essential for metabolism in humans and animals (Pechova and Pavlata, 2007). It is inert and non-volatile, whereas Cr(IV) is very unstable, toxic, and carcinogenic to a wide range of species (Bremer Neto et al., 2005; Younan et al., 2016). Hexavalent chromium is the most toxic form of the mineral and is rapidly absorbed by living cells; its effect depends on the dosage and exposure route over time. Cr(IV) is an effective oxidizing agent. In the lungs, liver, kidneys, and GIT, chromium induces detrimental consequences. Other metabolic disorders include metabolic acidosis, acute tubular acidosis, necrosis, failure of the renal system, and eventually death.

The patient's clinical presentations following acute exposure to high doses of Cr(IV) were similar regardless of the form of Cr(IV) ingested. Some symptoms include pain in the belly, nausea, vomiting, hematemesis and bloody diarrhea, caustic burning of the mouth, pharynx, esophagus, stomach, and duodenum, and bleeding from GIT. Anemia, irregular erythrocytes, intravascular hemolysis, hepatic toxicity, renal failure, cyanosis, metabolic acidosis, hypotension, and shock are caused by Cr sub-chronic and chronic exposure. Tissue biopsy data, including liver fat degeneration and necrosis of the renal tubes, have also been reported. Moreover, epidemiological studies proved that oral exposure to hexavalent chromium could be associated with cancer (Beaumont et al., 2008; Kerger et al., 2009).

Inhalation is the primary route of Cr ingestion. The physico-chemical properties of Cr particles affect the absorption of inhaled Cr. Cr^{6+} is absorbed at a faster rate than Cr^{3+} in the bloodstream. A significant amount of Cr, which is not absorbed through the lungs, enters the GIT. It is primarily absorbed in the jejunum. Following oral exposure, inorganic Cr^{6+} is more absorbed than Cr^{3+}. Cr is distributed to spleen, bone marrow, lungs, lymph nodes, liver, and kidney, where the Cr^{6+} is reduced to Cr^{3+}. The metal is excreted mostly through urine but is also eliminated by bile excretion and smaller quantities in nails, hair, milk, and sweat (Valko et al., 2005).

The Cr-induced carcinogenicity is mediated by the intermediates formed during the intracellular reduction of Cr^{6+} and Cr^{3+} along with other oxidative reactions. Further, Cr^{3+} forms toxic covalent adducts with various macromolecules such as DNA, RNA, proteins and other intracellular signaling molecules (Valko et al., 2005). Cr^{6+} is more toxic than Cr^{3+} due to its higher redox potential and higher potential to enter the cells. The free radicals formed during the chromium reduction ($Cr^{6+} > Cr^{5+} > Cr^{4+} > Cr^{3+}$) process were noted to induce DNA damage in various ways, leading to the production of DNA adducts, changes in replication sister chromatid exchanges, chromosomal aberrations, and transcription of DNA as seen in vivo and in vitro studies (Engwa et al., 2019). Further, oxygen radical species lead to oxidative stress, which may cause various deleterious effects such as lipid peroxidation, disruption of signaling pathways, modifications in cellular communications, activation apoptosis, and cell cycle arrest pathways (Barnhart, 1997).

11.4.2.1 Cr and GMB

The intestinal microbes are the body's first line of protection against intoxicants if ingested orally as they convert the more toxic Cr (VI) to less toxic Cr (III), suggesting that bacterial tolerance leads to host defense in animals supplemented with Cr (Upreti et al., 2004). Through the consumption of contaminated food and water, the microbes of the GIT are exposed to Cr. The human intake of Cr (VI) from drinking water at levels of 1–10 ppm is reportedly safe due to the excellent capacity of microorganisms in the GIT to reduce Cr (VI) to Cr(III) (Shrivastava et al., 2005). The three critical bacterial metal-binding mechanisms in the bacterial cell walls are as follows:

1) Exchange of ions with the peptidoglycan and teichoic acidic
2) Precipitation due to the core formation reactions
3) Compounding of ligands of nitrogen and oxygen

In addition, biosorption, oxidation, and enzymatic/bioaccumulation reduction are mechanisms by which toxic metals interact with microorganisms. The enzymatic decrease of Cr(VI) includes the binding of chromate reductase to the cell membrane during anaerobic respiration or the binding under aerobic conditions of cytosolic-soluble chromate reductase, the activity of which is an enzyme co-factor, boosted by GSH and NADH.

Cr is introduced to the bacterial flora in the GIT by the intake of contaminated water and food. Human ingestion of Cr (VI) in drinking water at levels of 1 to 10 ppm has been confirmed to be safe due to the GIT's high capacity to convert Cr (VI) to Cr (III). The effects of chronic ingestion of potassium dichromate [Cr (VI)] on the resident GMB of Wistar rats were investigated in a report. Cr-stressed rats were given 10 ppm Cr (VI) in their drinking water, while the other group was given plain water. *Lactobacillus, Pseudomonas* spp., and *E. coli* were isolated from the rat's caecum after 10 weeks. The promotion of facultative gut bacterial growth in Cr-stressed rats and the substantial increase in growth even at lower Cr concentrations were the study's major outcomes. As a result, Cr was concluded to be acting as a prebiotic. Furthermore, the capacity to reduce Cr (VI) to Cr (III) was substantially reduced, which was correlated with the growing antibiotic and HM resistance, as the transmissible plasmids containing the resistance towards these xenobiotics. *Pseudomonas, Aeromonas, Enterobacter, Escherichia, Bacillus, Streptomyces*, etc., have shown their ability to reduce Cr (VI) to Cr (III) (Upreti et al., 2004).

11.4.3 CADMIUM (Cd)

Cd is a pervading toxic metal that can arise from a variety of sources, such as batteries, paints, electroplating, plastics, and fertilizers. Due to the frequent use of metal, toxicity caused by Cd is inevitable. It has been reported that multiple cardiovascular disorders, hepatotoxicity, renal impairment, oxidative stress, osteoporosis, aberrant immunity, and tumorigenesis are linked to Cd-related toxicity (Fazeli et al., 2011). Cd toxicity mainly affects the GIT. Increased proinflammatory molecules such as lipopolysaccharide (LPS) are produced when the gut barrier function is disrupted in conjunction with Cd-induced changes in the viability of GMB components, which can contribute to systemic inflammation (Richardson et al., 2018). Cd enters the system circulation mainly via the inhalation route. Cd absorption from the lungs depends on the particle size. Smaller particles reach the alveoli and are absorbed according to the solubility, whereas the large particles are eliminated from the nasopharyngeal spaces by the mucociliary processes. The evidence also suggests that intestinal absorption of Cd involves the transporter proteins. Following the exposure, the Cd accumulates mostly in the liver and kidneys, and is eliminated through fecal and urinary excretion. Cd promotes apoptosis, oxidative stress, methylation of DNA, and DNA damage (Engwa et al., 2019). Cd is known as a potent carcinogen, which affects the kidneys, lungs, pancreas, bones, and prostate. Studies have shown that Cd causes apoptosis in several organs primarily via deactivation of a vital DNA repair activity (Valko et al., 2005). Chronic Cd exposure can cause nephrotoxicity (i.e., damage to the proximal tubular cells, proteinuria, aminoaciduria, glycosuria, polyuria, enzymuria, interstitial fibrosis, and a decrease in absorption of phosphate) in laboratory animals as well as in humans. Long-term Cd exposure is associated with osteoclastic activity leading to an increase in bone resorption.

11.4.3.1 Cd and GMB

In a study conducted by Fazeli et al. (2011), mice were treated with $CdCl_2$ in a dose of 23–50 mg/kg for 45 consecutive days. That chronic Cd exposure caused a significant decline in the colonies of all microbes in the fluid contents, and biopsy samples from all intestinal regions of the small intestine showed a higher toxic profile than in the colon and rectum, suggesting a regional preference for Cd toxicity in the gut. In addition, Gram (+) bacteria (*Bacillus cereus, Enterococcus spp.*) were more prone to Cd-related effects than Gram (-) bacteria (*E. coli, Klebsiella spp.*), probably due to different metal ion uptake capabilities (Fazeli et al., 2011). In another study, mice were exposed to a subchronic low dose of Cd (i.e., 10 mg/L of $CdCl_2$ for 10 weeks) to test the effects on the GMB. Microbial abundance assessed in caecum and feces by quantitative PCR and 16S rRNA gene amplicon sequencing analysis showed that the Cd exposure increased the population density of *Bacteriodetes* (*Bacteroidaceae, Paraprevotellaceae*) and reduced population densities of *Firmicutes* (*Lachnospiraceae, Ruminococcaceae, Clostidiaceae, Streptococcaceae*) and *Alpha-Proteobacteria* (*Bifidobacterium longum*). These bacterial modifications were associated with increased production of LPS, which could lead to the development of chronic liver diseases such as cirrhosis (Zhang et al., 2015).

Furthermore, to determine the effects of early-life Cd exposure on GMB and metabolism in adulthood, C57BL/6J mice were chronically treated with a 100 nM dose of Cd at an early stage of life. Early treatment with Cd reversed adiposity-induced dysfunction of hepatic lipid metabolism with increased serum triglyceride (TG), free fatty acids, and hepatic TG levels, especially in male mice compared to females (Ba et al. 2017). Compared to *Bacteroidetes*, Cd-mediated intestinal microbiota alterations at 8 weeks reflected a decreased abundance of *Firmicutes* and lower numbers of *Bifidobacterium* and *Prevotella*. This shows that the male *gut microbiome* was most susceptible to early exposure to Cd. This finding may be connected to the accumulation of fat followed by metabolic dysfunction in the latter part of life (Cox et al., 2014; Rodríguez et al., 2015). In this analysis, the changes in intestinal microbiota in two groups of mice were examined after they were given 20 and 100 mg/kg $CdCl_2$ for three weeks, while control group received $CdCl_2$-free water. According to this report, the gut barrier was weakened. TNF levels in the colon were also significantly elevated after exposure to Cd. On the other hand, Cd therapy delayed the growth of the GMB and decreased the abundance of total intestinal bacteria in mice. *Bacteroidetes*' growth was considerably slowed, while *Firmicutes*' growth was unaffected. Growth of *Lactobacillus* and *Bifidobacterium*, among other probiotics, was substantially inhibited. It was also found that Cd-treated groups had fewer copies of main genes involved in converting carbohydrates to short-chain fatty acids (SCFAs) than control groups. As a result, the concentrations of SCFAs in the colon shrank drastically. Thus, we can say that this report sheds light on the impact of Cd on the GMB of mice (Liu et al., 2014).

In another study, the mice were given 10 mg/L Cd in their drinking water for 10 weeks. They had higher levels of hepatic triacylglycerol, serum free fatty acid (FFA), and TG. The mRNA levels of several primary genes involved in *de novo* FFA synthesis and transport pathways and TG synthesis in the liver increased significantly in Cd-treated mice, suggesting that changes in these genes may be a potential molecular mechanism to explain subchronic Cd-induced hepatic toxicity. Firmicutes and proteobacteria concentrations in the feces decreased significantly after 4 weeks of Cd exposure. The quantity of Firmicutes in the cecum contents decreased significantly after 10 weeks of Cd exposure. Furthermore, 16S rRNA gene sequencing showed that Cd exposure altered the structure and richness of the gut microflora at the family and genus levels. Changes in gut microbiome composition may trigger an increase in serum lipopolysaccharide (LPS) and hepatic inflammation, which could cause energy homeostasis to be disrupted after Cd exposure. According to the findings, subchronic Cd exposure induced dysregulation of energy metabolism and altered the gut microbiome in mice (Zhang et al., 2015). Changes in the intestinal microflora population in healthy mice were studied after exposure to Cd in an experimental setting. For 45 days, five experimental groups received 23 to 50 mg/kg Cd in their drinking water, while the control group received Cd-free water. Bacterial counts were conducted after aseptically collecting intestinal contents and biopsy samples. *Bacillus cereus, Lactobacillus spp., Clostridium spp., Ecoli, Klebsiella spp., Pseudomonas spp., Enterococcus spp., and Proteus spp.* dominated the intestine microflora in the control community. A sharp decrease in the population of all microbial species in the intestine was observed due to dysbiosis caused by the introduction of Cd. The adverse effects of Cd tended to be less in the large intestine and rectum than in the small intestine, indicating that Cd has a site-specific effect. Because of their distinct ability to uptake metal ions, the gram-negative bacteria studied were less vulnerable to Cd than the gram-positive bacteria (Fazeli et al., 2011).

The gut microbiota is thought to be a key player in the toxic effects of HMs like Cd (Jin et al., 2017). Breton *et al.* discovered a substantial increase in blood, liver, kidney, and spleen Cd content and MT1 and MT2 expression in the duodenum and colon in germ-free mice treated with Cd (Breton et al. 2013a). In laboratory animals of various ages, chronic (and subchronic) and short-term Cd exposure has been shown to affect gut microbiome physiology. A study was conducted to see if chronic ingestion of low-dose Cd could affect allergic sensitization and, as a result, influence the development of allergic diseases. The bacterial diversity of conventional C57BL/6 mice was significantly reduced. The Firmicutes-to-Bacteriodetes ratio was altered after they were given low doses of Cd in their drinking water for 28 days. This treatment also promoted proinflammatory cytokine and antimicrobial responses in the gut by activating both the canonical and non-canonical NF-kβ pathways. Since these germ-free C57BL/6 mice were exposed to the same treatment, they formed the exact similar profile of responses. However, to a lesser extent, the effects of Cd were at least partially independent of the gut microbiome. Finally, after oral sensitization with ovalbumin (OVA) and cholera toxin as an adjuvant, traditional mice chronically treated with low doses of Cd produced higher antigen-specific IgE responses. Furthermore, Cd-treated mice developed higher airway allergic responses in response to nasal antigen, as evidenced by elevated IL-17 and Th1 responses compared to control mice. In conclusion, the environmental pollutant Cd has the potential to be a significant regulator of gut immune homeostasis and a source of increased allergic responses at distant mucosal sites (Kim et al., 2015).

11.4.4 ARSENIC (As)

As is a ubiquitous harmful environmental contaminant, present in inorganic (As^{3+} and As^{5+}) and organic (methylarsinic acid (MMA) and dimethylarsinic acid (DMA)) forms. Human exposure is primarily via infected food sources (e.g., rice, pork, and seafood) grown in As-contaminated soil and water. As is a known carcinogen, and its exposure has been associated with lungs, liver, kidney, and bladder tumors (Wang and Mulligan, 2008; Tsiaoussis et al., 2019). As poisoning can cause carcinogenesis and other health problems in a variety of organ systems. Water supplies, for example, are almost completely polluted with inorganic As (Arsenite and Arsenate), whereas food can contain both inorganic and organic As species. The trivalent arsenite, especially its methylated species, monomethylarsonous acid, is more toxic than the pentavalent forms (Richardson et al., 2018).

The primary use of As is in alloys of Pb (e.g., in car batteries and ammunition). As is a common n-type dopant in semiconductor electronic devices. It is also a component of the III-V compound semiconductor gallium arsenide. As and its compounds, especially trioxide, are used to produce pesticides, treated wood products, herbicides, and insecticides. These applications are declining with the increasing recognition of the toxicity of As and its compounds.

More than 75% of the orally ingested As is absorbed through the GIT. However, As can enter the system via oral, inhalation, and dermal routes. Humans and mice have been noted to absorb As through passive diffusion. The As and its metabolites are widely distributed to various organs in the body. The metal and its metabolites are excreted mostly through the urine and feces and minimally through the nails and hair.

Dose and duration of exposure are the main factors in As toxicity. Doses of 2 mg/kg of As and higher have caused fatalities in humans. Doses taken orally of 0.001 mg/kg of As taken long-term daily have been seen to cause skin diseases and cancer of the skin, bladder, liver, and kidney. Lung cancer is seen when As is inhaled at low doses of 0.05–0.07 mg/m^3. As is capable of generating many free radicals, such as superoxide O_2 -, nitric oxide NO, singlet oxygen $1O_2$, hydrogen peroxide H_2O_2, peroxyl radical ROO, dimethylarsinic radical $(CH_3)_2As$, and dimethylarsinic peroxyl radical $(CH_3)_2AsOO$ (Engwa et al., 2019). Toxicity and carcinogenicity caused by As have been noted to be associated with metabolic processes. Following the absorption, pentavalent As (As^{5+}) gets reduced to a more absorbable form, trivalent As (As^{3+}). Recent animal studies have shown carcinogenic activity in the skin, liver, urinary bladder, prostate, and lung, which is similar to As-induced cancer seen in humans. Carcinogenic mechanisms of As include epigenetic alterations, generation of ROS, and damage to the maintenance system of DNA (Engwa et al., 2019).

11.4.4.1 As and GMB

A direct effect of As exposure has been studied in various animal models suggesting gut microbial perturbations upon its exposure. In C57/BL6 mice, 10 ppm As dose for four weeks caused prominent deduction in the Firmicutes (*Clostridiales, Catabacteriaceae, Clostridiaceae*) and Tenericutes (*Erysipelotrichaceae*) abundances (Tsiaoussis et al., 2019). It, in turn, induced changes in the metabolism of isoflavone (daidzein), indole-containing (indoleacetic acid), and bile acid derivatives (Lu et al.,

2014b). These alterations of the GMB were associated with a significant decrease in urine dimethylated arsenicals and increased monomethylated arsenical species along with inorganic As, suggesting compromised detoxification of As by the GMB. Changes in the gut microbial profiles are greatly impacted by the biometabolism as well (Guo et al., 2014; Lu et al., 2014a; Dheer et al., 2015) showed that ppb level dose of inorganic As can cause structural changes in host gut lining and compositional alteration in microbial populations. They further suggested that the changes in microbiota composition were caused by disruptions in the microbial biofilm adjacent to the colonic mucosa. Increased levels of nitrate and nitrite in the colon are likely to be associated with nitrite reductase upregulation as well as increased toxic arginine derivatives formation, indicating stimulation of the metabolism of the host nitrogen and amino acid as a result of exposure to As (Dheer et al., 2015).

As exposure affects the carbohydrate, lipid, and amino acid metabolism of gut bacteria (Wang et al., 2020). *Bacteroides vulgatus* ATCC 8482, a common human gut symbiont, has shown high resistance to inorganic arsenics due to highly abundant arsenic resistance genes arsR and arsDABC (Li, 2017). The role of GMB in assimilation of this common environmental toxin was suggested by another study by Chi et al., where GMB-altered mice showed higher proportion of organic arsenicals species (mono- and di-methylated arsenic) in fecal samples, which suggests the importance of gut-microbial ecosystem in detoxification of such toxicants (Chi et al., 2019). HM binding ability of various probiotic strains has been documented, which suggests the potential of GMB in detoxification and bioremediation of HM exposure (Coryell et al., 2018; Feng et al., 2019; Duan et al., 2020). As is an anionic/negatively charged species, unlike the other HMs; this is troublesome for bacterial metal-binding interactions, since it is thought that the significant amount of metal consumed by microbes is due to charge attractions between the net negative bacterial cell and the positively charged metal (Richardson et al., 2018). Halttunen et al. attempted to solve the charge problem of As and bacterial surfaces by methylating a variety of *Lactobacilli* in order to neutralize surface negative charges, allowing positively charged amino groups on the cell wall to attract negatively charged metals. *Lactobacilli* cultures were lyophilized, incubated with As(III) or As(IV), and metal reduction was observed (Halttunen et al., 2007).

11.4.5 Mercury (Hg)

Hg has a long history of use in human applications, but due to its high toxicity, it has been phased out of many products. Hg is a potent water contaminant. This metal exists in both inorganic and organic forms, with the latter being the most toxic. Followed by the bacterial biotransformation of Hg to methylmercury (MeHg), it accumulates in bacterial cells, which increases the toxic load of aquatic ecosystem. Organic mercury is fat-soluble and is thus easily absorbed through the intestinal epithelium and bio-accumulative. This occurs most frequently in fish, especially large species near the top of the food chain, such as sharks and tuna; this bioaccumulation poses a danger to humans who eat seafood regularly (Zhao et al., 2020). The gut bacteria detoxify the organic Hg by converting the methylated Hg (highly absorbed in GIT) to inorganic Hg^{2+} (not easily absorbed in the GIT), thereby reducing the organic

Hg's toxic impact on health. Inorganic Hg is actively imported by the cell through mercury-specific transporters, while organic Hg is passively internalized. The chance of it being reabsorbed by the intestinal epithelium is reduced as a result of this isolation (Rowland et al., 1984). Inorganic sources of Hg, such as corrosive esophagitis and hematochezia, can cause gastrointestinal problems. Digestion and absorption are affected by a damaged GIT (Ruan et al., 2019).

Hg is absorbed mostly through inhalation, followed by ingestion. It passes easily through the blood-brain barrier and placenta by adhering to the red blood cells. Some Hg is taken up by the central nervous system though most of it is oxidized rapidly to mercuric Hg. Hg is deposited in many organs, including the thyroid, myocardium, breast, muscles, liver, adrenals, kidneys, skin, pancreas, sweat glands, salivary glands, lungs, enterocytes, prostate, testes, and breast milk. It also binds highly to T cell surfaces and sulphydryl groups, influencing the T cell function. Metallic Hg is excreted mostly as mercuric Hg. The excretion of organic and inorganic Hg mainly occurs through urine and feces. Some of the metal is excreted through sweat, tears, saliva, and breast milk (Bernhoft, 2012). Hg^{2+} is seen to deplete mitochondrial glutathione (GSH) at low concentrations, and in diminished respiratory chain electron transport, it enhances the hydrogen peroxide formation in kidney mitochondria. The increased formation of H_2O_2 can lead to oxidative tissue damage, seen in mercury-induced nephrotoxicity. Hg is seen to enhance lipid peroxidation, especially in the kidneys, brain, epididymis, and testes (Valko et al., 2005). Food-borne Hg is known to cause an excess risk of cardiovascular complications such as myocardial infarction and death caused by coronary heart disease. This increased risk is probably due to the lipid peroxidation caused by Hg (Valko et al., 2005). Some of the key effects of the abovementioned selected HMs and their impact on resident gut flora are listed in Table 11.4.

11.5 FUTURE PROSPECTIVES AND CONCLUSION

HM exposure has known causation with the precipitation of multiple pathological complications, including neurological and neurobehavioral aberrations, metabolic impairments, carcinogenesis, and teratogenic effects. The gut microbial ecosystem or the resident gut flora has been identified as an important player in maintaining physiological homeostasis by multiple mechanisms, which are some of the key producers of secondary metabolites, essential vitamins, hormones, and neurotransmitters. A bidirectional interaction with environmental toxicants has also been identified between the microbial ecosystem and such toxicants. For limiting their bioburden on physiological systems, this microbial diversity may play a major role and provide efficient host defense. Also, for a better understanding of the pathological basis of some of the abovementioned target organ toxicities upon HM exposure, their impact on the gut microbial diversity needs to be studied in more detail. Such studies may hold the key to understanding a complex and deep microbial-rooted phenomenon.

Financial support: No specific financial support was granted for this book chapter.
Conflict of interest: None to declare.

TABLE 11.4
Effects of Heavy Metals Exposure on Gut Microbial Ecosystem and Possible Mode of Their Actions

Heavy Metal	Model System	Route/Dose/Duration	Effects	Mechanism/Key Findings	References
As	C57BL/6 mice	10ppm As mixed with drinking water for 4 weeks		Metabolomics profiling revealed a synchronous effect, metabolites that are related to gut flora perturbed in multiple biological matrices.	(Lu et al., 2014a)
	C57BL/6 mice and GF mice	25 or 100ppm Na_3AsO_4 (Sodium Arsanate) mixed with drinking water for 14 days	The microbiome especially Faecalibacterium mitigates host exposure and uptake of ingested Na_3AsO_4 through the fecal excretion.	In the mice under study in contrast to WT (Wild Type) mice, methyltransferase-deficient mice (As3mt-KO) are dysfunctional in arsenic methylation, and tend to accumulate more intracellular inclusions in urothelial tissue during arsenic exposure. This results in increased overall arsenic-induced cytotoxicity.	(Coryell et al., 2018)
	Female CD-1 mice	As(III) dosing was determined for two life stages: (i) Adult animals: 0.05 mg/kg body weight [b.w.] single gavage, low dose 0.1 mg/kg b.w., medium dose 0.2 mg/kg b.w., high dose followed by repeated exposure at 1 mg/liter for 8 days and (ii) Postnatal day 10: PND10 and PND21 after single gavage (0.05 mg/kg b.w.)	Firmicutes ↑ Actinobacteria ↑ Bacteriodetes ↑ Proteobacteria ↑ Deferribacteres ↑	A change in the bacterial population to less reactive trivalent methylated species with an excess of arsenic resistance genes and indications of host arsenite metabolism. High levels of CC chemokine and proinflammatory and anti-inflammatory cytokine secretion in the intestine were caused by arsenic exposure in adult animals. The exposure of arsenic to PND21 resulted in the creation of distinct populations of bacteria.	(Gokulan et al., 2018)

(continued)

TABLE 11.4 (Continued)
Effects of Heavy Metals Exposure on Gut Microbial Ecosystem and Possible Mode of Their Actions

Heavy Metal	Model System	Route/Dose/Duration	Effects	Mechanism/Key Findings	References
	C57BL/6 mice	Concentrations of 0,10 or 250 ppb iAsIII for 2,5, or 10 weeks respectively	Bacteroidetes ↑ Porphyromonadaceae↓ Firmicutes↓	The altered metabolites of tryptophan primarily contribute to disrupting the cascade of neurological signaling. The upregulated sphingolipids and downregulated levels of acylcarnitine are also correlated with nervous system disruption.	(Dheer et al., 2015)
	C57BL/6 mice	100 ppb arsenic for 13 weeks	Alpha and beta diversities were altered, and genes involved in carbohydrate metabolism, especially pyruvate fermentation, SCFA synthesis, and starch utilization, were significantly changed, lipopolysaccharide biosynthesis genes and DNA repair genes were significantly increased in As-exposed mice, vitamins involved in producing folic acid, B6, B12, and K2 were enriched in As-treated mice.	(Chi et al., 2017)

	C57BL/6	10 ppm Arsenic metal for 4 weeks.	**Firmicute:** Clostridiales↓ Catabacteriaceae↓ Clostridiaceae↓ Catabacteriaceae↓ Clostridiaceae↓ Ruminococcacea↓ Lachnospiraceae↓ **Bacteroidetes**↑ **Tenericute:** Erysipelotrichaceae↓ Bacteroidetes↑ Verrucomicrobia ↑ Erysipelotrichaceae↓ These alterations of the gut microbiota were associated with a significant decrease in urine, DMAsV, and increased MMAsV and iAsV, suggesting a compromised detoxification of As by the gut microbiome. Changes in the gut microbial profiles are greatly impacted by the biometabolism as well.	The changes of the microbes, in turn, induced changes in the metabolism of isoflavone (daidzein),indole containing (Indoleacetic acid) and bile acid derivatives.	(Guo et al., 2014)
Pb	Kunming mice	1.34g/L PbCl$_2$ in drinking water for 8 weeks		Protective effect of CGA against Pb induced nephrotoxicity and hepatotoxicity expression of NF-κB, Bcl-2, Bax, cytochrome C, caspase-9	(Cheng et al., 2019)

(continued)

TABLE 11.4 (Continued)
Effects of Heavy Metals Exposure on Gut Microbial Ecosystem and Possible Mode of Their Actions

Heavy Metal	Model System	Route/Dose/Duration	Effects	Mechanism/Key Findings	References
	A(vy) strain wild-type non-agouti (a/a)	32 ppm in drinking water	Perinatal Pb exposure was associated with increased adult bodyweight in male but not in female offspring	Weighted Unfrac and AMOVA analysis, RDA analysis	(Wu et al., 2016)
	Balb/C female mice	100 or 500 ppm- mg/L of $PbCl_2$, 20 or 100 ppm- mg/L of $CdCl_2$ in drinking water for 8 weeks	Lachnospiracea↓ Relative amounts of Lactobacillaceae and Erysipelotrichaeceacae↑ due to alteration in Turicibacter spp	Excessive local and systemic oxidative stress or deregulation of immune responses	(Breton et al., 2013b)
	C57Bl/6 female mice	10 ppm PbCl2 in drinking water for 13 weeks	Vitamin E, bile acids, nitrogen metabolism, energy metabolism, oxidative stress, and the defense/detox mechanism were significantly disturbed by lead exposure.	Pb Exposure Reduced Vitamin E and Bile Acids in the Gut Microbiome: Cholic acid(CA) ↓, UDCA ↓, DCA ↓, α-tocopherol and γ-tocopherol ↓ Lead Exposure Altered the Nitrogen Metabolism of Gut Bacteria: UreE gene, hydroxylamine ↓ Pb Induced Oxidative Stress and Activated the Defense/Detoxification Mechanism in Gut Bacteria: CoADR MutT, tyrosine recombinase gene XerC.	(Gao et al., 2017)
	ICR male mice	0.1 mg/L Pb in drinking water for 15 weeks **Control group A** Group B: 0.01 mg/L Group C: 0.03 mg/L Group D: 0.1 mg/L	The levels of some genes related to lipid metabolism in the liver increased in Pb-treated mice. Exposure to lead further disturbed the hepatic lipid metabolism in mice Firmicutes↓ Bacteroidetes↑ Proteobacteria↑	After exposure to 0.1 mg/L, TG and TCH levels increased in the serum significantly. The levels of hepatic pyruvate increased in a dose-dependent manner and substantial changes were observed for over 15 weeks at 0.03 and 0.1 mg/L Pb. Following exposure to 0.1 mg/L Pb for 15 weeks, the amount of hepatic TG increased significantly.	(Xia et al., 2018)

Cd	Female Balb/c mice	CdCl$_2$ (Cadmium chloride) dosing in drinking water for 21 days Control group 1: Treated with water free from cadmium chloride Group 2: 20 mg/kg Group 3: 100 mg/kg	Elevation in levels of TNF-$\langle$ Ratio of Firmicutes to Bacteriodetes↓ SCFA↓ Lactobacillus↓ Lactobacillus and Bifidobacterium: Activity inhibited	The consumption of Cd resulted in a decrease in bacteria producing butyrate, resulting in a rise in cecal pH and a decrease in fecal SCFAs.	(Liu et al., 2014)
	Mice	10 mg/L Cd supplied in drinking water for 10 weeks	Triacylglycerol synthesis in the liver increased significantly. Cd exposure induced hepatic toxicity at a molecular level Firmicutes and γ-proteobacteria↓ significantly in the feces after 4 weeks Firmicutes↓significantly in the cecum content after 10 weeks ↑in serum lipopolysaccharide (LPS) and induce hepatic inflammation Disturbed energy meta	LPS from the gut microbiome interacts with several receptors or proteins on the host, such as TLR4, CD14, and MD-2 on the host cells and plays a role in producing Cytokines, which ultimately disturbs the metabolic functions of the liver. In this study, serum LPS levels were observed to have increased after exposure to Cd. This could result to a rise in Bacteroidaceae. Simultaneously, the transcriptional levels of TLR4, CD14, and MD-2 were also increased in the liver, leading to a rise in hepatic IL1β levels and the transcription of IL1β, IL6, and TNF.	(Zhang et al., 2015)

(continued)

TABLE 11.4 (Continued)
Effects of Heavy Metals Exposure on Gut Microbial Ecosystem and Possible Mode of Their Actions

Heavy Metal	Model System	Route/Dose/Duration	Effects	Mechanism/Key Findings	References
	C57BL/6J mice	100nM CdCl$_2$ in drinking water for 1 week	Dysfunction of hepatic lipid metabolism with increased serum TG, free fatty acids, and hepatic TG levels, is seen especially in males compared to female mice. In comparison to Bacteroidetes, Cd-mediated alterations at 8 weeks reflected a decreased abundance of Firmicutes and lower numbers of Bifidobacterium and Prevotella.	In male mice, LDC early exposure significantly increased plasma TG and TC levels. This facilitates a general impact on blood lipid levels. In addition, LDC increased the levels of leptin, a hormone secreted by adipocytes, and HDL, which transports fat molecules to the liver; while the levels of VLDL and LDL were unchanged, indicating that exposure to LDC decreases the metabolism related to body adiposity. LDC resulted in a rise of liver TG in male mice and lipid deposits in liver cells with respect to the accumulation of fat in liver tissue.	(Ba et al., 2017)
	C57BL/6 mice	Control Group A: pure water Group B: dose of 10 mg/L CdCl2 Group C: dose of 50 mg/L CdCl2 In pure water for 20 weeks	Actinobacteria↓ Clostridiales↓ Erysipelotrichales↑ Bacteroidales S247group↑ Erysipelotrichaceae↑ Bifidobacteriaceae↑ Bacteroidaceae↓ Lachnospiraceae↓ Ruminococcaceae↓	Following chronic Cd exposure, the gut SCFA-producing bacteria undergoes significant inhibition, resulting in reduced production of SCFAs in the gut. Significant metabolic functional changes associated with intestinal microbiota are affected. This is characterized by significant inhibitions of gene pathways associated with amino acid, carbohydrate, and energy metabolism, as well as promotions of metabolic pathways such as metabolism of glutathione and degradation of aminobenzoate.	(He et al., 2020)

Impact of Heavy Metals on Gut Microbial Ecosystem

	Female C57BL/6J mice	**Control Group A control**: pure drinking water. **Group B**: 10 mg/L CdCl$_2$ in drinking water **Group C**: Cd + antibiotic treatment. All doses lasted for 52 weeks	**control group**: Firmicutes (30%) Bacteroidetes (60%) **Low Cd treatment**: Firmicutes (48%), Bacteroidetes (30%) Proteobacteria (15%)	From the Transcriptomic analyses it came to light that 162 genes were differentially expressed. These included 59 upregulated and 103 downregulated in Cd treatment. These genes were proven to be involved in several important pathways, eventually increasing the intestinal permeability.	(Liu et al., 2020)
Hg	Kunming mice	Twelve subjects were divided into two groups ($n = 6$) Group-1: 0 mg/L Group 2: 80 mg/L HgCl$_2$ in drinking water for 90 day.	Mercury-exposure interfered with the growth and development of mice. It was shown to cause intestinal microbiota dysbiosis and metabolic disorder. It also aggravated apoptosis in mice. *Coprococcus*↑, *Oscillospira* ↑ *Helicobacter* ↑ *Lgnatzschineria* ↓ *Salinicoccus* ↓ *Bacillus*↓.	When exposed to mercury-the expression of pro-apoptotic genes including Bax, JNK, ASK1, caspase3, and TNF-α drastically increased whereas the dosing significantly decreased the expression of the anti-apoptotic gene Bcl-2.	(Zhao et al., 2020)
	Kunming mice	Twelve subjects were divided into two groups ($n = 6$) Administered with distilled water containing HgCl$_2$ Group-1: 0 mg/L HgCl$_2$ for 3 days Group-2:160 mg/L HgCl$_2$ for 3 days	The growth of mice was reduced, causing the gut flora concentration to be altered. *Clostridiales*↑, *Lactobacillus*↑ *Treponema* ↑, *Oscillospira* ↑, *Desulfovibrio* ↑ S24↓, *Acinetobacter*↓, *Staphylococcus*↓ An increase in the oxidative stress levels was observed.	Increase in the malondialdehyde (MDA) and reduction in the superoxide dismutase (SOD) and glutathione peroxidase (GSH) concentration resulted into the production of oxidative stress.	(Zhao et al., 2021)

(continued)

TABLE 11.4 (Continued)
Effects of Heavy Metals Exposure on Gut Microbial Ecosystem and Possible Mode of Their Actions

Heavy Metal	Model System	Route/Dose/Duration	Effects	Mechanism/Key Findings	References
	Kunming mice	Group-1: control group (CCk group), Group-2: Cu group (CCu group), Group-3: Hg group (CHg group), Group-4: Cu + Hg group (CCH group) for 90 days	**CCu group** *Rikenella↓*, *Jeotgailcoccus↓*, *Staphylococcus↓*, *Corynebacterium↑*. **CHg group and CCH group**. *Sporosarcina*, *Jeotgailcoccus*, *Staphylococcus↓* **CCH group**. *Anaeroplasma↑* Mercury caused decrease in the body weight and further causing histopathological lesions in the cecum.		(Ruan et al., 2019)
	Male Sprague Dawley rats	0.4 mg/L (equivalent to 10 μg/kg bw) MeHg	It is proven here that MeHg damages the intestinal integrity by causing inflammation, resulting into immunological and neurological responses to be generated.	MeHg induces gut dysbiosis structure, which further results in neuronal activity changes of the intestinal neurotransmitters and metabolites that are responsible for regulating this activity. This whole chain is further supported by the increased levels of BDNF.	(Lin et al., 2020)

Kunming mice	HgS (α-HgS, 30 mg/kg), Zuotai (β-HgS, 30 mg/kg), HgCl$_2$ (33.6 mg/kg, equivalent Hg as HgS), or MeHg (3.1 mg/kg, 1/10 Hg as HgS) for 7 days	Accumulation of Hg in the duodenum and ileum after HgCl$_2$ (30–40 fold) and MeHg (10–15 fold) was higher than HgS and Zuotai (~2-fold). **HgCl$_2$** *Bacteroidetes*↑, *Cyanobacteria* ↑ *Firmicutes* ↓ *Odoribacteraceae* ↑ **HgS** *Firmicutes* ↑, *Proteobacteria Rikenellaceae*↑ *Lactobacillaceae*↑, *Helicobacteracea* *Prevotellaceae* ↓, **HgCl$_2$** *Odoribacteraceae* ↑	HgCl$_2$ and MeHg decreased the levels of intestinal intake peptide transporter-1 and Ost-β, whereas it increased the ileal bile acid-binding protein and equilibrative nucleoside transporter-1. The efflux transporters ATP-binding cassette sub-family C member-4 (Abcc4), Abcg2, Abcg5/8, and Abcb1b were increased by HgCl$_2$, and to a lesser extent by MeHg, while HgS and Zuotai had minimal effect.	(Zhang et al., 2019)

REFERENCES

Ahmed E, and Holmström SJM (2014) Siderophores in environmental research: roles and applications. *Microb Biotechnol* 7:196–208.

Alava P, Du Laing G, Tack F, et al. (2015) Westernized diets lower arsenic gastrointestinal bioaccessibility but increase microbial arsenic speciation changes in the colon. *Chemosphere* 119:757–7762.

Assefa S, and Köhler G (2020) Intestinal microbiome and metal toxicity. *Curr. Opin. Toxicol.* 19:21–27.

Ba Q, Li M, Chen P, et al. (2017) Sex-dependent effects of cadmium exposure in early life on gut microbiota and fat accumulation in mice. *Environ Health Perspect* 125:437–446.

Barnhart J (1997) Occurrences, uses, and properties of chromium. *Regul Toxicol Pharmacol* 26:S3–S7. https://doi.org/10.1006/rtph.1997.1132.

Beaumont JJ, Sedman RM, Reynolds SD, et al. (2008) Cancer mortality in a Chinese population exposed to hexavalent chromium in drinking water. *Epidemiology* 19: 12–23.

Bernhoft RA (2012) Mercury toxicity and treatment: a review of the literature. *J Environ Public Health* 2012. https://doi.org/10.1155/2012/460508.

Beyersmann D, and Hartwig A (2008) Carcinogenic metal compounds: recent insight into molecular and cellular mechanisms. *Arch Toxicol* 82:493–512.

Bisanz JE, Enos MK, Mwanga JR, et al. (2014) Randomized open-label pilot study of the influence of probiotics and the gut microbiome on toxic metal levels in Tanzanian pregnant women and school children. *MBio* 5:e01580–14.

Bremer Neto H, Graner CAF, Pezzato LE, and Padovani CR (2005) The spectrophotometric method on the routine of 1, 5-diphenylcarbazide was adjusted on chromium determination in feces, after its utilization as a biological marker as chromium (III) oxide. *Ciência Rural* 35:691–697.

Breton J, Daniel C, Dewulf J, et al. (2013a) Gut microbiota limits heavy metals burden caused by chronic oral exposure. *Toxicol Lett* 222:132–138.

Breton J, Massart S, Vandamme P, et al. (2013b) Ecotoxicology inside the gut: impact of heavy metals on the mouse microbiome. *BMC Pharmacol Toxicol* 14:1–11.

Bridou R, Monperrus M, Gonzalez PR, et al. (2011) Simultaneous determination of mercury methylation and demethylation capacities of various sulfate-reducing bacteria using species-specific isotopic tracers. *Environ Toxicol Chem* 30:337–344.

Byrne CS, Chambers ES, Morrison DJ, and Frost G (2015) The role of short chain fatty acids in appetite regulation and energy homeostasis. *Int J Obes (Lond)* 39 :1331–1338. https://doi.org/doi:10.1038/ijo.2015.84. Epub 2015 May 14.

Cai L, Liu G, Rensing C, and Wang G (2009) Genes involved in arsenic transformation and resistance associated with different levels of arsenic-contaminated soils. *BMC Microbiol* 9:1–11.

Cani PD, Neyrinck AM, Fava F, et al. (2007) Selective increases of bifidobacteria in gut microflora improve high-fat-diet-induced diabetes in mice through a mechanism associated with endotoxaemia. *Diabetologia* 50:2374–2383.

Chang LW, Magos L, and Suzuki T (1996) *Toxicology of Metals*. CRC Boca Raton, FL

Chary NS, Kamala CT, and Raj DSS (2008) Assessing risk of heavy metals from consuming food grown on sewage irrigated soils and food chain transfer. *Ecotoxicol Environ Saf* 69:513–524.

Chatterjee S, Mukherjee A, Sarkar A, and Roy P (2012) Bioremediation of lead by lead-resistant microorganisms, isolated from industrial sample. *Advances in Bioscience and Biotechnology* 3: 290–295.

Chellaiah ER (2018) Cadmium (heavy metals) bioremediation by Pseudomonas aeruginosa: a minireview. *Appl Water Sci* 8:1–10.

Cheng D, Li H, Zhou J, and Wang S (2019) Chlorogenic acid relieves lead-induced cognitive impairments and hepato-renal damage via regulating the dysbiosis of the gut microbiota in mice. *Food Funct* 10:681–690.

Chi L, Bian X, Gao B, et al. (2017) The effects of an environmentally relevant level of arsenic on the gut microbiome and its functional metagenome. *Toxicol Sci* 160:193–204.

Chi L, Xue J, Tu P, et al. (2019) Gut microbiome disruption altered the biotransformation and liver toxicity of arsenic in mice. *Arch Toxicol* 93:25–35.

Choudhary M, Kumar R, Datta A, et al. (2017) Bioremediation of heavy metals by microbes. In: Arora S, Singh AK, Singh YP (eds), *Bioremediation of Salt Affected Soils: An Indian Perspective*. Springer, pp 233–255.

Coryell M, McAlpine M, Pinkham NV, et al. (2018) The gut microbiome is required for full protection against acute arsenic toxicity in mouse models. *Nat Commun* 9:1–9.

Cox LM, Yamanishi S, Sohn J, et al. (2014) Altering the intestinal microbiota during a critical developmental window has lasting metabolic consequences. *Cell* 158:705–721.

Cruz Viggi C, Pagnanelli F, Cibati A, et al. (2009) Sulphate reducing bacteria for the treatment of heavy metals contaminated waters in permeable reactive barriers. In: *Advanced Materials Research*. Trans Tech Publ, pp 565–568.

Cryan JF, O'Riordan KJ, Sandhu K, et al. (2020) The gut microbiome in neurological disorders. *Lancet Neurol* 19:179–194.

DC Rubin SSC, Alava P, and Zekker I, et al. (2014) Arsenic thiolation and the role of sulfate-reducing bacteria from the human intestinal tract. *Environ Health Perspect* 122:817–822.

Dethlefsen L, Huse S, Sogin ML, and Relman DA (2008) The pervasive effects of an antibiotic on the human gut microbiota, as revealed by deep 16S rRNA sequencing. *PLoS Biol* 6:e280.

Dethlefsen L, and Relman DA (2011) Incomplete recovery and individualized responses of the human distal gut microbiota to repeated antibiotic perturbation. *Proc Natl Acad Sci* 108:4554–4561.

Dheer R, Patterson J, Dudash M, et al. (2015) Arsenic induces structural and compositional colonic microbiome change and promotes host nitrogen and amino acid metabolism. *Toxicol Appl Pharmacol* 289:397–408.

Dimkpa CO, Merten D, Svatoš A, et al. (2009) Siderophores mediate reduced and increased uptake of cadmium by Streptomyces tendae F4 and sunflower (Helianthus annuus), respectively. *J Appl Microbiol* 107:1687–1696.

Ding R, Goh W-R, Wu R, et al. (2019) Revisit gut microbiota and its impact on human health and disease. *J Food Drug Anal* 27:623–631.

Duan H, Yu L, Tian F, et al. (2020) Gut microbiota: a target for heavy metal toxicity and a probiotic protective strategy. *Sci. Total Environ.* 742:140429.

Duca F, Gerard P, Covasa M, and Lepage P (2014) Metabolic interplay between gut bacteria and their host. In: Delhanty PJD and van der Lely AJ (eds), *How Gut and Brain Control Metabolism*. Karger Publishers, pp 73–82.

Duffus JH (2002) "Heavy metals" a meaningless term? (IUPAC Technical Report). *Pure Appl Chem* 74:793–807.

Eggers S, Safdar N, Sethi AK, et al. (2019) Urinary lead concentration and composition of the adult gut microbiota in a cross-sectional population-based sample. *Environ Int* 133:105122.

Engwa GA, Ferdinand PU, Nwalo FN, and Unachukwu MN (2019) Mechanism and health effects of heavy metal toxicity in humans. In: Karcioglu O and Arslan B (eds), *Poisoning Mod World—New Tricks an Old Dog?* London: IntechOpen, DOI: 10.5772/intechopen.82511

Faith JJ, Guruge JL, Charbonneau M, et al. (2013) The long-term stability of the human gut microbiota. *Science* 341:1237439.

Fan Y, and Pedersen O (2021) Gut microbiota in human metabolic health and disease. *Nat Rev Microbiol* 19:55–71.

Fazeli M, Hassanzadeh P, and Alaei S (2011) Cadmium chloride exhibits a profound toxic effect on bacterial microflora of the mice gastrointestinal tract. *Hum Exp Toxicol* 30:152–159.

Feng P, Ye Z, Kakade A, et al. (2019) A review on gut remediation of selected environmental contaminants: possible roles of probiotics and gut microbiota. *Nutrients* 11:22.

Forsythe P, and Kunze WA (2013) Voices from within: gut microbes and the CNS. *Cell Mol Life Sci* 70:55–69.

Fung TC, Olson CA, and Hsiao EY (2017) Interactions between the microbiota, immune and nervous systems in health and disease. *Nat Neurosci* 20:145–155.

Gao B, Chi L, Mahbub R, et al. (2017) Multi-omics reveals that lead exposure disturbs gut microbiome development, key metabolites, and metabolic pathways. *Chem Res Toxicol* 30:996–1005.

Gergen I, and Harmanescu M (2012) Application of principal component analysis in the pollution assessment with heavy metals of vegetable food chain in the old mining areas. *Chem Cent J* 6:1–13.

Gilmour CC, Podar M, Bullock AL, et al. (2013) Mercury methylation by novel microorganisms from new environments. *Environ Sci Technol* 47:11810–11820.

Gokulan K, Arnold MG, Jensen J, et al. (2018) Exposure to arsenite in CD-1 mice during juvenile and adult stages: effects on intestinal microbiota and gut-associated immune status. *MBio* 9: e01418–18.

Guo X, Liu S, Wang Z, et al. (2014) Metagenomic profiles and antibiotic resistance genes in gut microbiota of mice exposed to arsenic and iron. *Chemosphere* 112:1–8.

Halttunen T, Finell M, and Salminen S (2007) Arsenic removal by native and chemically modified lactic acid bacteria. *Int J Food Microbiol* 120:173–178.

He X, Qi Z, Hou H, et al. (2020) Structural and functional alterations of gut microbiome in mice induced by chronic cadmium exposure. *Chemosphere* 246:125747.

Heianza Y, Sun D, Li X, et al. (2019) Gut microbiota metabolites, amino acid metabolites and improvements in insulin sensitivity and glucose metabolism: the POUNDS Lost trial. *Gut* 68:263–270.

Ingerman L, Jones DG, Keith S, and Rosemond ZA (2002) Toxicological profile for aluminum. In: *ATSDR's Toxicological Profiles*. Agency for Toxic Substances and Disease Registry, Department of Health and Human Services, Division of Toxicology and Environmental Medicine, Atlanta, Georgia, EUA.

Jin Y, Wu S, Zeng Z, and Fu Z (2017) Effects of environmental pollutants on gut microbiota. *Environ Pollut* 222:1–9.

Kerger BD, Butler WJ, Paustenbach DJ, et al. (2009) Cancer mortality in Chinese populations surrounding an alloy plant with chromium smelting operations. *J Toxicol Environ Heal Part A* 72:329–344.

Khan MF, and Wang H (2020) Environmental exposures and autoimmune diseases: contribution of gut microbiome. *Front Immunol* 10:3094.

Kim E, Xu X, Steiner H, et al. (2015) Chronic ingestion of low doses of cadmium alters the gut microbiome and immune homeostasis to enhance allergic sensitization (MUC9P. 743). *J Immunol*, 194 (1 Supplement) 205: 7.

Kim Y-K, and Shin C (2018) The microbiota-gut-brain axis in neuropsychiatric disorders: pathophysiological mechanisms and novel treatments. *Curr Neuropharmacol* 16:559–573.

Koenig JE, Spor A, Scalfone N, et al. (2011) Succession of microbial consortia in the developing infant gut microbiome. *Proc Natl Acad Sci U S A* 108:4578–85. https://doi.org/doi:10.1073/pnas.1000081107. Epub 2010 Jul 28.

Konepudi KK, Singh DP, Podili K, et al. (2015) Role of Probiotics and Prebiotics in the Management of Obesity. In: Khardori N (Ed), *Food Microbiology*. CRC Press, pp 157–186.

LeBlanc JG, Milani C, De Giori GS, et al. (2013) Bacteria as vitamin suppliers to their host: a gut microbiota perspective. *Curr Opin Biotechnol* 24:160–168.

Li J (2017) *Arsenic Biotransformations in Microbes and Humans, and Catalytic Properties of Human AS3MT Variants*.

Lin T-H, and Pan T-M (2019) Characterization of an antimicrobial substance produced by Lactobacillus plantarum NTU 102. *J Microbiol Immunol Infect* 52:409–417.

Lin X, Zhao J, Zhang W, et al. (2020) Acute oral methylmercury exposure perturbs the gut microbiome and alters gut-brain axis related metabolites in rats. *Ecotoxicol Environ Saf* 190:110130.

Lind PM, and Lind L (2018) Endocrine-disrupting chemicals and risk of diabetes: an evidence-based review. *Diabetologia* 61:1495–1502.

Liu Y, Li Y, Liu K, and Shen J (2014) Exposing to cadmium stress cause profound toxic effect on microbiota of the mice intestinal tract. *PLoS One* 9:e85323.

Liu Y, Li Y, Xia Y, et al. (2020) The dysbiosis of gut microbiota caused by low-dose cadmium aggravate the injury of mice liver through increasing intestinal permeability. *Microorganisms* 8:211.

Lu K, Abo RP, Schlieper KA, et al. (2014a) Arsenic exposure perturbs the gut microbiome and its metabolic profile in mice: an integrated metagenomics and metabolomics analysis. *Environ Health Perspect* 122:284–291.

Lu K, Mahbub R, Cable PH, et al. (2014b) Gut microbiome phenotypes driven by host genetics affect arsenic metabolism. *Chem Res Toxicol* 27:172–174.

Mahurpawar M (2015) Effects of heavy metals on human health. *Int J Res* 3:1–7.

Martin S, Griswold Wendy. (2009) Human Health Effects of Heavy Metals. *Environ Sci Technol Briefs Citizens* 15: 1–6.

Martin YE, and Johnson EA (2012) Biogeosciences survey: studying interactions of the biosphere with the lithosphere, hydrosphere and atmosphere. *Prog Phys Geogr* 36:833–852. https://doi.org/10.1177/0309133312457107.

Masindi V, and Muedi KL (2018) Environmental contamination by heavy metals. In: Saleh HE M, and Aglan RF (eds)., *Heavy Metals* IntechOpen. pp. 115–133.

Matsumoto M, Kibe R, Ooga T, et al. (2012) Impact of intestinal microbiota on intestinal luminal metabolome. *Sci Rep* 2:233. https://doi.org/10.1038/srep00233.

Mazmanian SK, Round JL, and Kasper DL (2008) A microbial symbiosis factor prevents intestinal inflammatory disease. *Nature* 453:620–625.

McDermott TR, Stolz JF, and Oremland RS (2020) Arsenic and the gastrointestinal tract microbiome. *Environ Microbiol Rep* 12:136–159.

Mitra A, Chatterjee S, and Gupta DK (2017) Potential role of microbes in bioremediation of arsenic. In: Gupta DK, and Chatterjee S (eds.), *Arsenic Contamination in the Environment*. Springer, pp 195–213.

Mörkl S, Butler MI, Holl A, et al. (2020) Probiotics and the microbiota-gut-brain axis: focus on psychiatry. *Curr Nutr Rep* 9:171–182.

Muccioli GG, Naslain D, Bäckhed F, et al. (2010) The endocannabinoid system links gut microbiota to adipogenesis. *Mol Syst Biol* 6:392.

Nadal A, Quesada I, Tudurí E, et al. (2017) Endocrine-disrupting chemicals and the regulation of energy balance. *Nat Rev Endocrinol* 13:536–546.

Nicholson JK, Holmes E, Kinross J, et al. (2012) Host-gut microbiota metabolic interactions. *Science* 336:1262–1267.

O'Hara AM, and Shanahan F (2006) The gut flora as a forgotten organ. *EMBO Rep* 7:688–693.

Ojekunle O, Banwo K, and Sanni AI (2017) In vitro and in vivo evaluation of Weissella cibaria and Lactobacillus plantarum for their protective effect against cadmium and lead toxicities. *Lett Appl Microbiol* 64:379–385.

Pan X, Chen Z, Li L, et al. (2017) Microbial strategy for potential lead remediation: a review study. *World J Microbiol Biotechnol* 33:35.

Parkar SG, Trower TM, and Stevenson DE (2013) Fecal microbial metabolism of polyphenols and its effects on human gut microbiota. *Anaerobe* 23:12–19.

Pechova A, and Pavlata L (2007) Chromium as an essential nutrient: a review. *Vet Med (Praha)* 52:1.

Prakash S, Rodes L, Coussa-Charley M, and Tomaro-Duchesneau C (2011) Gut microbiota: next frontier in understanding human health and development of biotherapeutics. *Biologics* 5:71–86.

Rakoff-Nahoum S, Paglino J, Eslami-Varzaneh F, et al. (2004) Recognition of commensal microflora by toll-like receptors is required for intestinal homeostasis. *Cell* 118:229–241.

Rehman K, Fatima F, Waheed I, and Akash MSH (2018) Prevalence of exposure of heavy metals and their impact on health consequences. *J Cell Biochem* 119:157–184.

Richardson JB, Dancy BCR, Horton CL, et al. (2018) Exposure to toxic metals triggers unique responses from the rat gut microbiota. *Sci Rep* 8:1–12.

Ridlon JM, Kang DJ, and Hylemon PB (2006) Bile salt biotransformations by human intestinal bacteria. *J Lipid Res* 47:241–59. Epub 2005 Nov 18.

Rinninella E, Raoul P, Cintoni M, et al. (2019) What is the healthy gut microbiota composition? A changing ecosystem across age, environment, diet, and diseases. *Microorganisms*. 7: 14 https://doi.org/10.3390/microorganisms7010014.

Rodríguez JM, Murphy K, Stanton C, et al. (2015) The composition of the gut microbiota throughout life, with an emphasis on early life. *Microb Ecol Health Dis* 26:26050.

Rosenfeld CS (2017) Gut dysbiosis in animals due to environmental chemical exposures. *Front Cell Infect Microbiol* 7:396.

Rowland IR, Robinson RD, and Doherty RA (1984) Effects of diet on mercury metabolism and excretion in mice given methylmercury: role of gut flora. *Arch Environ Heal An Int J* 39:401–408.

Ruan Y, Wu C, Guo X, et al. (2019) High doses of copper and mercury changed cecal microbiota in female mice. *Biol Trace Elem Res* 189:134–144.

Samuel BS, Shaito A, Motoike T, et al. (2008) Effects of the gut microbiota on host adiposity are modulated by the short-chain fatty-acid binding G protein-coupled receptor, *Gpr41*. *Proc Natl Acad Sci* 105:16767–16772.

Serino M, Luche E, Gres S, et al. (2012) Metabolic adaptation to a high-fat diet is associated with a change in the gut microbiota. *Gut* 61:543–553. https://doi.org/doi:10.1136/gutjnl-2011-301012. Epub 2011 Nov 22.

Shrivastava R, Kannan A, Upreti RK, and Chaturvedi UC (2005) Effects of chromium on the resident gut bacteria of rat. *Toxicol Mech Methods* 15:211–218.

Sobolev D, and Begonia M (2008) Effects of heavy metal contamination upon soil microbes: lead-induced changes in general and denitrifying microbial communities as evidenced by molecular markers. *Int J Environ Res Public Health* 5:450–456.

Stacchiotti V, Rezzi S, Eggersdorfer M, and Galli F (2020) Metabolic and functional interplay between gut microbiota and fat-soluble vitamins. *Crit Rev Food Sci Nutr* 61: 1–22.

Swann JR, Want EJ, Geier FM, et al. (2011) Systemic gut microbial modulation of bile acid metabolism in host tissue compartments. *Proc Natl Acad Sci U S A* 108:4523–4530. https://doi.org/10.1073/pnas.1006734107.

Tchounwou PB, Yedjou CG, Patlolla AK, and Sutton DJ (2012) Heavy metal toxicity and the environment. *EXS* 101:133–164.
Thursby E, and Juge N (2017) Introduction to the human gut microbiota. *Biochem J* 474:1823–1836.
Tinkov AA, Gritsenko VA, Skalnaya MG, et al. (2018) Gut as a target for cadmium toxicity. *Environ Pollut* 235:429–434.
Tsiaoussis J, Antoniou MN, Koliarakis I, et al. (2019) Effects of single and combined toxic exposures on the gut microbiome: current knowledge and future directions. *Toxicol Lett* 312:72–97.
Turroni F, Ventura M, Buttó LF, et al. (2014) Molecular dialogue between the human gut microbiota and the host: a Lactobacillus and Bifidobacterium perspective. *Cell Mol Life Sci* 71:183–203.
Upreti RK, Shrivastava R, and Chaturvedi UC (2004) Gut microflora & toxic metals: chromium as a model. *Indian J Med Res* 119:49–59.
Valko M, Morris H, and Cronin MTD (2005) Metals, toxicity and oxidative stress. *Curr Med Chem* 12:1161–1208. https://doi.org/10.2174/0929867053764635.
Wagner-Döbler I (2013) *Bioremediation of Mercury*. Caister Academic Press.
Wang J, Hu W, Yang H, et al. (2020) Arsenic concentrations, diversity and co-occurrence patterns of bacterial and fungal communities in the feces of mice under sub-chronic arsenic exposure through food. *Environ Int* 138:105600.
Wang S, and Mulligan CN (2008) Speciation and surface structure of inorganic arsenic in solid phases: A review. *Environ Int* 34:867–879.
WangX S, and Shi X (2001) Molecular mechanisms of metal toxicity and carcinogenesis. *Mol Cell Biochem* 222:3–9.
Wang Z, Klipfell E, Bennett BJ, et al. (2011) Gut flora metabolism of phosphatidylcholine promotes cardiovascular disease. *Nature* 472:57–63. https://doi.org/doi:10.1038/nature09922.
Wong JMW, de Souza R, Kendall CWC, et al. (2006) Colonic health: fermentation and short chain fatty acids. *J Clin Gastroenterol* 40:235–243.
Wu J, Wen XW, Faulk C, et al. (2016) Perinatal lead exposure alters gut microbiota composition and results in sex-specific bodyweight increases in adult mice. *Toxicol Sci* 151:324–333.
Xia J, Jin C, Pan Z, et al. (2018) Chronic exposure to low concentrations of lead induces metabolic disorder and dysbiosis of the gut microbiota in mice. *Sci Total Environ* 631:439–448.
Yatsunenko T, Rey FE, Manary MJ, et al. (2012) Human gut microbiome viewed across age and geography. *Nature* 486:222–227.
Yilmaz B, Terekeci H, Sandal S, and Kelestimur F (2020) Endocrine disrupting chemicals: exposure, effects on human health, mechanism of action, models for testing and strategies for prevention. *Rev Endocr Metab Disord* 21:127–147.
Yin K, Wang Q, Lv M, and Chen L (2019) Microorganism remediation strategies towards heavy metals. *Chem Eng J* 360:1553–1563. https://doi.org/10.1016/j.cej.2018.10.226.
Younan S, Sakita GZ, Albuquerque TR, et al. (2016) Chromium (VI) bioremediation by probiotics. *J Sci Food Agric* 96:3977–3982.
Zawadzka AM, Paszczynski AJ, and Crawford RL (2009) Transformations of toxic metals and metalloids by Pseudomonas stutzeri strain KC and its siderophore pyridine-2, 6-bis (thiocarboxylic acid). In: *Advances in Applied Bioremediation*. Springer, pp 221–238.
Zhai Q, Wang G, Zhao J, et al. (2014) Protective effects of Lactobacillus plantarum CCFM8610 against chronic cadmium toxicity in mice indicate routes of protection besides intestinal sequestration. *Appl Environ Microbiol* 80:4063–4071.

Zhang B-B, Liu Y-M, Hu A-L, et al. (2019) HgS and Zuotai differ from $HgCl_2$ and methyl mercury in intestinal Hg absorption, transporter expression and gut microbiome in mice. *Toxicol Appl Pharmacol* 379:114615.

Zhang C, Zhang M, Wang S, et al. (2010) Interactions between gut microbiota, host genetics and diet relevant to development of metabolic syndromes in mice. *ISME J* 4:232–241.

Zhang S, Jin Y, Zeng Z, et al. (2015) Subchronic exposure of mice to cadmium perturbs their hepatic energy metabolism and gut microbiome. *Chem Res Toxicol* 28:2000–2009.

Zhao J, Jing Y, Zhang J, et al. (2019) Aged refuse enhances anaerobic fermentation of food waste to produce short-chain fatty acids. *Bioresour Technol* 289:121547.

Zhao Y, Zhou C, Guo X, et al. (2021) Exposed to mercury-induced oxidative stress, changes of intestinal microflora, and association between them in mice. *Biol Trace Elem Res* 199:1900–1907.

Zhao Y, Zhou C, Wu C, et al. (2020) Subchronic oral mercury caused intestinal injury and changed gut microbiota in mice. *Sci Total Environ* 721:137639.

Zheng D, Liwinski T, and Elinav E (2020) Interaction between microbiota and immunity in health and disease. *Cell Res* 30:492–506.

Zheng X, Xie G, Zhao A, et al. (2011) The footprints of gut microbial-mammalian co-metabolism. *J Proteome Res* 10:5512–5522. doi: 10.1021/pr2007945. Epub 2011 Oct 24.

Zhou B, Yuan Y, Zhang S, et al. (2020) Intestinal flora and disease mutually shape the regional immune system in the intestinal tract. *Front Immunol* 11:575.

12 Roadmaps for Environmental Protection and Sustainable Development

Suman Mallick and Ratna Ghosal

CONTENTS

- 12.1 Introduction 268
- 12.2 Types of Freshwater Ecosystems and Their Ecological Roles....................268
  - 12.2.1 Lotic Ecosystems..268
  - 12.2.2 Lentic Ecosystems 269
  - 12.2.3 Wetland Ecosystems 270
  - 12.2.4 Ground Water Table and Aquifers 271
- 12.3 Impacts on Freshwater Ecosystems 271
  - 12.3.1 Urbanization 273
  - 12.3.2 Climate Change 273
  - 12.3.3 Introduction of Exotic Species 274
- 12.4 Roadmaps Towards Sustainable Development 275
  - 12.4.1 Monitoring..276
    - 12.4.1.1 Monitoring Water Quality..276
    - 12.4.1.2 Monitoring Biodiversity 277
    - 12.4.1.3 Monitoring Toxicity..277
    - 12.4.1.4 Monitoring Using Molecular Biomarkers 278
  - 12.4.2 Restoration..278
    - 12.4.2.1 Restoration of Lotic Systems...278
    - 12.4.2.2 Restoration of Lentic Systems 279
  - 12.4.3 Protection and Sustainable Development 280
- 12.5 Future Goals and Conclusion...281
- Author Contributions 281
- References...282

12.1 INTRODUCTION

With an increase in urbanization, and overexploitation of natural resources, the sustenance of most ecosystems is at risk. One such ecosystem that supports a wide variety of flora and fauna, including humans as well, are the freshwater ecosystems. Freshwater ecosystems make up only about 0.01% of the world's water bodies and cover around 0.8% of the earth's surface (Dudgeon et al., 2006). Freshwater resources include ponds, lakes, streams, wetlands, and underground aquifers, and support a great range of biodiversity. Freshwater ecosystems support about 10,000 species of fish (Lundberg et al., 2000), which make up about 40% of global fish diversity. Apart from ichthyofaunal diversity, freshwater also consists of several other vertebrate species (Strayer et al., 2010). Aquatic reptiles (crocodiles, turtles, and snakes) and mammals (otters, river dolphins, and platypus) are prominent creatures that add to the faunal diversity of freshwaters (Dudgeon et al., 2006). Larval stages of mst of the amphibians develop in freshwater (Balian et al., 2007), and out of 5828 amphibian species prevalent on the earth's surface, 4117 have been reported to dwell in freshwater systems. Similarly, 5% of the so far known snake species inhabits freshwater (Pauwels et al., 2007). Apart from vertebrates, invertebrates also play a vital role in transforming detritus from sedimentary substratum into dissolved nutrients that influence the primary productivity (Covich et al., 2004) of the freshwater ecosystems. The invertebrate diversity in freshwater ecosystems comprise about 175,000 species (Lévêque, 2001) that have, so far, been discovered while actual number of species is certainly much higher than this. A few prominent species in the invertebrate category include dipteran insects, nematodes, and crustaceans (Palmer et al., 1997). Overall, freshwater presents a great resource for most of the biotic life on the earth's surface. Additionally, the human population heavily depends on freshwater resources for their day-to-day activities that include harvesting water for cooking, bathing, animal rearing, irrigation, aquaculture, and using freshwater habitats as a sewage discharge medium, as well. Overall, the freshwater reservoirs not only serve as a habitat to a great variety of biotic organisms but also are heavily exploited by humans for agricultural, industrial, and daily household needs. Thus, it is important to conserve the freshwater ecosystems to help maintain the global ecological balance (Darwall et al., 2005). However, to conserve an ecosystem, we need to understand it better. Thus, in this chapter, we outline the types and ecological roles of different freshwater habitats, the impact of anthropogenic disturbance on such habitats, and then present possible roadmaps for building a sustainable future for our freshwater systems.

12.2 TYPES OF FRESHWATER ECOSYSTEMS AND THEIR ECOLOGICAL ROLES

Broadly, the freshwater habitats can be classified into lotic, lentic, wetlands, and aquifers types.

12.2.1 Lotic Ecosystems

The most common examples of lotic habitats are rivers, waterfalls, and rapids. A river can be termed as any natural water body either having a perennial or an intermittent flow along a natural channel (Taylor et al., 2005). On the other hand, rapids are shallow area with fast flowing water (Torrente-Vilara et al., 2011), while a waterfall is an isolated landscape characterized by an abrupt fall in topography and strong flow rate of water (Clayton and Pearson, 2016). A waterfall can be classified into two zones: a high flow vertical zone and a spray zone that supports various microhabitats (Clayton and Pearson 2016). Waterfall habitats are highly dynamic with variable flow rate, having different depths of water columns, and thus, support species with widely different ecological needs. For example, the spray zone in a waterfall area is the safest place for larvae of various invertebrates, amphibians, and fish, as the thin films of water over rock surface nourish them with nutrients and limit predator evasion. Many freshwater gobies (*Sicyopterus stimpsoni, Lentipes concolor, Awaous guamensis*) are adapted to live within the fast flowing parts of the waterfall (Schoenfuss and Blob, 2003) whereas adult Hawaiian stream fish (*Stenogobius hawaiiensis*) can successfully climb waterfalls with the help of their ventral suckers (Blob et al., 2008). In addition to the waterfall and the streams, rivers play a major role in facilitating fish migration (Myers et al., 2006). Anadromous fishes (rainbow trout, *Onchorhyncus mykiss*, Atlantic salmon, *Salmo salar*) spend the majority of their lifecycle at sea, but move into the freshwater streams during spawning (Gunter et al., 1942). On the other hand, catadromous fishes (European eel, *Anguilla*, bumblebee goby, *Brachygobius xanthomelas*, mountain mullet, *Agonostomus monticola*) spend their entire lives in freshwater and migrate to the sea for the purpose of breeding (Norman et al., 1930). Thus, rivers are the navigation channels for most fish species and higher vertebrates as well.

12.2.2 LENTIC ECOSYSTEMS

A system consisting of standing water can be categorized as a lentic ecosystem. Ponds and lakes can be defined as isolated freshwater systems with shallow water depth that support a few unique life forms (Stanley, 1975). Groups of free-floating microalgae termed as phytoplankton are the predominant producer of freshwater ponds. Apart from primary producer, phytoplankton is also regarded as the bioindicator of health and productivity for the lentic ecosystems (Ekhalak et al., 2013). Phytoplanktons are also available within the lotic ecosystems but the planktons have greater diversity within the lentic systems (Hutorowicz et al., 2008). Grabowska et al. (2014) concluded lower abundance of phytoplankton in rivers (0.98 x 10^6 individual/liter) in comparison to floodplain lakes (1.15–3.58 x 10^6 individual/liter) in North Eastern Poland. Green algae (*Chlorophycea*), Desmids (*Desmidiaceae*), Euglenophytes (*Euglenophyceae*), Cynobacteria (*Cyanophyceae*), and Diatoms (*Bacilariophyceae*) are prominent phytoplankton found mostly in ponds and lakes (Carvalho et al., 2013). Apart from maintaining plankton diversity, ponds and lakes are known to recharge the groundwater. The impounded water in the ponds and lake infiltrate through the soil and percolate to the aquifer (Winter et al., 1999). India witnessed a reduction in national per capita annual availability of water in the form of lakes and ponds, which reduced from 1816 cubic meters in 2001 to 1544 cubic meters in 2011 due to anthropogenic activities and irrigation pressure. Consequently, the impact was reflected in net availability of groundwater that decreased from 433 billion cubic meters/year to 398 billion cubic meters/year (Suhag, 2016). While most lakes and ponds have diverse environmental functions, they represent isolated ecosystems with low immigration and emigration rate of individuals. Thus, the species inhabiting such ecosystems are specialists, having unique ecological needs and are highly adapted for that particular ecosystem. American bullfrog (*Lithobates catesbeianus*) is one such specialist species native to Canada, which prefers to inhabit in freshwater ponds, swamps, lakes, reservoirs, and irrigation ditches throughout its life stages (Gascon, 2007). Moreover, inland freshwater systems are crucial drivers of global carbon dynamics (Abhizova et al., 2012). Holgerson (2015) reported relatively high levels of carbondioxide (CO_2) and methane (CH_4) saturation in forested ponds of Connecticut, USA when compared to atmosphere. Ponds and lakes also play a major role in enhancing resistance to flood and drought (Goyal et al., 2021). Poland is a group of countries with low water resources and high frequency of extreme phenomena like floods and drought (Mioduszewski, 1998). During the dry years of 2018–20, Poland faced severe drought with rainfall deficit reaching a record value of 412.7 mm (Pinskwar et al., 2020). However, due to construction of water reservoirs and ponds, the water retention capacity of Poland soils increased (by 4%) and thus helped in recharging the groundwater in such a water-deficit country (Mioduszewski, 1998).

12.2.3 Wetland Ecosystems

Cowardin et al. (1979) defined wetland as a transitional land between terrestrial and aquatic systems where the water table is close to the land surface. Such lands are covered by shallow water and undrained hydric soil (Denny, 1985). Inland wetlands can be further classified into freshwater marshes, peatlands, and swamps (Semeniuk and Semeniuk, 1997). The wetlands that have standing water for most of the time are ideal sites for shrubs and trees, and are often termed as swamps. Wetlands play a significant role in regulating hydrological cycles thereby improving water quality (Wigham et al., 1988). For example, the riparian wetlands in Europe have been reported to reduce excess nutrients in flow-through water by removing nitrate and phosphorous from surface and sub-surface runoff (Verhoevin et al., 2006). Patrick and Khalid (1974) concluded that most wetlands act as sinks of different elements; for example, phosphorous, organic carbon that accumulates in the sediments and is removed by burial within the wetlands. Removal of trace elements by wetland vegetation can be promising in sequestering heavy metals from freshwater systems (Dunbabin and Bowmer, 1992). Several wetlands have been artificially constructed in treating drainage waters from urea mines (Overall and Parry, 2004), gold mines (Bishay and Kadlec, 2005), and various metal ore mines (Sobolweski, 1996). Constructed wetlands in Czech Republic are efficiently used in removing cadmium, zinc, and copper from various wastewater sources like municipal sewage, textile industry, pulp, and paper production units (Vyamazal et al., 2009). Apart from this, wetlands also serve as an excellent nursery for aquatic fauna as well. Freshwater Alaskan wetlands include bogs, fens, tundra, marshes, and meadows that support millions of shorebirds and waterfowls by providing suitable habitat for resting, feeding, and nesting (Fretwell, 1996). Wetland ecosystems are extremely crucial habitats for migratory birds. Migratory birds are equipped with sublime navigation and long distance flight abilities, where they utilize wetlands as stopovers during the long flight between wintering and breeding grounds (Downs and Horner, 2008). Loss and degradation of wetlands have been shown to threaten the migratory bird population by impacting their migratory performance as well as reproduction (Merken et al., 2015). Loss of 60% of all wetlands in Greece during 1950 and 1985 resulted in 40% decline in migratory pattern of Afro-Palearctic migrant species (Vickery et al., 2014). Wetlands also play an important role in mitigation of flood due to high water retention properties of sediments (Brody et al., 2007). Padmanabhan and Bengston (2001) found that restoration of wetlands in the Maple river watershed in North Dakota, USA gave satisfactory result in preventing flood events due to improvement in the water retention capability of sediment soil within the wetland ecosystems.

12.2.4 Ground Water Table and Aquifers

Groundwater is a vital source of renewable resource, and is replenished naturally by various methods including precipitation and infiltration from various freshwater regimes (ponds, lakes, and rivers) (Jinwal and Dixit 2008). An aquifer is a body of porous rock or sediment saturated with groundwater. Calligaris et al. (2018) demonstrated that aquifers provided up to 75% of drinking water in European union, where available lentic and lotic resources were scarce. Jinwal and Dixit (2008) focused on the importance of conserving the groundwater, as it is a crucial factor for rock formation that impacts geological structures and geomorphological and hydrological conditions as well. Aquifers also support unique biodiversity, for example, Stygobites, a group of strictly subterranean, aquatic animals are adapted to thrive in the aphotic, subsurface environment (for example, the aquifers) (Maurice and Bloomfield, 2012). In particular, the Stygobites live in spaces between porous aquifers and graze over bacterial and fungal biofilms. Groundwater plays a vital role in soil fertility. Vertisols are the cracking clayey soil found in alluvial low lands, where groundwater infiltrates into aquifers. Such soils are regarded as fertile due to transport of water that are enriched with solutes and gas complexes through the voids of the clayey lowland soil (Kurtzman et al., 2016).

12.3 IMPACTS ON FRESHWATER ECOSYSTEMS

Our earth has a great diversity and presents rich freshwater resources. However, most freshwater ecosystems are under peril by multiple stressors (Table 12.1). Table 12.1 provides a few examples on the ecological and environmental status of different ecosystems around the globe that suffer from various kinds of threats and challenges. Humans have been intensifying the impacts on freshwater ecosystems through construction of dams, overharvesting of freshwater fishes and other aquatic organisms, artificially modifying the course of the river, and closing down lakes and ponds through landfilling (Walmsley et al., 2001). Famiglietti et al. (2014) reported that globally, groundwater depletion has imposed a great threat to global water security, and is pumped at greater rates than it can be naturally replenished. In this part of the chapter, we will discuss the important stressors that are causing a huge negative impact on global freshwater resources.

TABLE 12.1
A Few Ecosystems Representing a Severe Degraded Status and the Potential Cause and Threat of Such Degradation

Name of Ecosystem	Type	Threat	Status	Citation
Ganga-Brahmaputra river basin, India	Lotic	Rise in temperature	Global rise in water temperature (0.05°C/ year) and elevated carbon dioxide content (280 ppm in 1700 to 380 ppm in 2005 resulted in acidification of riverine system contributing to extinction of 20–30% of native species.	Mitra et al., 2009
Western Ghats, India	Aquifers	Depletion of groundwater	Overexploitation of groundwater (230 km^3 /year) for urban water supply and agriculture depleted the aquifers endangering the existence of endemic subterranean fish population.	Raghavan et al., 2021
Rogen lake, Norway	Lentic	Acidification of lakes due to extreme events of acid rain.	Since 1930s, acid rain decreased the pH by up to 1.8 pH units and caused local extinction of Arctic char (*Salvelinus alpinus*), a species belonging to the Salmonidae family	Schofield, 1976
Greek wetlands, Greece	Wetlands	Dam construction, land clearing, overfishing, and draining of wetlands	Rapid urbanization led to depletion of natural resources, alteration of water quality in lakes (48% change in hydrological parameters) and rivers (75% change in hydrological parameters), and 63% of the wetlands deteriorated due to anthropogenic disturbances.	Zalidis, 1997
Hai river basin, North China	Lotic	Excessive irrigation for agricultural purpose and rapid urbanization	Long term water table decline (0.5 m / year) that lead to average recharge deficit of 40–90 mm/ year.	Foster and Chilton, 2003
Tunggak River, Malaysia	Lotic	Tremendous industrial discharge	46% Malaysian rivers have been contaminated by heavy metal elements lowering dissolved oxygen concentration (<1.5mg/L) and leading to anoxic conditions for aquatic biota.	Sajaul et al., 2013
Prairie Pothole Region, North America	Wetlands	Drying of wetlands	Habitat loss and degradation of waterfowl population due to drying of wetlands. Threatening 45% its population, and impacting nesting propensity and clutch size of the birds.	Niemuth et al., 2014 Steen et al., 2014

12.3.1 Urbanization

Our world is urbanizing at an unusual pace as the global proportion of urban population has dramatically increased from 28.3% in 1950 to 50% in 2010 (Wang et al., 2012). One of the characteristic features of urbanization is the establishment and growth of industries and factories (Kim, 2005). The environmental damage due to industrialization is quite evident within the freshwater ecosystems. For example, the Ulhas river estuary in Maharashtra, India is one of the highly polluted rivers owing to the heavy load of industrial pollutant and untreated sewage discharges (Menon et al., 2010). Mercury discharge has grown exponentially in the Ulhas river estuary from 15 tons a^{-1} in 2001 to 41 tons a^{-1} in 2020 (Chakroborty et al., 2013). Apart from water pollution due to industrial and sewage discharge, other pollution types, for example, air and soil, can also impact freshwater ecosystems. Heathwaite (2010) demonstrated that fragile freshwater systems are subjected to multiple stressors such as changing climate, increase in land use pattern, changing demand of water resources, and nutrient cycles that will intensify the impact over freshwater availability in the coming decades. All ecosystems are interconnected and toxicity released a particular system can further magnify along the food chain, thus resulting in a greater threat to human health. Soofi et al. (2013) reported that excessive release of iron caused iron clog in the gills of fishes, and in turn, affected human health due to consumption of infected fish species. Freshwater ecosystems witness discharge of contaminants from various sources, such as release of pathogens and fecal matters from untreated sewage (Banks et al., 1997; Baron and Poff, 2004), petroleum slicks (Mushtaq et al., 2020) and fly ash from nuclear and thermal power plants (Thorat Prerana and Charde, 2013; Walia and Mehra, 1998), pesticides and fertilizers from agricultural lands (Sankhla et al., 2018), and trace elements (mercury, lead, chromium) from industrial discharges (Owa 2013; Arora and Kakkar, 2017). These pollutants increase bacterial load, making the ecosystem anoxic, and increase the toxicity exposure for aquatic organisms. Plastic contamination is one of the prominent environmental issue in freshwater systems (Thompsen et al., 2015). Plastic pieces in small size scale (<5 mm) termed as microplastics have reached high densities (10,00,000 microplastics/m^3) within freshwater habitats and are interacting with aquatic species as well as with environment, thus causing detrimental impact on the ecosystem (Eerkes-Medrano et al., 2015). Chronic exposure of female zebrafish (*Danio rario*) to 25 ng/l of Di-(2-ethylhexyl)-phthalate (DEHP) have severely impaired ovulation and embryonic development of the species (Corradetti et al., 2013). DEHP, the building blocks of plastic, is released in freshwater bodies through breakdown of plastic waste (bags and bottles).

12.3.2 Climate Change

Climate change is another potential stressor of freshwater ecosystems that affects survival of aquatic taxa to a great extent (Hassan et al., 2020). Human activities have contributed to an increase in greenhouse gas emission, mainly carbon dioxide, which is responsible for warming of the atmosphere and the aquatic habitats (Huntington 2006). The released greenhouse gas threatens to cause a global rise in thermal regime, and the predicted increase may happen from 1.5 to 5.8°C during the next decade (Houghton et al., 2001). Being poikilothermic, fish are extremely sensitive to high temperature and are easily susceptible to bacterial and fungal infections. Besides fish, other poikilotherms, particularly amphibians and aquatic reptiles, are also highly vulnerable to temperature fluctuations within the freshwater habitats. Atkinson et al. (1995) reported that temperature affects morphology of aquatic animals. For example, elevated rearing temperature will lead to 90% decrease in body size for freshwater fishes. Conover (1984) showed the effect of temperature on sex determination of amphibians through his experiment on painted turtles (*Chrysemys picta*). He showed that 2°C rise in surface temperature would result in skewed sex ratio and 4°C rise will eliminate males from the turtle population (Conover 1984). Temperature has adverse effects on food chain and energy transmission from one trophic level to another. Besides aquatic organisms, climate change has an adverse effect on the physical characteristics of a freshwater habitat. Australia's Murray–Darling Basin was impacted by change in climatic conditions such as increased temperature, salinity, evaporation, and reduced rainfall (Pittock and Finlayson 2011). Natural flow conditions of the Murray Darling river basin were estimated to be 12,233 GL/year. However, since 2002, the dry period began due to change in climatic conditions and the water volume was reduced to 4733 GL/year. Such reduction in water flow imposed a severe challenge to the aquatic community inhabiting the river basin due to deterioration of water quality (Kingsford et al., 2011), including increased salinity (115μS/cm to 800 μS/cm) (Goss 2003), change in temperature (Pittock et al., 2008), and increased acidification (Hall et al., 2006). Bowling and Baker (1996) also reported appearance of a toxic cyanobacterial bloom due to deterioration of water quality of the Darling River under altered climatic conditions.

12.3.3 Introduction of Exotic Species

Exotic species can be termed as any species that is introduced outside its natural distribution range (Mills et al., 1996). Most exotic species were imported outside their native range as they act as biocontrol agents for several aquaculture farms, and are often used as ornamental pets for recreation (Simberloff et al., 1997). For example, silver carp (*Hypophthalmichthys molitrix*) native to Eastern Asia were imported to the United States during the early 1970s. The carp were introduced in the aquaculture ponds to improve water quality as they feed over large cladocerans (*Daphnia*) (Wu et al., 1997). Similarly, sailfin catfish (*Pterigoplichthyes pardalis*) native to the Amazon Basin were introduced to aquaculture ponds in India (Murugan et al., 2015), Bangladesh, Indonesia (Muchlisin, 2012), United States (Orfinger and Goodding 2018), and several other Eurasian countries (Hossain et al., 2018) for its voracious grazing habit over the fine blanket of algae and thus, preventing formation of algal bloom (Hoover 2004). However, due to accidental release from aquaculture ponds, several exotic species were introduced into the wild waters (Mills et al., 1993). In the absence of a natural predator, most exotics become voracious feeders, which resulted in huge population growth (Savidge, 1987). Exotic species become invasive when they impact the native species and the native ecosystems by competing for food and space (Simberloff et al., 2013). For example, sea lamprey (*Petromyzon marinus*), a parasitic fish native to the Atlantic Ocean, entered into the Great Lakes of North America through man-made shipping canals (Siefkes et al., 2013). Lamprey severely impacted the native population of lake trout (*Salvelinus namaycush*). The annual harvest rate of lake trout was around 6.8 million kg before the invasion of sea lamprey, however, post-invasion the catch rate tremendously reduced to 136.07 kg due to mass mortality of trout because of lamprey infection (Lawrie, 1970). In the mission to control the lamprey spread and invasion, the United States invested $19.4 million (Dettmers et al., 2012). Deliberate release of exotic species in freshwater systems has tremendous effect on indigenous species. The African catfish (*Clarias gariepinus*) is another prominent example that invaded several freshwater systems in West Bengal and Assam states of India, and imposed severe threat to native, economically important carp species, rohu (*Labeo rohita*) and catla (*Labeo Catla*). The catfish has a high level of niche overlap (92%) with the native carp species, and thus triggered a strong competition for similar resources (Baruah et al., 1999, Pannikar et al., 2021). Besides fish, aquatic plants (e.g., duckweeds commonly known as water hyacinth (*Eichhornia crassipes*) native to Amazon basins have turned into a highly invasive species. Globally, water hyacinth was introduced to various freshwater systems for bioremediation of persistent pollutants (Saha et al., 2017). However, the plant soon turned out to be an invasive species, and impacted the growth of native plants. Further, water hyacinth led to an alteration of physical properties of the freshwater ecosystems by lowering primary productivity, and by reducing dissolved oxygen levels (Chapungu et al., 2018). De Groote et al. (2003) reported that water hyacinth (*Eichhornia crassipes*) of South American origin impacted the economy of Southern Benin, a country in South Africa, by impacting the livelihood of 200,000 fishermen. Annual income of the fishermen community reduced by US$ 84 million because of the decline (by 43%) in native fish population due to widespread invasion of the water hyacinth (De Groote et al., 2003).

12.4 ROADMAPS TOWARDS SUSTAINABLE DEVELOPMENT

In order to sustain the freshwater resources, there is an urgent need to conduct long-term monitoring of the status of the freshwater ecosystems. Freshwater species constitute 9.5% of known animal species on earth. A declining trend in the global population of freshwater species (annually 3.9% since 1970), which is significantly higher than the terrestrial species (annually 1.1%), has become a matter of concern (Dudgeon, 2019). Globally, freshwater species populations are impacted by multiple stressors, for example, habitat modification, fragmentation, destruction, invasive species, overfishing, pollution, diseases, and climate change (Oy et al., 2018). Thus, proper management of freshwater systems needs three key steps, as outlined in the following three sections: monitoring, restoration, and protection and sustainable development of the freshwater habitats to safeguard the valuable resources of the biosphere. A schematic diagram depicting the path towards building a sustainable freshwater system is represented in Figure 12.1.

FIGURE 12.1 Path towards environmental sustainability.

12.4.1 Monitoring

Conservation of freshwater ecosystems and associated ecosystem services has become a great concern for better sustenance of aquatic biota and human beings, as well. The first and most crucial stage in sustainable ecosystem management is evaluation of the current status of a target ecosystem. There is a need for a long-term, multipronged approach to monitor the health of an ecosystem on a temporal and a spatial scale. Long-term monitoring of ecosystems helps us to interpret ecosystem processes, driving mechanisms, adaptations, interactions with environmental changes, and global anthropogenic pressure (Beaugrand et al., 2019, Luo et al., 2015; Seehausen 2008). Here, we describe several monitoring approaches ranging from traditional to advanced molecular methods that are currently in practice to monitor the health of a freshwater ecosystem.

12.4.1.1 Monitoring Water Quality

Water quality is a strong indicator of health of an aquatic ecosystem. A healthy water body is essential to sustain a high biodiversity of species (Dudgeon, 2019). Parameters to evaluate water quality can be classified as physical, chemical, and biological. One of the biological attributes of water parameters includes measurement of phytoplankton, which is responsible for primary production and maintenance of energy flow within trophic levels through food chain. The physical category of water parameters includes temperature. A small rise in temperature can impact the survivality of aquatic species like fishes since they are poikilothermic (Persson et al., 1986). Similarly, chemical parameters are crucial for maintaining the balance of biogeochemical cycles that play a vital role in availability of nutrients to the biotic community (Prasad et al., 2021). All three categories of water parameters are measured concurrently to determine and assess the quality of water in the freshwater systems. Traditionally, most of these measurements either relies on tedious, qualitative methods or longer analytical processes in the laboratory. For example, turbidity is an important parameter giving an estimation of light scattering by particles (for example, clay, silt, organic matter, phytoplankton, and other microorganisms) that are present in water (Srivastva et al., 2003). Higher turbidity denotes excessive amounts of suspended solid in a water body that can impact the primary productivity of the pond due to low penetration of sunlight (Suber et al., 1953). In earlier days, turbidity measurements were carried out by secchi disc, with alternating black and white quadrant with a long rope bearing measuring scale (Pilgrim et al., 1987). However, modern estimation of turbidity involves Nepthalo Turbidity Meter (NTU), which uses sensor-based probes (Verma et al., 2012) for rapid on-field estimations. Turbidity is only a representative example. Likewise, most methods to measure water parameters now mostly rely on novel chemical sensors and low-cost calorimetric probes for accurate and precise on-site estimations. Nonetheless, measuring a cohort of different water parameters to assess the health of an ecosystem is still in practice.

12.4.1.2 Monitoring Biodiversity

The presence of a variety of plants and animals in an ecosystem indicates good health and viability of the system. The protocols for measuring the diverse groups of flora and fauna ranges from analytical methods, such as spectrophotometric analysis of chlorophyll concentration (Ediger et al., 2006) to analyze phytoplankton abundance and biomass, to laborious, physical methods, such as the mark-recapture method using gill net sampling for stock assessment of freshwater fish species. In the case of several freshwater species, particularly the burrowing organisms and the benthic invertebrates, traditional methods are not feasible to measure their richness and abundance. Thus, advanced, computational methods have been developed to reliably estimate the population of several freshwater organisms. One such computational technique is deep learning, which has witnessed a rapid growth of attention in many domains, including ecology due to improved computational power and for efficient management of large data sets (Hinton et al., 2015). For example, Schneider and Zhuang (2020) used a deep learning algorithm and SONAR (Sound Navigation and Ranging) images of the endangered population of Amazon river dolphin (*Inia geoffrensis*) to identify the species. Such computational tools greatly contribute towards the conservation of river dolphin, which suffers from high poaching rate.

12.4.1.3 Monitoring Toxicity

Bioaccumulation of contaminants in fish tissue needs investigations as it has high implications for human and ecological health (MacCarty and Mackay, 1992). Most aquatic organisms acquire contaminants and concentrate in their tissues by profuse water uptake or by dietary intake (Branson et al., 1985). Stahl et al., 2009 depicted the detection of dioxins and furans in 81% of predatory fish like largemouth bass (*Micropterus salmoides*), walleye (*Sander vitreus*), northern pike (*Esox lucius*), and 99% of bottom dwellers such as common carp (*Cyprinus carpio*) and channel catfish (*Ictalurus punctatus*). Toxicity estimation involves various methods such as quantification of toxin in fish tissue (Papadimitriou et al., 2010), and biochemical analysis of toxic agents released in open waters (Stahl et al., 2009). For example, Nyamah and Torgbor (1986) showed the use of colorimetric method to estimate arsenic contamination in water samples collected from several different rivers and lakes in Ghana,

Africa. In another study using tissues of Prussian carp (*Carassius gibello*) from several lakes in Greece, accumulation of microcystin (a toxicant produced by cyanobacteria) was measured by a spectrophotometric assay method and the study reported a higher health risk of consuming the fish species due to increased concentration of the toxicant present in their bodies (Papadimitrou et al., 2009).

12.4.1.4 Monitoring Using Molecular Biomarkers

Molecular biomarkers are emerging tools for monitoring biodiversity and detection of exotic species in freshwater ecosystems. For example, metagenomics, a study of the total genomic content of a microbial community, has gained significant attention for understanding the dynamics of microbial community in environmental systems. Targeted metagenomics is usually used to investigate phylogenetic diversity and relative abundance of a particular gene (Oh et al., 2011). With restricted access to protected organisms and difficulty in locating/capturing cryptic species, modern research has focused on non-invasive methods of sampling. Environmental DNA (eDNA) is one such molecular tool extensively used in monitoring, management, and conservation practices of ecological systems. Aquatic animals such as fish often release their genetic material through scales, mucus, feces, and gametes in the habitat, which can be used as an excellent clue for their detection and surveillance (Thomsen and Willerslev, 2015). Spear et al. (2021) demonstrated the use of eDNA tool in estimating abundance of the walleye (*Sander vitreus*), an economically important sportfish of Northern Wisconsin, USA. Similarly, Chucholl et al. (2021) monitored invasive crayfish species (*Pacifastacus leniusculus, Faxonius limosus, Faxonius immunes*) population with the help of eDNA detection in Central Europe.

12.4.2 Restoration

Besides monitoring and evaluating the status of an ecosystem, there is an urgent need to develop corrective measures for aquatic regimes in order to sustain freshwater resources. Thus, restoration of freshwater systems has become a major focus in the field of ecosystem conservation. Restoration can be defined as a holistic process aiming at re-establishment of pre-disturbed aquatic functions and related physical, chemical, and biological characteristics (Kentula et al., 1992). Wendel (1991) suggested that restoration requires re-construction of physical conditions that were originally present in a system, adjustment of chemical constituents of soil and water, and biological intervention in the form of re-introduction of native fauna and flora. However, the process of restoration is widely different between lentic and lotic ecosystems, and here we provide a few representative examples for the same.

12.4.2.1 Restoration of Lotic Systems

Large river ecology has evolved through time from recognizing rivers as large streams to accepting rivers as complex ecosystems (McCain, 2013). Rivers suffer a high rate of degradation through various anthropogenic activities; for example, habitat alteration, pollution due to industrial discharge, and altered flow and flooding regime due to dam constructions. The ecological integrity of river systems is heavily dependent upon the degree of connectivity between main channel and floodplain, and on disturbances (annual flood, seasonal droughts, channel modification) impacting the resilience of the ecosystem (Lubinski, 1998). The river restoration program works on the basis of a few commonly stated goals, water quality enhancement, management of riparian zones, improvement of stream habitat, facilitating fish passages, and bank stabilization. Each of the stated goals can be achieved using a variety of different methods (for example, the quality of water in river systems can be improved by using chemical methods). For example, treatment of rivers with polyaluminium chloride promotes removal of suspended solids and algae by flocculation, precipitation, and oxidation (Geng et al., 2021). Similarly, aquatic animals like clam, snail, silver carp and grass carp, and other filter feeding fish are also known to improve the quality of river water by increasing the utilization of nutrient uptake (Li et al., 2018; Ateia et al., 2016). Beechie et al. (2010) proposed three different models to project restoration or recovery of the river ecosystems. The first one is the rubberband model that signifies limiting the source of disturbance and can help the ecosystem to recover fast. The second model is hysteresis or broken leg model, which supports the idea of allowing an ecosystem to recover slowly on its own. The third one, the Humpty-Dumpty model, denotes that restoration may end up following a different trajectory leading to the formation of a stable ecosystem, which may be widely different from the original system. For example, Lake et al. (2007) proposed restoration of an ecosystem using the Humpty-Dumpty model pathway, which led to the formation of a new community with an entirely different species composition than the original system, but the species richness was equivalent to the pre-disturbed state.

12.4.2.2 Restoration of Lentic Systems

Lentic habitats, for example, lakes and ponds, being closed systems require different recovery methods compared to rivers and streams. For example, natural chemical properties in a contaminated stream can be restored by eliminating the source of contamination and further rely on self-cleansing properties of the flowing waters. However, such recovery method is not feasible for lakes, particularly smaller ones, which require within-lake manipulation due to the stagnancy of the water body. Biomanipulation is an age-old technique generally performed to improvise ecological state of eutrophic lakes by reducing abundance of planktivorous fish or by addition of piscivorous fish or through manual removal of undesired fish species (Drenner and Hambright, 1999). The biomanipulation of Lake Væng, a shallow lake of Denmark, Europe was shown to give positive response by shifting from turbid, phytoplankton dominated state to clear, macrophyte dominated state by removing two of the fish species, common roach (*Rutilus rutilus*) and bream (*Abramis brama*) (Sondergaard et al., 2017). Restoration of deep, stratified lakes is conducted by introducing oxygen into hypolimnetic waters, which helps in increasing redox potential and further reducing phosphorous release. Besides chemicals, manipulation of biological communities is also carried out to restore a lentic system. charophytes are a group of green algae that are profusely used in lake restoration programs (Van den Berg et al., 1999). *Chara* and *Nitella* are aquatic algae belonging to the group of Charophytes that have gained attention in the context of restoration of lentic community. Both the algae form meadows of vegetation that reduce resuspension of sediments (Van den Berg et al., 1999) Moreover, the vegetation meadows are considered as a great habitat for littoral invertebrates (Rosine, 1955, Quade, 1969; Allanson, 1973; Hargeby et al., 1994), fish (Fassett, 1953; Leslie et al., 1994), and are a major food source for herbivorous water birds (Hargeby et al., 1994; Van den Berg et al., 1999).

12.4.3 Protection and Sustainable Development

Sustainable development is defined as development that meets the needs of present without compromising future generations to meet their own needs. In the context of freshwater, one of the resources that is heavily exploited and is also linked with the economy of a country is the freshwater fish species. To control overharvesting and provide sustainable management of fish stocks, the United States established a series of small reserves, where fishing is prohibited during the monsoon season, which is the breeding time for most fish species. Such controlled fishing was implemented to protect the viable breeding population of overexploited fishes, and the policy helped in contributing to an annual increase in target species' population (Roberts and Polunin, 1991). Individual Transferable Quota (ITQ) is a well-established method to reduce overfishing that provides fisherman a permissible limit to harvest fish, referred to as the Total Allowable Catch (TAC). TAC limits are determined after evaluating the health of a particular ecosystem and assessing the stock population available for a target species (Costello et al., 2008). Science-based policymaking offered great success to the Fraser River salmon management program. The restriction to build the Moran Dam on the mainstream of Fraser River, Canada due to migratory passage of sockeye salmon (*Oncorhynchus nerka*) resulted in salmon population increase (3.3 million to 7.8 million fish on average per year (Patterson et al., 2016). The recent census of global freshwater species suggests 22.5% (6000 out of 26,400 species) are at severe risk due to environmental pollution and require urgent attention for conservation and management (Gozlan and Britton, 2015). The issues of sustainability in terrestrial and marine ecosystems have been resolved by creation of protected areas, such as national parks, biosphere reserves, and sanctuaries. However, similar protection is not provided to freshwater species, and thus there is an urgent need to design freshwater protected areas in order to safeguard the declining species, maximize long-term sustainability of biodiversity, and support recovery of vulnerable and endangered species (Gozlan and Britton, 2015). Physical alteration of river course/structure is another major problem that contributes towards threatening the health of a river ecosystem. However, recently, several projects have moved towards adopting a sustainable approach that will maximize benefits for people and nature as well. For example, The Nature Conservancy (TNC) and US Army Corps of Engineers collaboratively launched the Sustainable River Project (SRP) in 2002 in order to manage river infrastructure in a better way (Warner et al., 2014). The focus of SRP included determination of unique flow requirements of rivers and creating operative plan for dams that achieve environmental flows (timing, quantity, and quality of water flow) to ensure sustenance of critical ecological function and habitat of indigenous species. The project also proved to be beneficial in mitigating flood impacts in the United States and played a crucial role in restoring the ecological factors of rivers.

12.5 FUTURE GOALS AND CONCLUSION

In addition, scientific methods applied to restore the ecological integrity of the rivers, science education, awareness, and outreach have contributed significantly in reducing the rate of global depletion of freshwater resources. Environmental education was promoted to reveal the connection between the changing concerns of the environment and its associated effects (Tilbury, 1995). The origin of environmental education was in Britain during the 1970s. Since then many government and international organizations have developed a vocabulary for environmental education with the concept of 'sustainability' in mind. The World Conservation strategy is one such organization that focuses on essential ecological processes, life support systems, genetic diversity as well as sustainable use of natural resources. Apart from this, they also highlight the role of education in bringing about changes in sustainable lifestyle. The World Conservation strategy later developed into various subunits, namely IUCN (International Union of Conservation of Nature), UNEP (United Nations Environment Programme), and WWF (Worldwide Fund for nature). An environmental education involves changes in attitudes and practices of many people that contribute towards sustainable living. For example, keeping fish as pets in aquariums is a common practice, and sustainable education may help pet owners to classify them as exotic and native species and thus, contribute towards environmental awareness.

It is high time to prioritize freshwater ecosystems as global biodiversity hotspots in order to safeguard the sentinel aquatic life of the biosphere. While climate change is becoming a gradual threat over the fate of freshwater ecosystems, we should not underestimate the immediate impacts of dam construction, extreme land use change, and pollution, which can intensify the rate of fish extinction far higher than the present rate. There is an urgent need of action to conserve freshwater biodiversity in order to enhance the ecosystem resilience that contributes to sustenance of freshwater biodiversity. In this chapter, we addressed the potential stressors of freshwater ecosystems and discussed various strategies to monitor their health. In order to mitigate the fragile freshwater ecosystem and constrain its stability, a fully integrated approach related to global initiatives is needed. For example, sustainable development goals that account for social and economic concerns and necessary changes in the water governance need to be undertaken.

AUTHOR CONTRIBUTIONS

Biological and Life Sciences, School of Arts and Sciences, Ahmedabad University, Gujarat 380009, India.
SM and RG conceptualized the articles, and SM wrote the manuscript. RG modified and edited the content of the manuscript.
SM acknowledged Department of Science and Technology (DST) – Science and Engineering Research Board (SERB), Government of India, for supporting his fellowship during this period.

REFERENCES

Ateia, M., Yoshimura, C. and Nasr, M., 2016. In-situ biological water treatment technologies for environmental remediation: a review. *Journal of Bioremediation and Biodegradation*, 7(348), p. 2.

Balian, E.V., Segers, H., Martens, K. and Lévéque, C., 2007. The freshwater animal diversity assessment: an overview of the results. *Freshwater Animal Diversity Assessment*, 595, pp. 627–637.

Beechie, T.J., Sear, D.A., Olden, J.D., Pess, G.R., Buffington, J.M., Moir, H., Roni, P. and Pollock, M.M., 2010. Process-based principles for restoring river ecosystems. *BioScience*, 60(3), pp. 209–222.

Blob, R.W., Bridges, W.C., Ptacek, M.B., Maie, T., Cediel, R.A., Bertolas, M.M., Julius, M.L. and Schoenfuss, H.L., 2008. Morphological selection in an extreme flow environment: body shape and waterfall-climbing success in the Hawaiian stream fish *Sicyopterus stimpsoni*. *Integrative and comparative Biology*, 48(6), pp. 734–749.

Brody, S.D., Highfield, W.E., Ryu, H.C. and Spanel-Weber, L., 2007. Examining the relationship between wetland alteration and watershed flooding in Texas and Florida. *Natural Hazards*, 40(2), pp. 413–428.

Calligaris, C., Mezga, K., Slejko, F.F., Urbanc, J. and Zini, L., 2018. Groundwater characterization by means of conservative ($\delta 18O$ and $\delta 2H$) and non-conservative (87Sr/86Sr) isotopic values: The classical karst region aquifer case (Italy–Slovenia). *Geosciences*, 8(9), p. 321.

Chucholl, F., Fiolka, F., Segelbacher, G. and Epp, L.S., 2021. eDNA detection of native and invasive crayfish species allows for year-round monitoring and large-scale screening of lotic systems. *Frontiers in Environmental Science*, 9, p. 23.

Clayton, P.D. and Pearson, R.G., 2016. Harsh habitats? Waterfalls and their faunal dynamics in tropical Australia. *Hydrobiologia*, 775(1), pp. 123–137.

Corradetti, B., Stronati, A., Tosti, L., Manicardi, G., Carnevali, O. and Bizzaro, D., 2013. Bis-(2-ethylexhyl) phthalate impairs spermatogenesis in zebrafish (*Danio rerio*). *Reproductive Biology*, 13(3), pp. 195–202.

Costello, C., Gaines, S.D. and Lynham, J., 2008. Can catch shares prevent fisheries collapse? *Science*, 321(5896), pp. 1678–1681.

Covich, A.P., Austen, M.C., Bärlocher, F., Chauvet, E., Cardinale, B.J., Biles, C.L., Inchausti, P., Dangles, O., Solan, M., Gessner, M.O. and Statzner, B., 2004. The role of biodiversity in the functioning of freshwater and marine benthic ecosystems. *BioScience*, 54(8), pp. 767–775.

Darwall, W., Smith, K., Lowe, T. and Vié, J.C., 2005. *The Status and Distribution of Freshwater Biodiversity in Eastern Africa*. Occasional Paper of the IUCN Species Survival Commission No. 31.

Drenner, R.W. and Hambright, K.D., 1999. Biomanipulation of fish assemblages as a lake restoration technique. *Archiv für Hydrobiologie*, 146(2), pp. 129–165.

Dudgeon, D., Arthington, A.H., Gessner, M.O., Kawabata, Z.I., Knowler, D.J., Lévêque, C., Naiman, R.J., Prieur-Richard, A.H., Soto, D., Stiassny, M.L. and Sullivan, C.A., 2006. Freshwater biodiversity: importance, threats, status and conservation challenges. *Biological Reviews*, 81(2), pp. 163–182.

Dunbabin, J.S. and Bowmer, K.H., 1992. Potential use of constructed wetlands for treatment of industrial wastewaters containing metals. *Science of the Total Environment*, 111(2–3), pp. 151–168.

Foster, S.S.D. and Chilton, P.J., 2003. Groundwater: the processes and global significance of aquifer degradation. *Philosophical Transactions of the Royal Society of London: Series B: Biological Sciences*, 358(1440), pp. 1957–1972.

Geng, M., Wang, K., Yang, N., Li, F., Zou, Y., Chen, X., Deng, Z. and Xie, Y., 2021. Spatiotemporal water quality variations and their relationship with hydrological conditions in Dongting Lake after the operation of the Three Gorges Dam, China. *Journal of Cleaner Production, 283*, p. 124644.

Goyal, V.C., Singh, O., Singh, R., Chhoden, K., Kumar, J., Yadav, S., Singh, N., Shrivastava, N.G. and Carvalho, L., 2021. Ecological health and water quality of village ponds in the subtropics limiting their use for water supply and groundwater recharge. *Journal of Environmental Management, 277*, p. 111450.

Gozlan, R.E. and Britton, J.R., 2015. Sustainable freshwater fisheries: the search for workable solutions. *Freshwater Fisheries Ecology, 48*(3), pp. 616–621.

Holgerson, M.A., 2015. Drivers of carbon dioxide and methane supersaturation in small, temporary ponds. *Biogeochemistry, 124*(1), pp. 305–318.

Kentula, M.E., Brooks, R.P., Gwin, S.E., Holland, C.C. and Sherman, A.D., 1992. *Approach to Improving Decision Making in Wetland Restoration and Creation* (No. PB-92-217306/XAB; EPA-600/R-92/150). Environmental Protection Agency, Corvallis, OR (United States). Environmental Research Lab.

Kurtzman, D., Baram, S. and Dahan, O., 2016. Soil–aquifer phenomena affecting groundwater under vertisols: a review. *Hydrology and Earth System Sciences, 20*(1), pp. 1–12.

Lake, P.S., Bond, N. and Reich, P., 2007. Linking ecological theory with stream restoration. *Freshwater Biology, 52*(4), pp. 597–615.

Li, D., Pi, J., Zhang, T., Tan, X. and Fraser, D.J., 2018. Evaluating a 5-year metal contamination remediation and the biomonitoring potential of a freshwater gastropod along the Xiangjiang River, China. *Environmental Science and Pollution Research, 25*(21), pp. 21127–21137.

Lubinski, K., 1998. Floodplain river ecology and the concept of river ecological health. *Ecological Status and Trends of the Upper Mississippi River System*. In, K.Lubinski and C.Theiling (Eds.) US Geological Survey, Wisconsin, pp.1–12.

Lundberg, J.G., Kottelat, M., Smith, G.R., Stiassny, M.L. and Gill, A.C., 2000. So many fishes, so little time: an overview of recent ichthyological discovery in continental waters. *Annals of the Missouri Botanical Garden*, 87(1), pp. 26–62.

McCain, K.N., 2013. Moving large river ecology from past theories to future actions: a review. *Reviews in Fisheries Science, 21*(1), pp. 39–48.

McCarty, L.S., Dixon, D.G., MacKay, D., Smith, A.D. and Ozburn, G.W., 1992. Residue-based interpretation of toxicity and bioconcentration QSARs from aquatic bioassays: neutral narcotic organics. *Environmental Toxicology and Chemistry: An International Journal, 11*(7), pp. 917–930.

Menon, J.S. and Mahajan, S.V., 2010. Site-wise mercury levels in Ulhas River Estuary and Thane Creek near Mumbai, India and its relation to water parameters. *Our Nature, 8*(1), pp. 170–179.

Mills, E.L., Leach, J.H., Carlton, J.T. and Secor, C.L., 1993. Exotic species in the Great Lakes: a history of biotic crises and anthropogenic introductions. *Journal of Great Lakes Research, 19*(1), pp. 1–54.

Mitra, A., Gangopadhyay, A., Dube, A., Schmidt, A.C. and Banerjee, K., 2009. Observed changes in water mass properties in the Indian Sundarbans (northwestern Bay of Bengal) during 1980–2007. *Current Science*, 97(10), pp. 1445–1452.

Myers, J.M., Busack, C.A., Rawding, D., Marshall, A.R., Teel, D.J., Van Doornik, D.M. and Maher, M.T., 2006. *Historical Population Structure of Pacific Salmonids in the Willamette River and the Lower Columbia River Basins*. U.S. department of commerce, National Oceanic and Atmospheric Administration, National Marine Fisheries Service.

Niemuth, N.D., Fleming, K.K. and Reynolds, R.E., 2014. Waterfowl conservation in the US Prairie Pothole Region: confronting the complexities of climate change. *PloS One*, *9*(6), p. e100034.

Nyamah, D. and Torgbor, J.O., 1986. Colorimetric method for the determination of arsenic in potable water. *Water Research*, *20*(11), pp. 1341–1344.

Oh, S., Caro-Quintero, A., Tsementzi, D., DeLeon-Rodriguez, N., Luo, C., Poretsky, R. and Konstantinidis, K.T., 2011. Metagenomic insights into the evolution, function, and complexity of the planktonic microbial community of Lake Lanier, a temperate freshwater ecosystem. *Applied and Environmental Microbiology*, *77*(17), pp. 6000–6011.

Palmer, I.M., 1997. Biodiversity and ecosystem processes. *Ambio*, *26*(8), pp.571-577.

Papadimitriou, T., Kagalou, I., Bacopoulos, V. and Leonardos, I.D., 2010. Accumulation of microcystins in water and fish tissues: an estimation of risks associated with microcystins in most of the Greek Lakes. *Environmental Toxicology*, *25*(4), pp. 418–427.

Pauwels, O.S., Wallach, V. and David, P., 2007. Global diversity of snakes (Serpentes; Reptilia) in freshwater. *Freshwater Animal Diversity Assessment*, 595, pp. 599–605.

Persson, L., 1986. Temperature-induced shift in foraging ability in two fish species, roach (Rutilus rutilus) and perch (Perca fluviatilis): implications for coexistence between poikilotherms. *The Journal of Animal Ecology*, *55*(3), pp. 829–839.

Pilgrim, D.A., 1987. Measurement and estimation of the extinction coefficient in turbid estuarine waters. *Continental Shelf Research*, *7*(11–12), pp. 1425–1428.

Pittock, J. and Finlayson, C.M., 2011. Australia's Murray–Darling Basin: freshwater ecosystem conservation options in an era of climate change. *Marine and Freshwater Research*, *62*(3), pp. 232–243.

Raghavan, R., Britz, R. and Dahanukar, N., 2021. Poor groundwater governance threatens ancient subterranean fishes. *Trends in Ecology & Evolution*, 36(10), pp.875-878.

Roberts, C.M. and Polunin, N.V., 1991. Are marine reserves effective in management of reef fisheries? *Reviews in Fish biology and Fisheries*, *1*(1), pp. 65–91.

Schneider, S. and Zhuang, A., 2020. *Counting Fish and Dolphins in Sonar Images Using Deep Learning*. arXiv preprint arXiv:2007.12808.

Schoenfuss, H.L. and Blob, R.W., 2003. Kinematics of waterfall climbing in Hawaiian freshwater fishes (Gobiidae): vertical propulsion at the aquatic–terrestrial interface. *Journal of Zoology*, *261*(2), pp. 191–205.

Schofield, C.L., 1976. Acid precipitation: effects on fish. *Ambio*, 5(5–6) pp. 228–230.

Siefkes, M.J., Steeves, T.B., Sullivan, W.P., Twohey, M.B. and Li, W., 2013. Sea lamprey control: past, present, and future. In William W. Taylor (Ed.), *Great Lakes Fisheries Policy and Management*, In Michigan State University Press, East Lansing. pp. 651–704.

Søndergaard, M., Lauridsen, T.L., Johansson, L.S. and Jeppesen, E., 2017. Repeated fish removal to restore lakes: case study of Lake Væng, Denmark—two biomanipulations during 30 years of monitoring. *Water*, *9*(1), p. 43.

Spear, M.J., Embke, H.S., Krysan, P.J. and Vander Zanden, M.J., 2021. Application of eDNA as a tool for assessing fish population abundance. *Environmental DNA*, *3*(1), pp. 83–91.

Stahl, L.L., Snyder, B.D., Olsen, A.R. and Pitt, J.L., 2009. Contaminants in fish tissue from US lakes and reservoirs: a national probabilistic study. *Environmental Monitoring and Assessment*, *150*(1), pp. 3–19.

Steen, V., Skagen, S.K. and Noon, B.R., 2014. Vulnerability of breeding waterbirds to climate change in the Prairie Pothole Region, USA. *PloS One*, *9*(6), p. e96747.

Strayer, D.L. and Dudgeon, D., 2010. Freshwater biodiversity conservation: recent progress and future challenges. *Journal of the North American Benthological Society*, *29*(1), pp. 344–358.

Taylor, M.P. and Stokes, R., 2005. When is a river not a river? Consideration of the legal definition of a river for geomorphologists practising in New South Wales, Australia. *Australian Geographer*, *36*(2), pp. 183–200.

Thomsen, P.F. and Willerslev, E., 2015. Environmental DNA–An emerging tool in conservation for monitoring past and present biodiversity. *Biological Conservation*, *183*, pp. 4–18.

Tilbury, D., 1995. Environmental education for sustainability: defining the new focus of environmental education in the 1990s. *Environmental Education Research*, *1*(2), pp. 195–212.

Torrente-Vilara, G., Zuanon, J., Leprieur, F., Oberdorff, T. and Tedesco, P.A., 2011. Effects of natural rapids and waterfalls on fish assemblage structure in the Madeira River (Amazon Basin). *Ecology of Freshwater Fish*, *20*(4), pp. 588–597.

Van den Berg, M.S., Scheffer, M., Van Nes, E. and Coops, H., 1999. Dynamics and stability of Chara sp. and *Potamogeton pectinatus* in a shallow lake changing in eutrophication level. *Hydrobiologia*, *408*, pp. 335–342.

Warner, A.T., Bach, L.B. and Hickey, J.T., 2014. Restoring environmental flows through adaptive reservoir management: planning, science, and implementation through the Sustainable Rivers Project. *Hydrological Sciences Journal*, *59*(3–4), pp. 770–785.

Zalidis, G.C., Mantzavelas, A.L. and Gourvelou, E., 1997. Environmental impacts on Greek wetlands. *Wetlands*, *17*(3), pp. 339–345.

Zedler, J.B., Williams, G.D. and Desmond, J.S., 1997. Wetland mitigation: can fishes distinguish between natural and constructed wetlands? *Fisheries*, *22*(3), pp. 26–28.

Index

Note: Figures are indicated by *italics*. Tables are indicated by **bold**.

A

abiotic 63
absorbance 217–19
absorption 64, 72
accumulation evidences 188, 193, 194, 198, 202
acetylcholinesterase 45
ACToR 217
adsorbed 60, 65
adsorption 63–4
adverse effects of pesticides 168
aggregation 60, 64–5
air 123
Aldrins dieldrins 192, 194
alkylphenols 87, 106
anthropogenic activities 269, 278
aquatic environment 62–3, 68
arsenic (As) 37–8, 42, 140, 142, 248–9
aryl hydrocarbon receptor (AhR) 168
azo compounds 192, 193

B

benzoyl compounds 180
bidirectional relationship to heavy metals 234–5
bioaccumulation 6–7, 62, 64, 67–9, 142, 145–8, 150, 153, **238**
bioaccumulation factor (BAF) 153, 154
bioavailability 60, *63*, 64–5, 149, 152, 179, 181–3, 185–7, 189, 191, 193, 195–7, 199–201, 203–4
bioconcentration 179, 181–3, 185–7, 189, 191, 193–5, 197, 199, 201–4
bioconcentration factor (BCF) 182
biodegradability 6
biodiversity 2–3
biomagnification 3, 60, 62–4, 67–8, 154, 179, 181–3, 185–7, 189, 191, 193, 195–7, 199–201, 204
biotic 63–4
biotransformation 236, **238**
bisphenol A (BPA) 87
Brownian diffusion 22
bulk material 13, 16

C

cadmium (Cd) 37–8, 41–2, 140, 245–7
carbamate insecticides 45
carbon based nanomaterials (nps) 46, 200, 203
carcinogenic and genotoxic effects of pesticides 45
carcinoma cells 22, 23
cerium 154
ChEMBL 217
chemicals of emerging concern (CEC) 86–8, 92, 94, 106, 108, 109
chromium (Cr) 140, 141, 243–5
clathrin-/caveolin-mediated endocytosis 15
climate change 273–4, 281
confocal Raman micro spectroscopy (CRM) 220–1
conservations 2–3
constitutive androgen receptor (CAR) 168
contaminants 7, 36–7, 44
coral reefs 148, 149
cytochrome P450 (CYPs) 167, 168, 171, 173
cytokines 48, 50
cytotoxicity 36, 41, 48, 50

D

dendrimers 46
detection 128
dichlorodiphenyltrichloroethane (DDT) 44, 87, 96, 187, 191–5
dispersive solid phase (micro) extraction (DSPE/DSPME) 88–9
DNA damage 20, 40–2, 44, 46, 50
dosage 4
dose-response assessment 4, 6
drug-metabolizing enzymes (DMEs) 167

E

ecological risk 2–4
ecological risk assessment 2–4
ecosystem 2–3, 5–6
ecosystem services 275
ecotoxicity 215, 217, 221, 224–5
effects on gut microbiota 243, 244, 246–9, **251–6**
effects on gut microbiota and mechanism of actions **257–9**
electrochemical 130
electrostatic interaction 60, 64
emerging contaminants 86, 87, 89–91, 96, 98, 100, 102
emerging pollutants 85, 86, 88–90
endocrine disrupting compounds (EDC) 86, 87, 94, 172

287

endosulfan 45
engineered nanomaterials (ENMs) 60, 63–4, 67–8, 217
environmental education 281
environmental laws 4
environmental remediation by microbes 236
environmental risk 2, 4–5, 8
environmental science 2–3
environmental toxicology 36–7
EU Commission 104
European Union 4–5
evaluation 2–3, 5
exposure assessment 4–7
exposure to pesticide 169
extraction 85, 88–94, 96, 98, 100, 102

F

fertilizers 3
fluorescence 217–19
focused ion beam-scanning electron microscopy (FIB-SEM) 221
food chain 3, 140, 142, 147, 152, 154, 155
forest ecosystem 143, 144
Fourier-transform ion cyclotron resonance (FT-ICR) 225
fresh water ecosystem 145
fullerenes 201, 202, 204
fullerenols 203, 204
fungicide 4

G

gas chromatography (GC) 85, 93–4, 97–9, 101, 103, 106
genotoxicity 19, 20, 22
global warming 2
glutathione s-transferase (GST) 167, 170, 171
graphene 202
grassland ecosystem 144, 145
greenhouse gases 121
gut-barrier function 235
gut microbiota (GMB) 234, 239

H

harmful 1–3, 6
hazard identification 4, 6
hazardous 2, 4, 217, 223–4
heavy metal pollution 186
heavy metals 132, 233–5
HeLa cells 15
hexachloro benzenes (HCBs) 193
high performance liquid chromatography (HPLC) 93, 94, 106
high resolutiom mass spectrometry (HRMS) 93, 99, 108

high resolution gas chromatography (HRGC) 93
high throughput screening (HTS) 216–17
HPLC with tandem mass spectrometry (HPLC-MS/MS) 94
human health 62–3, 67
humic acid 66–7

I

indigenous species 275, 280
industries 3, 5
inorganic pollutants 186, 192
insecticide 4
interaction with bacterial species **240**
internalization 16
international union for conservation of nature (IUCN) 195, 196
invasive species 274–5
ion-trap mass spectrometry (IT-MS) 93

L

lead (Pb) 37–9, 42, 140, 141, 143, 149, 150, 242–3
limit of quantitation (LOQ) 93–4, 97
liquid chromatography high resolution mass spectrometry (LC-HRMS) 94–5, 105
liquid chromatography time-of-flight mass spectrometry (LC-ToF-MS) 93
liquid liquid extraction (LLE) 88, 90, 92
living systems 4
luminescence 217–18

M

mangrove ecosystem 141, 146–8
matrices: soil 87; sediment 87; food 87
mechanism of action 39, 43–4, 47
MeHg 40–1
mercury (Hg) 37–8, 40–1, 140, 142, 152, 249–50
metabolism of pesticides 165, 169
metal ions 12, 36–40, 48
metal oxide NPs 16, 18
metallic nanoparticles 199–203
microalgae 146, 147, 155
microarray 217–18
microplastics 48–50, 179, 180, 182, 186, 195–9
modulation of xenobiotics 239
mutagenic pollutants 172

N

n-acetyltransferase (NAT) 170
nanomaterials (NMs) 36–7, 46–8, 50, 199–201, 203, 204, 215, 217
nanoparticles (NPs) 224
nanoplastics 49
nanotechnology 59, 68

Index

nanotoxicity 14, 18
nanotoxicity 47–8
natural organic matter (NOM) 60, 64–7
nucleotide sequencing 225

O

octachlorobenzo-p-dioxin (OCDD) 193
organic carbon 7
organic pollutants 191–4, 197, 204
organophosphorus pesticides 45
oxidative stress 14, 18, 19, 21, 23

P

paper-based 131
parabens 87
per-fluorinated compounds (PFC) 86, 92–4, 108–9
perfluorononanoate (PFNA) 92
perfluorooctanesulfonic acid (PFOS) 92–4
perfluorooctanoic acid (PFOA) 92
persistent organic pollutants (POPs) 166, 191, 192, 194, 196, 197
personal care products (PCP) 86, 87, 94
pesticides 3, 36–7, 43–6, 50, 168–73
pharmaceutical 4
phenols 191, 192, 194
phthalate esters 180, 191, 192, 197
phthalates 87, 106
physicochemical properties **237**
phytoplankton 155
pollutant 62–3, 67, 122
polyamide 198
polyaromatic hydrocarbons (PAHs) 180, 191, 193
polybrominated diphenyl ethers (PBDE) 87, 98, 195
polychlorinated biphenyls (PCBs) 87, 98, 102, 192, 193, 195
polycyclic aromatic hydrocarbons (PAH) 88, 98, 166, 173
polyester 198
polyethylene 198
polyfluoroalkyl substances (PFAS) 182
polystyrene 198
polyvinyl chloride (PVC) 198
porosity 7
pregnane X receptor (PXR) 168
problem formulation 5
producers 155
PubChem 217

Q

quadrupole time-of-flight mass spectrometry (QToF-MS) 93, 94, 99
quantum dots 12

R

radioactive wastes 179, 190
reactive nitrogen species 40, 43
reactive oxygen species (ROS) 14, 18–21, 23, 24, 39–40, 220, 226
reproduction 2
resilience of ecosystem 279, 281
restoration 278–80
risk assessment 2–8
risk characterization 4–6
rodenticide 4
role in brain function 234
routes of exposure 37–8, 43, 46, 49

S

sample preparation 87–90, 92, 94–5
scintillation 217
sedimentation 63–4
separation 127
silver NPs 48
single particle ICP-MS 219
single-walled carbon nanotubes (SWCNTs) 14
sociobehavioural 8
socioeconomic 8
solid phase extraction (SPE) 88–90, 92, 94, 96, 98, 100, 102, 106
solvation 64–5
sources and toxicities 249
sources 236, 243, 245, 248
sources, bio-distribution, chronic exposure effects 242
spectroscopy 126
stressors 2–3
sulfotransferases (SULT) 170–2
superoxide dismutase (SOD) 225
sustainability 276, 280–1
synthesis of important secondary metabolites **241–2**

T

tandem mass spectrometry (MSn) 93
time-of-flight secondary ion mass spectrometry (ToF-SIM) 220–1
TiO_2 NPs 48
ToxCast program 217, 222
toxicities 234, **238**
toxicities, bio-distribution and metabolism, physicochemical properties 244, 246, 248, **237**
toxicity 3, 6–7
toxicology 35–7
transfer factor (TF) 190, 199
transparency 8
triple quadrupole 93, 108

triple quadrupole/linear ion trap mass spectrometer (QTRAP/QLIT) 108
triple quadrupole mass spectrometry (QqQ-MS) 93
trophic levels 180, 181, 184–6, 190, 194, 199, 203
tropic transfer 147, 154

U

UDP-glucuronosyltransferases (UGT) 170, 171
ultra high performance liquid chromatography (U-HPLC) 106
uncertainty 3, 7
United States Environmental Protection Agency 13
uptake 142, 144, 145, 152, 153
urbanization 273

US Environmental Protection Agency (US EPA) 217

V

volatile organic compounds (VOCs) 88, 192

W

WHO 145–8, 155

X

xenobiotics 166, 167, 170, 173

Z

zeta potential 13, 19
zinc nanoparticles (ZnO NPs) 48